Gas Chromatography
Mass Spectrometry
Applications in
Microbiology

Gas Chromatography
Mass Spectrometry
Applications in
Microbiology

Edited by

Göran Odham
Lennart Larsson
and
Per-Anders Mårdh

University of Lund
Lund, Sweden

PLENUM PRESS • NEW YORK AND LONDON

Library of Congress Cataloging in Publication Data

Main entry under title:

Gas chromatography/mass spectrometry applications in microbiology.

Bibliography: p.
Includes index.
1. Gas chromatography. 2. Mass spectrometry. 3. Microbiology—Technique. I.
Odham, Göran, 1939–   . II. Larsson, Lennart, 1949–   . III. Mårdh,
Per-Anders.
QR69.G27G37 1983   576   83-16102
ISBN 0-306-41314-0

Printed in the United States of America

CONTRIBUTORS

Cécile Asselineau *Centre de Recherche de Biochimie et Génétique Cellulaires du CNRS, 118 route de Narbonne, 31062 Toulouse Cedex, France*

Jean Asselineau *Centre de Recherche de Biochimie et Génétique Cellulaires du CNRS, 118 route de Narbonne, 31062 Toulouse Cedex, France*

Göran Bengtsson *Laboratory of Ecological Chemistry, University of Lund, Ecology Building, Helgonavägen 5, S-223 62 Lund, Sweden*

Johan Haverkamp *Department of Biomolecular Physics, FOM-Institute for Atomic and Molecular Physics, Kruislaan 407, 1098 SJ Amsterdam, The Netherlands*

Nancy J. Hayward *Bacteriology Department, Alfred Hospital, Commercial Road, Prahran, Melbourne 3182, Victoria, Australia*

Erik Jantzen *National Institute of Public Health, Postuttak, Oslo 1, Norway*

Lennart Larsson *Department of Medical Microbiology, University of Lund, Sölvegatan 23, S-223 62 Lund, Sweden*

Samuel L. MacKenzie *Prairie Regional Laboratory, National Research Council of Canada, 110 Gymnasium Road, Saskatoon, Saskatchewan S7N 0W9, Canada*

Per-Anders Mårdh *Department of Medical Microbiology, University of Lund, Sölvegatan 23, S-223 62 Lund, Sweden*

Henk L. C. Meuzelaar *Biomaterials Profiling Center, University of Utah, 391 South Chipewa Way, Research Park, Salt Lake City, Utah*

v

Göran Odham *Laboratory of Ecological Chemistry, University of Lund, Ecology Building, Helgonavägen 5, S-223 62 Lund, Sweden*

Danielle Patouraux-Promé *Centre de Recherche de Biochimie et Génétique Cellulaires du CNRS, Université Paul Sabatier, 118 route de Narbonne, 31062 Toulouse Cedex, France*

Jean-Claude Promé *Centre de Recherche de Biochimie et Génétique Cellulaires du CNRS, Université Paul Sabatier, 118 route de Narbonne, 31062 Toulouse Cedex, France*

Jean K. Whelan *Chemistry Department, Clark Building, Woods Hole Oceanographic Institution, Woods Hole, Massachusetts*

Gerhan Wieten *Department of Biomolecular Physics, FOM-Institute for Atomic and Molecular Physics, Kruislaan 407, 1098 SJ Amsterdam, The Netherlands*

PREFACE

During recent years there has been increasing interest in the value of a number of chemical and physical–chemical analytical methods for the detection and characterization of microorganisms. Furthermore, such methods are currently used in studies on microbial metabolic processes, on the role of microorganisms in the turnover of inorganic and organic compounds, and on the impact on environmental changes by microbial activity. Moreover, the introduction of some of these methods not only shortens the analytical time period compared to "traditional" techniques, but also improves the analytical quality.

Mass spectrometry (MS) combined with chromatographic inlet systems, particularly gas chromatography (GC), belongs to those methods which during recent years have established their value for the above-mentioned purposes.

The present volume starts with basic chapters on the principles for MS and common inlet systems, particulary GC. It discusses applications of these techniques to a number of microbiological disciplines, e.g., ecological and medical microbiology. Emphasis is laid on organic compound classes of special relevance to microbiology, e.g., volatiles, lipids, amino acids, peptides and carbohydrates. Some compound classes of a more general biochemical rather than specific microbiological importance, e.g., steroids and nucleotides, are dealt with briefly.

The editors wish to thank all those who have contributed to this book. We hope it will stimulate further research in this futuristic field and will be of practical value.

Lund, Sweden   G. Odham
                                                        L. Larsson
                                                        P.-A. Mårdh

CONTENTS

Chapter 2

Mass Spectrometry
Göran Odham and Lennart Larsson

II. ANALYSIS OF MICROBIAL CONSTITUENTS

Chapter 5
Amino Acids and Peptides
Samuel L. MacKenzie

III. APPLICATIONS

Chapter 6
Analysis of Volatile Metabolites in Indentification of Microbes and Diagnosis of Infectious Diseases
Lennart Larsson, Per-Anders Mårdh, and Göran Odham

Chapter 7
Head-Space/Gas–Liquid Chromatography in Clinical Microbiology with Special Reference to the Laboratory Diagnosis of Urinary Tract Infections
Nancy J. Hayward

Chapter 10
Analytical Pyrolysis in Clinical and Pharmaceutical Microbiology
Gerhan Wieten, Henk L. C. Meuzelaar, and Johan Haverkamp

Chapter 11
Volatile C^1–C^8 Compounds in Marine Sediments
Jean K. Whelan

Chapter 12
Mass Spectrometry of Nitrogen Compounds in Ecological Microbiology
Göran Bengtsson

PART I

INTRODUCTION TO GAS CHROMATOGRAPHY/ MASS SPECTROMETRY AND APPLICATIONS IN MICROBIOLOGY

INTRODUCTION

Per-Anders Mårdh, Department of Medical
Microbiology, University of Lund, Sölvegatan 23,
S-223 62 Lund, Sweden

Since its introduction in the early fifties, gas chromatography (GC) has become a much used analytical technique in several disciplines. However, it was not until the last decade that GC has been routinely applied in microbiology.

The development during recent years of improved GC techniques employing, e.g., automated injectors, capillary columns working with split- and splitless injections, pulsed electron capture detectors, as well as computerized reading of chromatograms, has increased the use of GC as a useful analytical technique in microbiology.

GC combined with mass spectrometry (MS) had not been made use of in microbiology until recent years, but lately there have been an increasing number of applications in this field. In addition to GC/MS, systems making use of high-performance liquid chromatography (HPLC) as inlet devices have recently been introduced. The reduced need for elaborate derivatization procedures and the facility for measuring components of high molecular weight should make these systems particularly useful in microbial work.

The conventional continuous scanning of complete mass spectra using magnetic or quadrupole types of mass spectrometers has in several new instruments been supplemented by selected ion monitoring (SIM), which permits the measurement of a limited number of preselected ions. This technique has met a number of analytical demands in microbiology, where the high sensitivity and selectivity of the MS instruments have been found valuable. Thus, SIM allows search for and quantification of one or a few selected compounds. The extremely high selectivity of the SIM technique is of particular value when testing complex biological samples. In theory, it also allows the manufacture of cheaper mass spectrometers, which can be used for specific diagnostic tasks. Both electric (EI) and chemical (CI) ionization are currently used.

GC/MS studies of microbial cellular and extracellular compounds include compound classes such as hydrocarbons, alcohols, fatty acids, amines, aldehydes, and ketones. Macromolecules, i.e., proteins, lipids, polysaccharides, lipoproteins, and glycolipids, can be analyzed after degradation to monomers (amino acids, fatty acids, saccharides), followed by derivatization. Definite identification of these compounds requires tests by MS.

MS can be used to identify microbial compounds which may be of value in taxonomic studies and for establishing manuals for the differential diagnosis of microbes. MS has also been used to establish the identity of microbial metabolites in culture media. Through knowledge of the identity of such metabolites, specific components may be added to the medium, thereby increasing the differential diagnostic capacity of the GC analytical system.

Furthermore, MS can be used to identify the presence of specific microbial compounds in body fluids of infected hosts. Thus, by GC/MS operating in the SIM mode, compounds unique to a group of organisms— e.g., tuberculostearic acid occurring in microbes of the order of *Actinomycetales*, or C_{32}-mycocerosic acid, which occurs only in a restricted number of species of *Mycobacterium*, i.e., in *M. tuberculosis, M. bovis, M. africanus*, and *M. kansasi*—can be used as markers for these organisms.

Still another example of the use of GC/MS in clinical microbiology is the demonstration of endotoxin in the blood of patients with endotoxinemia (endotoxin shock), endotoxin being a cell wall constituent of gram-negative bacteria.

GC/MS has also found application in ecological studies. Good examples are the determination of minute amounts of amino acids of microbial origin on leaves in river waters and the study of volatiles in sediments from the oceans.

In the above-mentioned examples, the extremely high sensitivity and selectivity of the GC/MS analytical technique is utilized. This allows the detection of compounds in the low picogram range.

Current research also suggests that these unique properties can be used in the study of the chemical compositions at the interfaces between microorganisms and various hosts. Provided that suitable sampling techniques are developed, MS can open up possibilities for chemical studies of such small environments.

1

GAS CHROMATOGRAPHY

Lennart Larsson, Department of Medical Microbiology,
    University of Lund, Sölvegatan 23, S-223 62 Lund, Sweden
Göran Odham, Laboratory of Ecological Chemistry,
    University of Lund, Ecology Building,
    Helgonavägen 5, S-223 62 Lund, Sweden

1. INTRODUCTION

Gas chromatography (GC), which was introduced in 1952, is now one of the most extensively used instrumental procedures for separation of compounds. GC is employed mainly in analytical work, but it can be used also for preparative purposes. In principle, all covalent compounds of moderate molecular weight can be separated by means of GC. In addition, GC often permits tentative identification of the compounds studied.

The sample is introduced into a stream of heated carrier gas (the mobile phase), where its components are volatilized and swept through the chromatographic column, which contains a stationary phase (Figure 1). In the GC column, the components are selectively retarded according to their interactions with the stationary phase. The separated components enter a detector connected with a recorder or printer/plotter and are consecutively registered as chromatographic peaks. The area beneath each peak can be correlated to the concentration of the corresponding component in the injected sample.

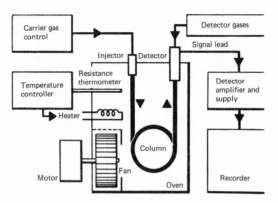

FIGURE 1. The principle of gas chromatography (Drucker, 1976).

Numerous textbooks on gas chromatography have been published. The present chapter is intended only as a brief introduction to the theory of GC separation, different types of columns, stationary phases, injection systems, and detectors. For more extensive accounts of the theory and practice of GC, specialized literature should be consulted (Grob, 1977; Jennings, 1980, 1981; Bertsch *et al.*, 1981).

2. GAS CHROMATOGRAPHIC SEPARATION THEORY

After introduction and vaporization in the injection port or directly onto the column, each of the sample components (solutes) is allowed to distribute between the stationary (liquid or solid) and the mobile (gaseous) phase. The distribution is governed by the partition coefficients (distribution constants) of the solutes between the two phases, viz.,

$$K = \frac{C_S}{C_M}$$

where K is the partition coefficient (distribution constant), C_S is the concentration of solute in the stationary phase, and C_M is the concentration of solute in the mobile phase.

The greater the solubility of a solute in the stationary phase, i.e., the higher the partition coefficient, the longer will be the time lapse from injection of the sample until the component reaches the detector. The factors determining this *retention time* include the type of column, the carrier-gas flow rate, and the temperature. Separation of two compounds by GC is possible only if they have different K values.

Provided that separation procedure and detector(s) are appropriate, the chromatograms present one peak corresponding to each of the separated components in the injected sample. The *adjusted retention time* of a solute can be defined as

$$t'_R = t_R - t_0$$

where t'_R is the adjusted retention time, t_R is the retention time, and t_0 is the retention time of inert gas, "dead time."

A compound's *relative retention* is its adjusted retention time in relation to that of an internal standard, viz.,

$$\alpha = t'_R / t'_{Ri}$$

where α is the relative retention and t'_{Ri} is the adjusted retention time of the internal standard.

As a rule, the internal standard is added to the sample before the analysis, in order to compensate for possible differences in the analytical conditions. The relative retention is frequently used to characterize a compound at GC. This value is unaffected by fluctuations in the carrier-gas flow rate and depends only on the temperature and polarity of the stationary phase, irrespective of the type and dimensions of the column. In addition, α is virtually constant within a limited temperature range.

The *efficiency* of a GC column can be expressed as the number of theoretical plates. This term was originally defined for distillation columns for separation. The ratio of the GC column length to the number of theoretical plates is designated the *height equivalent to a theoretical plate* (HETP), viz.,

$$\text{HETP} = L/n$$

where L is the length of the column and n is the number of theoretical plates.

HETP is basically a measure of peak broadening during passage of the sample through the column. The smaller the HETP, i.e., the sharper the chromatographic peaks, the more efficient is the GC column.

Column efficiency for a given compound in the prevailing chromatographic conditions can be calculated from the chromatogram (Figure 2), expressed as the relationship

$$n = 16\left(\frac{t_R}{w}\right)^2 = 5.54\left(\frac{t_R}{w_h}\right)^2$$

where w is the peak width, defined as the segment of the base line that is

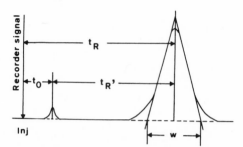

FIGURE 2. Chromatogram of a single solute.

cut by tangents drawn at the inflection points on either side of the peak, and w_h is the peak width at half peak height.

The *effective theoretical plate number* is

$$N = 16\left(\frac{t'_R}{w}\right)^2 = n\left(\frac{k}{1+k}\right)^2$$

where N is the effective theoretical plate number and k is the capacity ratio (partition ratio) = t'_R/t_0.

Specifications of column efficiency generally refer to compounds with relatively high k values ($k > 4$), i.e., solutes with reasonably long retention time.

For details regarding the derivation of the cited relationships the reader is referred to Grob (1977).

The relationship between HETP and the carrier-gas flow rate is expressed by van Deemter's equation. In its simplest form, the relationship is expressed as

$$\text{HETP} = A + B/\bar{\mu} + C\,\bar{\mu}$$

where $\bar{\mu}$ is the average linear velocity of carrier gas, and A, B, and C are complex factors representing the effect of multiplicity of gas paths through a packed column due to eddy diffusion (A), the longitudinal diffusion of the solute molecules in the gas phase (B), and the mass transfer resistance between the gaseous and stationary phases (C). The correlation is exemplified in Figure 3. To obtain the most efficient separation, i.e., the lowest HETP value, the carrier-gas flow rate obviously must be carefully optimized.

Whereas the efficiency, and thus the *performance*, of the column is a measure of the sharpness of single peaks, the *resolution* denotes the ability

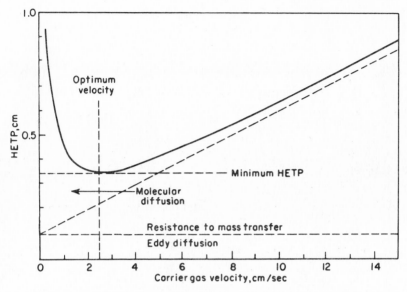

FIGURE 3. Visualization of van Deemter's equation (Willard, Merrit, and Dean, 1970).

of the column to separate two solutes eluting as two consecutive peaks (Figure 4):

$$R = \Delta t_R / w_m$$

where R is the resolution and

$$\Delta t_R = t_{R_2} - t_{R_1} = \Delta t'_R = t_{R'_2} - t_{R'_1}$$

$$w_m = (w_1 + w_2) \times \tfrac{1}{2}$$

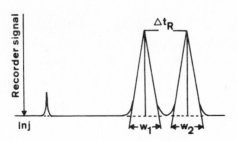

FIGURE 4. Resolution between two solutes.

The resolution rises with column efficiency. If $R = 0$, there is no separation between two peaks, while $R = 1.5$ means virtually (99.7%) complete separation ("base-line resolution"). $R = 1$ corresponds to 98% separation.

3. COLUMNS

3.1. Packed and Capillary Columns

The most widely used GC columns are stainless steel or glass tubes 1–4 m long and with inner diameter 2–4 mm, which are packed with granular support material coated with a stationary-phase film. Other tubing metals, such as nickel, copper, or aluminum, may also be employed. Glass columns are fragile, but have the advantages of inertness and transparency. They are essential when studying compounds which may decompose on contact with hot metal, due to catalytic action. This tendency is more pronounced with capillary than with packed columns. The inner surface of glass tubing can be deactivated by silylation, in order to remove polar groups which may reduce column performance.

Micropacked columns, generally with 1 mm inner diameter, are less commonly used. They are more efficient than ordinary packed columns. Because of the smallness of their support granules, the permeability of micropacked columns is low, which means that a high inlet pressure of the carrier gas is required.

Deposition of a thin (0.1–1.5 μm) film of liquid stationary phase directly onto the inner surface of a long, narrow-bore (capillary) tube was proposed as early as 1958 (Golay). However, such wall-coated open tubular (WCOT) columns of stainless steel or glass have not yet achieved the widespread use that would seem to be warranted by their superior efficiency as compared with packed columns. The reasons are the somewhat more complicated technical arrangements (inlet systems and makeup gas facilities) than those for packed columns, and also the brittleness of the glass capillaries. Use of nickel capillary tubing, which combines the chemical inertness of glass with the durability of metal, and of the recently introduced fused silica-glass capillary columns, may drastically change this situation. The fused silica capillaries have a thin outer coating of polyimide and are flexible and almost unbreakable. They are therefore suitable for use in routine procedures.

Recent studies with fused silica capillaries in microbiology have concerned cellular constituents (Moss et al., 1980) and metabolic end-products (Larsson and Holst, 1982) of microorganisms. The superior differential diagnostic capacity of fused silica capillaries as compared with packed columns in identification of bacteria was demonstrated in studies on cellular fatty acids of *Legionella* (Moss, 1981).

Currently used methods for pretreating glass capillary columns (soda lime glass, borosilicate glass, lead glass, fused quartz, or fused silica glass) include deactivation and roughening of the inner surface of the capillaries as well as static or dynamic coating (Jennings, 1980). Fused silica columns have very low surface energy, which implies rejection of high-polarity phases. Work is in progress to modify these surfaces, in order to give a wider range of stationary phases in these flexible columns. Extensive comparisons between conventional glass and fused silica-glass capillary columns were reported by Jennings (1981).

At optimum carrier-gas velocity, the number of theoretical plates per length unit is comparable for packed and open tubular column types (Table 1). The practical length of packed columns is limited by the large pressure drop, and thus high inlet pressure is required to effectuate an adequate flow rate for carrier gas. Because of the low resistance to flow in capillary columns—about 1% of that in packed columns—very long capillaries can be used, with a correspondingly high total number of plates. But because of the smaller sample capacity of capillary columns (Table 1) special injector systems (see Section 4) must as a rule be employed in order to avoid overloading and resultant decrease of efficiency.

Wide-bore (0.5 mm) WCOT capillaries, having a relatively thick layer of stationary phase, and such capillaries where support granules are being attached to the inner walls prior to coating [support-coated open tubular (SCOT) columns] are also available. The separating qualities of these columns are superior to those in packed columns, but less efficient than in narrow-bore capillaries (Table 1). SCOT and wide-bore WCOT columns can be used with inlet systems of the same types as for packed columns.

The *coating efficiency* (a percent value) is frequently used for assessing the performance of a WCOT column.

$$\text{Coating efficiency} = \frac{\text{HETP}}{\text{HETP}_{\text{min}}} \times 100$$

where HETP is calculated from the chromatogram (L/n) and HETR$_{\text{min}}$ is derived from the equation

$$\text{HETP}_{\text{min}} = r\left[\frac{1 + 6k + 11k^2}{3(1 + k)^2}\right]^{1/2}$$

where k is the capacity ratio and r is the radius of the column. An important point is that the coating efficiency can be used *only* in regard to WCOT values.

TABLE 1. Some Characteristics of Packed, Micropacked, Support-Coated, and Wall-Coated Gas Chromatographic Columns[a]

Type of column	WCOT	SCOT	Micropacked	Packed
Typical inside diameter	0.25 mm 0.50 mm	0.50 mm	1 mm	2 mm 4 mm
Typical length	10–100 m	10–100 m	1–6 m	1–4 m
Typical efficiency	1000–3000 plates/m	600–1200 plates/m	1000–3000 plates/m	500–1000 plates/m
Sample sizes	10–100 ng	10 ng–1 µg	10 ng–10 µg	10 ng–1 mg
Linear flow rate (optimum)	20–30 cm/s, H_2, He 10–15 cm/s, N_2, Ar	20–30 cm/s, H_2, He 10–15 cm/s, N_2, Ar	8–15 cm/s, H_2, He 3–10 cm/s, N_2, Ar	4–6 cm/s, H_2, He 2–5 cm/s, N_2, Ar
Volume flow rate (optimum)	1–5 ml/min, H_2, He 0.5–4 ml/min, N_2, Ar	2–8 ml/min, H_2, He 1–4 ml/min, N_2, Ar	2–6 ml/min, H_2, He 1–3 ml/min, N_2, Ar	20–60 ml/min, H_2, He 15–50 ml/min, N_2, Ar
Pressures required	Low	Low	Very high	High
Detector makeup gas	Usually required	Usually required	Usually required	Not required
Speed of analysis	Fast	Fast	Medium	Slow
Chemical inertness	Best			Poorest
Permeability	High	High	Low	Low

[a] Alltech Associates Chromatography Products Catalog, 1980.

In addition to offering separation qualities superior to those of packed columns within a given analysis time, capillary columns can be used to shorten considerably the analysis time at an efficiency level which is comparable with that when packed columns are used. We feel that the potential of this latter aspect of capillary GC has hitherto been disregarded.

3.2. Stationary Phases and Supports

The selection of *stationary phase* is one of the most important decisions in work with GC. The stationary phase must be a good differential solvent, in order to effect good separation of the solutes. It should also be thermostable, especially in analyses of compounds with high molecular weight, and it must not react irreversibly with any of the components to be separated. A good rule of thumb is that the selected stationary phase should have a chemical structure similar to the solute structure: a polar stationary phase should be used to separate polar compounds and a nonpolar for nonpolar compounds. Thus, free fatty acids can be analyzed with a polar acidic phase. Amines can be chromatographed with an alkaline polar stationary phase, while hydrocarbons are usually best separated on a nonpolar phase.

The liquid stationary phases that have been used in GC number several hundreds. Each of them can be characterized by Rorschneider's and McReynold's constants. These constants describe, as retention indices, the interaction between a given phase and certain test compounds with a wide range of polarity. Most suppliers of products for chromatography now provide McReynold's constants for the stationary phases on sale. Despite differences in name, many of the commercially available stationary phases are virtually identical, e.g., the polymethyl silicones SE-30, OV-101, OV-1, and SP 2100. Other examples are OV-202, OV-210 and SP-2401. In the selection of an appropriate stationary phase, therefore, consideration of the relevant constants (Rorschneider's or McReynold's), in addition to maximum operating temperature, is of great importance.

Ideally, the support granules upon which the liquid stationary phase is distributed in packed columns should play no part in the separation process. When the film of stationary phase is very thin in this gas–liquid chromatography (GLC), with weight corresponding to only 1%–2% of the support, the requirement for inertness of the support is particularly vital. Silylation is frequently employed to deactivate the surface of the granules.

Desirable qualities of the solid support, still predominantly of diatomaceous earth type, include large surface area and regular granular size. The latter is expressed as mesh range—the higher the mesh range, the smaller the granules. The selected mesh size is usually a compromise dictated by the wish to use small granules providing a large area of total surface for efficient separation and the associated increase of flow resistance.

The solid support as such occasionally affects separation, i.e., gas-solid chromatography (GSC). The materials that have been used for GSC, particularly for separation of solutes with relatively low molecular weight, include silica gels, molecular sieves, activated charcoal, graphitized carbon, and porous polymers. Water and gases are among these solutes (Chapter 6). Microbial acidic, alkaline, and neutral volatiles have also been separated by GSC (Larsson *et al.*, 1978).

Covalent bonding of liquid stationary phase to the solid support has been described. As a rule, a low concentration of stationary phase, typically 0.2%, is used. Stationary phase may also be bonded directly (cross-linked) to silanol groups on the inner glass surface of capillary columns (Jennings, 1981). Impurities in cross-linked stationary phases which may accumulate after repeated injections, e.g., of extracts from biological samples, can usually be removed by washing the column with organic solvents.

Guidelines concerning chromatographic conditions for the analysis of constituents and metabolic end-products of bacteria, including choice of stationary phase, are given in Chapters 3–13.

4. SAMPLE INJECTION

In ideal circumstances the sample is instantaneously vaporized at the beginning of the column to form a homogeneous narrow "plug" of sample mixed with the carrier gas. In practice this ideal is the exception rather than the rule. Poor injection technique results in peak broadening. The result of chromatography depends largely on the skill of the operator in injecting the sample. This section deals with injection techniques for samples in solution. In regard to analysis of gases and solid samples, reference is made to Chapters 6 and 10, respectively.

4.1. Injection onto Packed Columns

In the usual procedure, a 0.5–5 µl portion of solution is introduced into the column by means of a microsyringe which penetrates the rubber septum of the chromatograph. The solution is vaporized either in an empty insert liner, to which the column is attached, or directly in the column ("on-column injection"). The liner method is preferable if the sample contains‛appreciable amounts of nonvolatile compounds which can be trapped at the injection port. Such material may act as a strong adsorbent for solutes, which can be released when the next sample is injected, leading to "ghost" peaks. Removal of trapped compounds is easier from an insert liner than from a column, since in the latter case the first inches of the packing material in the column must be removed and renewed.

4.2. Injection onto Capillary Columns

Narrow-bore (0.25 mm) capillary columns are usually operated at a carrier-gas flow rate around 1 ml/min. Use of this low flow rate at injection of the sample would lead to appreciable peak broadening, because of the long time (minutes) required for the carrier gas to sweep the vaporized sample from the injection port to the column. To circumvent this obstacle, the sample can be injected at a relatively high flow rate of carrier gas, which is split into two unequal parts after evaporation. The smaller part is connected to the column, while the larger is vented out (Figure 5). Such *split injection* also means that the volume introduced into the column can be kept small enough to avoid overloading (Table 1). Care must be taken to ensure linearity of the splitting device, i.e., nondiscrimination between sample components according to high or low boiling point. The split ratios generally recommended to ensure satisfactory linearity are 1:50–1:100. This means that only 1%–2% of the injected material enters the column. To avoid decomposition of the sample, an all-glass splitter is preferable to the frequently employed metal splitters. The splitting device can be deactivated by silylation *in situ*.

Although inlet splitting of the sample is the most widely used injection technique in capillary GC, it may be unsuitable in some circumstances, viz., when only small amounts of a sample are available or when the

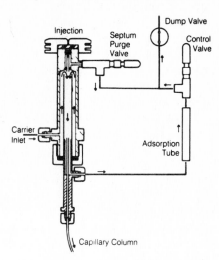

FIGURE 5. Inlet splitter for gas chromatographic analysis on a capillary column (SGE Catalog, 1980).

concentration of sample components is low. Other inlet systems have therefore been devised, though hitherto they are seldom used for analysis of microbial constituents and metabolites.

In the *cold trapping* technique the sample is injected at a column temperature at least 100°C below the boiling points of the solutes. In consequence, the solutes are more or less quantitatively trapped and concentrated at the beginning of the column. No splitting device is used. When the vaporization chamber has been freed from solvent by flushing carrier gas through the column, the temperature of the oven heating the column is slowly increased to release the solutes consecutively according to their boiling points. The chromatographic process then begins. Highly volatile compounds—even gases—can be trapped by cooling the oven, or the first part of the column, to subambient temperature, for instance with liquid nitrogen.

The *splitless injection* technique utilizes, instead of cold trapping, a solvent effect (Grob and Grob, 1981). The interaction between solvent and liquid stationary phase causes stronger retention at the front than at the rear of the sample plug, leading to a band narrowing. The sample thus is reconcentrated prior to entry onto the column. To obtain a good solvent effect, not less than 0.5 µl of the sample should be injected at a column temperature of at least 5–10°C below the boiling point of the solvent. For example, whereas hexane and isooctane will give good solvent effect, polar and aromatic solvents are not recommended.

In order to reduce the otherwise pronounced "tailing" of the solvent peak, splitless injection must include purging of the system from the last portion of the injected solvent. "Splitless," therefore, is valid only at the injection stage. Ideally, the purging system should be activated immediately after all the sample components have entered the column, leaving only a fraction of the solvent in the vaporization chamber. Too early activation of the purging system leads to loss of sample, while too late activation results in peak tailing. Jennings (1980) recommended the purge function to be activated when a carrier gas volume equivalent to 1.5 times the volume of the vaporization chamber has swept through that chamber.

Recent progress permits placing of the sample directly *on-column*, by using an extremely fine needle to connect the syringe directly with the capillary column. As such needles are too brittle to penetrate a rubber septum, a septum-free injection system must be used. Cross-linked stationary phases are preferred because these are not sensitive towards the injection of even relatively large amounts of organic solvents. On-column injection is less suitable for use on a routine basis but is advantageous in analyses of thermally unstable compounds. This injection technique usually decreases the discrimination of certain solutes, which sometimes can be observed in splitless injections.

4.3. Syringe Technique

The "hot needle" injection technique implies that the entire sample is drawn into the glass barrel of the syringe, leaving the needle empty. The needle pierces the septum and is allowed to warm up in the injection port for 3–4 s. The sample is then rapidly injected. This technique is preferable to retaining part of the sample in the needle, since sample components then may fractionate from the needle when the septum has been penetrated, before the injection as such has been made. This release leads to band broadening and reduced resolution.

For further details of various sample injection techniques in capillary GC, Jennings (1980) should be consulted.

Several systems for automatic injection of samples are commercially available. They permit multiple automated and reproducible injections even outside the laboratory's normal working hours, and thus increased capacity of the gas chromatograph.

5. DETECTORS

The GC detector should respond with high sensitivity to the components of interest in the sample but be insensitive to fluctuations of carrier-gas flow and temperature. It should have a wide linear dynamic range, i.e., a linear relation between solute concentration and detector response, and a small "dead volume." Further, it must react quickly to eluted components. There is no universally "best" GC detector for analytical purposes. The choice of detector must be related to the chemical character of the studied compounds.

The *thermal conductivity* (TC) detector utilizes differences in thermal conductivity, measured by a Wheatstone bridge, between a stream of carrier gas containing solutes and a reference carrier-gas stream. Disequilibrium of the bridge generates a signal which is amplified and registered. High thermal conductivity of the carrier gas is desirable for sensitivity, hydrogen and helium being superior in this respect. But because of the relatively high cost of helium and the fire hazards associated with hydrogen, nitrogen is sometimes used instead, despite its considerably poorer sensitivity. As compared with most other detectors used in GC, the TC type has low sensitivity. However, it is almost universally applicable, since it responds to all compounds, including permanent gases. The TC detector, moreover, is nondestructive and therefore suitable in, for instance, preparative work.

The *flame ionization* (FI) detector utilizes an air-hydrogen flame with temperature high enough to ionize eluting solutes. The resulting increase of the ion current is registered by two electrodes, situated across the flame.

The FI detector responds to all compounds containing a CH group and in general gives considerably higher sensitivity than the TC detector. The detection limit normally lies around 1 ng. Inorganic compounds such as permanent gases are not registered. The lack of FI detector response to air and water can be a favorable factor, e.g., in head-space analyses (Chapter 6).

Electron capture (EC) detectors measure a decrease of current signal rather than a positively produced current. A radioactive source (tritium, scandium, or ^{63}Ni) creates a constant current from the nitrogen carrier gas between two electrodes. The introduction of effluent containing sample molecules between the electrodes reduces the current to an extent which is dependent on the ability of the solutes to attract electrons. The EC detector is extremely sensitive to compounds possessing a high electron affinity, including those that contain halogen atoms. It is widely used in studies on pesticides. Its sensitivity to nonelectrophilic molecules is very low, however. The EC detector, being thus extremely selective, can be used for sensitive (picogram level) detection of a variety of compounds, including bacterial metabolites which have been derivatized to halogen-containing products.

Several other detectors, utilizing argon ionization, photoionization, flame photometry, or nitrogen–phosphorous detection, are available for use in GC. For detailed accounts of currently available detectors in GC, see Grob (1977).

In addition to identification of compounds separated by GC, the *mass spectrometer* can be employed as a GC detector when adapted for selected ion monitoring (SIM). This technique permits determinations of very high sensitivity (picogram range) and is even more selective than the EC detector. SIM is therefore particularly valuable for identifying and quantifying compounds present in trace amounts in complex (e.g., biological) environments. See Chapters 2 and 9 for details.

6. APPLYING GAS CHROMATOGRAPHY

Essentially all compounds with reasonably high vapor pressure can be directly analyzed with GC. The method can be used also for molecules with little or no volatility, if they are suitably derivatized. This process usually implies replacing of one or more hydrogen atoms bound to nitrogen or oxygen in the molecule by an atom or a group which reduces the polarity of the molecule, and consequently also its boiling point. Macromolecules must be degraded to smaller units, which in turn often require derivatization prior to GC analysis.

Low and medium molecular weight hydrocarbons, fatty acids, alcohols, amines, aldehydes, ketones, sulfur compounds, gases, and many other compounds thus can be chromatographed without prior derivatization (Chapter 6). Macromolecules such as proteins, complex lipids, polysaccharides, lipoproteins, and glycolipids—common cellular constituents of microorganisms—must first be degraded to monosaccharides, fatty acids, and amino acids, which are then derivatized (Chapter 8). Table 2 shows some common derivatives used in GC analysis of these monomers.

The *choice of gas chromatograph* is complicated by the negligible differences in price between instruments of different makes but with comparable performance. Many gas chromatographs are now fitted with a microprocesser by which a number of functions can be controlled. The microprocesser, however, does not add to the separation efficiency of the system.

When capillary columns are used, certain qualities are essential in the chromatograph. These include possibilities for attaching various injection devices and for septum purge and makeup gas (see below). Furthermore, it is vital to minimize the "dead volume" between column and detector, in order to avoid unnecessary broadening of peaks.

When using packed columns a mesh size of 80/100 (0.18–0.15 mm) or 100/120 (0.15–0.13 mm) for the *solid support* granules is generally suitable. Low concentration of the liquid stationary phase on the support means low column bleed and short retention time, but decreased resolution. A good compromise generally is to use 1%–5% (w/w) concentration of liquid phase on the support.

Thorough *conditioning* of the column is of cardinal importance for removal of moisture and impurities. Column conditioning is conveniently accomplished by connecting the column to the injector and thereafter flushing with the carrier gas at a temperature which is programmed (see below) to rise slowly from ambient to a level close to the limit of the stationary phase. This temperature level should be maintained for some hours before the column is connected to the detector.

TABLE 2. Some Derivatives of Common Bacterial Constituents for Gas Chromatographic Analysis[a]

Bacterial constituents	GC derivatives[a]
Carbohydrates	TFA (TMS) derivatized methyl glycosides
Lipids	Fatty acid methyl (propyl, butyl) esters
Proteins	N–HFB alkyl esters, N-TFA alkyl esters of amino acids

[a] TMS, trimethylsilyl; TFA, trifluoroacetyl; and HFB, heptafluorobutyryl.

Except when studying inorganic compounds, the *flame ionization detec-tor* (FID) can be recommended for a variety of uses, although its response to formic acid and formaldehyde, for instance, is poor. FID otherwise is highly sensitive and reliable, and, in contrast to the EC detector, it is not susceptible to overloading. Examples for application of *EC detector* are selective and sensitive analysis of metabolites in body fluids where alcohols, fatty acids, and amines are derivatized to halogen-containing molecules (Craven *et al.*, 1977). Such methods have been successful in the diagnosis of infectious diseases by direct GC analysis of body fluids (Figure 6).

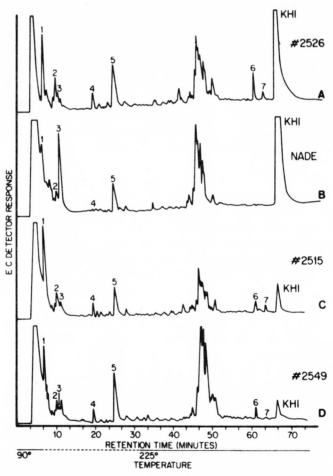

FIGURE 6. Chromatograms representing derivatized amines from spinal fluid of patients with meningitis due to *Mycobacterium tuberculosis*, using electron capture detection (Craven *et al.*, 1977).

When decisions have been made in regard to column, detector and sample preparation, the chromatographic conditions must be adjusted to give satisfactory separation. An isothermal analysis of a sample containing compounds with a very wide range of boiling points may result in a chromatogram where rapidly eluting peaks are poorly resolved, while solutes with higher boiling points have very long retention time and appear as excessively broad peaks. This drawback can be avoided by gradual, generally linear, increase of the column temperature during the chromatographic run. Such *temperature programming* means that the temperature is kept low enough during the first part of the analysis to allow ready separation of rapidly eluting compounds, but high enough at the end of the run to let larger molecules elute as narrow, well-resolved peaks within a reasonable time. The effect of temperature programming is illustrated in Figure 7.

When temperature programming is used, particularly in capillary GC, "ghost peaks", i.e., chromatographic peaks representing components not present in the injected sample, may occasionally appear. This phenomenon is usually attributable to impurities that have been trapped on the septum or in the injection port and are released when the temperature rises. To avoid ghost peaks, a small portion of the carrier gas can be used to sweep across the septum—septum purge—thus preventing its contact with the carrier gas sweeping through the column (Figure 5). Alternatively, a "septum swinger" may be used, so that the septum is in contact with the carrier gas only at the moment of injection. The quality of commercially available septa varies considerably. Unfortunately, septa which are suitable for high-temperature work usually are less serviceable for repeated injections.

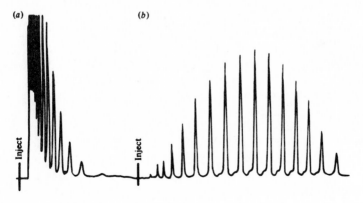

FIGURE 7. Comparison between isothermal analysis (a) and use of temperature programming (b) of the same sample (Andrews, 1970).

According to van Deemter's equation (Section 2), the flow rate of the carrier gas must be optimized for good separation. With packed columns, a flow of nitrogen carrier gas of 20–30 ml/min generally is satisfactory, whereas narrow-bore capillary columns work at considerably lower flow rates, viz., around 1 ml/min (Table 1). For optimal results, capillary columns also require an additional gas flow (makeup gas, 20–30 ml/min) at the detector to enhance its performance and to decrease band broadening of the eluates. Flame ionization detectors require access to air ($\approx$200 ml/min) and hydrogen (20–30 ml/min).

Compared to when using a carrier gas of relatively high density (e.g., nitrogen), gases of lower density such as helium or hydrogen can be used at considerably higher flow rates without much reduction of separation efficiency. This means a shorter analysis time. However, since nitrogen is inert and less expensive than the other gases mentioned, it is frequently used in connection with an FI or constant current EC detector. Hydrogen or helium are recommended when TC detection is being used or in cases when it is vital to shorten the analysis time as much as possible.

Peak identification can be tentatively accomplished from data on retention time. Preferably, comparisons should be made of relative retentions, with adjusted retention times of unknown solutes related to that of an appropriate internal standard (Section 2).

Several compounds have highly similar or identical retention times on a GC column, even if this is of capillary type. GC alone, therefore, does not give 100% accuracy in peak identification. For positive and unambiguous identification, mass spectrometry is the most sensitive method, particularly when it is used on-line with gas chromatography (Chapter 2). In many cases, comparisons of retention time data nevertheless can be useful for tentative identification. It is vital that chromatographic conditions are optimized to ensure maximum resolution. Addition of an authentic compound to an aliquot of the sample prior to injection into the column is advantageous. If the molecular structure of such a compound is identical with that of the unknown solute, its peak size should increase without appearance of extraneous peaks in the chromatogram.

Analysis of a sample on two columns having stationary phases of differing polarity can give additional information on the chemical structure of the unknown sample.

Quantification. The peak area is dependent upon the amount of the component that is injected. However, the *detector response* varies significantly according to the type of compound. The extreme selectivity of the EC detector in registering molecules containing an electron-attracting group in preference to compounds without such groups was mentioned in Section 5. The TC and FI detectors also respond differently to different types of molecules. The area of one peak, therefore, can be related only to the quantity of the corresponding compound and *not* to other solutes. *Measurements of peak areas* are preferably accomplished by using an electronic integrator. Other techniques include planimetry, weighing the paper of cut peaks, and triangulation. Measurement of peak heights instead of areas can be recommended only if the peak widths are of comparable size.

When *internal normalization* is used for quantification, possible differences in detector response to the various solutes are disregarded. Instead, the relative peak areas are assumed to express the proportions of corresponding components in the injected sample. This method assumes that all of the injected material is eluted from the column and registered as chromatographic peaks.

Internal standardization means that mixtures of known amounts of an internal standard and of the solute(s) are first analyzed. A calibration curve can then be constructed, with peak area ratios of solute to internal standard plotted against corresponding weight ratios. Addition of a known amount of the same internal standard to the actual sample will then give a chromatogram from which area ratios can be correlated to weight ratios, using the calibration curve. Being virtually independent of the reproducibility of the analysis, the internal standardization method is generally regarded as the most accurate means for quantification. The selected internal standard should have peak height and retention time close to these values in the solute(s) of interest.

In the *external standardization* method, peak areas representing known quantities of solutes are compared with the areas of the same solutes in the sample. In contrast to internal standardization, this method requires both highly reproducible sample injections and constant analytical conditions.

REFERENCES

Alltech Associate Chromatography Products Catalog, 1980, **35**:37.

Andrews, R. K., 1970, in *Introduction to Gas Chromatography*, Pye Unicam, Cambridge.

Bertsch, W., Jennings, W. G., and Kaiser, R. E. (eds.), 1981, *Recent Advances in Capillary Gas Chromatography*, Hütig Verlag, Heidelberg.

Craven, R. B., Brooks, J. B., Edman, D. C., Converse, J. D., Greenlee, J., Schlossberg, D., Furlow, T. L., Gwaltney, J. M., and Miner, W. F., 1977, Rapid diagnosis of lymphocytic meningitis by frequency-pulsed electron capture gas–liquid chromatography: Differentiation of tuberculous, cryptococcal and viral meningitis, *J. Clin. Microbiol.* **6**:27–32.

Drucker, D. B., 1976, Gas–liquid chromatographic chemotaxonomy, in *"Methods of Microbiology"* J. R. Norris, ed. **9**:52–125, Academic Press, London.

Golay, M. J. E., 1958, "Theory and Practice of Gas Liquid Partition Chromatography with Coated Capillaries," in *Gas Chromatography* (V. J. Coates, H. J. Noebles, and I. S. Fagerson, eds.) pp. 1–13, Academic Press, New York.

Grob, K., and Grob, J., Jr., 1981, "Splitless Injection and the Solvent Effect," in *Recent Advances in Capillary Gas Chromatography* (W. Bertsch, W. G. Jennings, and R. E. Kaiser, eds.) pp. 455–474, Hütig Verlag, Heidelberg.

Grob, R. L., 1977, in *Modern Practice of Gas Chromatography*, John Wiley and Sons, New York.

Jennings, W. G., 1980, in *Gas Chromatography with Glass Capillary Columns* (2nd ed), Academic Press, New York.

Jennings, W. G., 1981, *Comparisons of Fused Silica and Other Glass Columns in Gas Chromatography*, Hütig Verlag, Heidelberg.

Larsson, L., and Holst, E., 1982, Feasibility of automated head-space gas chromatography in identification of anaerobic bacteria, *Acta Pathol. Microbiol. Scand. Sect. B*, **90**:125.

Larsson, L., Mårdh, P.-A., and Odham, G., 1978, Analysis of amines and other bacterial products by head-space gas chromatography, *Acta Pathol. Microbiol. Scand. Sect B* **86**: 207–213.

Moss, C. W., 1981, Gas–liquid chromatography as an analytical tool in microbiology, *J. Chromatogr.* **203**; 337–347.

Moss, C. W., Dees, S. B., and Guerrant, G. O., 1980, Gas–liquid chromatography of bacterial fatty acids with a fused-silica capillary column, *J. Clin. Microbiol.* **12**; 127–130.

SGE Chromatography and Mass Spectrometry Products, 1980, p. 26.

Willard, H. H., Merritt, L. L., and Dean, J. A., 1970, "Gas Chromatography", in *Instrumental Methods of Analysis* (4th ed.), pp. 494–530, Van Nostrand Company, New York.

2

MASS SPECTROMETRY

Göran Odham, Laboratory of Ecological Chemistry,
University of Lund, Ecology Building,
Helgonavägen 5, S-223 62 Lund, Sweden
Lennart Larsson, Department of Medical Microbiology,
University of Lund, Sölvegatan 23, S-223 62
Lund, Sweden

1. INTRODUCTION

The basic principles of mass spectrometry (MS) have long been known. In 1912, for instance, Thomson used MS to demonstrate the existence of two neon isotopes. But it is only in the past two decades that MS has evolved into one of the most sensitive and reliable techniques for structural analysis of organic compounds. Progress in electronics, resulting in greater and more constant accuracy of instrumentation, together with proof that ionized organic compounds in the gaseous phase fragment uniquely according to their chemical structure, have led organic chemists to adopt and further develop the MS technique.

Some of the most important spectrometric improvements in recent years are sophisticated inlet systems employing capillary gas chromatography and high-pressure liquid chromatography on-line to the mass spectrometer, various methods of ionization, fast scanning of full spectra, heightened sensitivity, and computers specially designed for handling MS data. Modern MS systems have many uses. These include studies on the fragmentation of various types of ions, determination of the structure of organic molecules, qualitative and quantitative determination of components in mixtures, and turnover of stable isotopes, such as ^{13}C and ^{15}N in biological processes. In these and other fields, MS is commonly employed as an outstanding, and often a necessary, analytical tool.

In this chapter we do not propose to deal in detail with the various instrumental designs of mass spectrometers, or with the interpretation of mass spectra of different types of molecules. Such information is available in textbooks, among the more recent and comprehensive of which those by Budzikiewicz *et al.* (*Mass Spectrometry of Organic Compounds*, 1967) and Beynon *et al.* (*The Mass Spectra of Organic Molecules*, 1968) merit mention. Of particular value to the biologist are the collective volumes edited by Waller (*Biochemical Applications of Mass Spectrometry*, 1972) and Waller and Dermer (*Biochemical Applications of Mass Spectrometry*, first supplementary volume, 1980).

We hope that this introductory chapter on MS will assist understanding of the subsequent chapters and will give the reader some insight into the potentialities of MS in microbiological research.

2. METHODS OF IONIZATION

2.1. Electron Impact

The basic principle of MS is separation and registration of atomic masses. In the study of organic compounds, this implies ionization of the molecule prior to the analysis and recording of its behavior. The ionization procedure takes place in the *ion source* of the instrument and can be accomplished in several ways. In an electron impact (EI) ion source held at high vacuum ($\sim 10^{-7}$ Torr) the vaporized sample is bombarded with a beam of electrons (Figure 1), the energy of which can be varied from about 10 to 100 eV. The spectrometer is usually operated at 70 eV, a value which yields spectra of good reproducibility and high diagnostic value.

For an organic compound, M, the electron bombardment may lead to a number of different reactions in the ion source, reactions which can be illustrated as follows:

$$M + e \rightarrow M + e \tag{1}$$

$$M + e \rightarrow [M]^{\ddagger} + 2e \tag{2}$$

$$M + e \rightarrow [M]^{2+} + 3e \tag{3}$$

$$M + e \rightarrow [M]^{\bar{\cdot}} \tag{4}$$

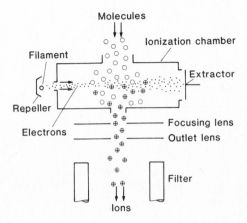

FIGURE 1. Mass spectrometer ion source (reprinted with permission from Nermag S.A., Paris, France).

Despite the high energy of the electrons, Eq. (1) represents the most common event in the ion source, i.e., the situation in which no ionization occurs. Thus, it has been estimated that only about 1 molecule per thousand is ionized. Most of the ions formed will be singly positively charged [Eq. (2)], while significant amounts of doubly charged ions [Eq. (3)] are observed only for a limited number of organic molecule types, such as fused aromatics, containing several rings. The negatively charged portion [Eq. (4)] will be very small. A noteworthy point is that singly charged molecule ions, being either positive or negative, contain an odd number of electrons and consequently are radicals.

The amount of energy required to remove an electron from M (the ionization potential) is typically in the range 7–20 eV. As mentioned, the electron energy used in MS with EI usually is considerably higher (70 eV), and this evokes further reactions in the ion source. If the sample pressure is kept low ($\sim 10^{-7}$ Torr) these reactions are predominantly unimolecular, while bimolecular (ion–molecule) or other collision reactions are negligible. Thus, molecular ion radicals of a wide energy range are primarily formed, of which those with low energy can subsequently be collected as M^{+} ions. If sufficiently excited, however, the M^{+} ions decompose to form a variety of charged and neutral species. The situation may be illustrated as follows:

$$ABCD \rightarrow e \rightarrow ABCD^{+} + 2e$$

$$ABCD^{+} \rightarrow A^{+} \rightarrow BCD^{.}$$

$$\rightarrow A^{\cdot} + BCD^{+}$$
$$\rightarrow AB^{+} + CD^{\cdot}$$
etc.

where ABCD symbolizes the general formula of an organic molecule.

Reactions involving only a simple cleavage of the molecular ion radical (ABCD‡), or a successive series of such simple cleavages, are often accompanied by reactions that involve cleavage of two bonds followed by rearrangements. The overall behavioral result of electron impact on the molecules in the ion source is conveniently illustrated by a graph (mass spectrum) in which the abscissa indicates the ratio of mass to charge (m/z) and the ordinate indicates the relative abundance of the ions produced.

As an example, the mass spectrum of the methyl ester of stearic acid obtained at electron energy 70 eV is illustrated in Figure 2 (Ryhage and Stenhagen, 1960). The molecular weight (molecular ion radicals) is indicated by the peak at m/z 298 (about 25%). The spectrum also shows a series of ions resulting from (a) fission α to the ester carbonyl group, (b) fission β to the carbonyl group, (c) fission at the hydrocarbon chain with charge retention on the oxygen-containing fragments, and (d) the same type of fission as in (c) but with charge retention on the hydrocarbon fragments.

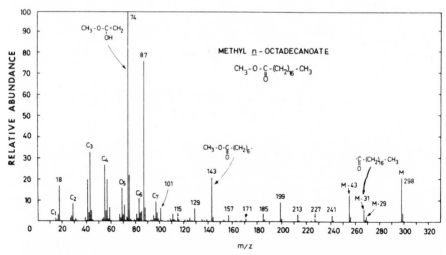

FIGURE 2. Mass spectrum of methyl n-octadecanoate (methyl stearate) 200°C, 70 eV EI (Ryhage and Stenhagen, 1960).

Fission of the molecule

$$R-\overset{\overset{\displaystyle O}{\|}}{C}-OCH_3$$

according to (a) can occur in four ways, resulting in the ions

$$R^+ \qquad R-C\equiv O^+ + O\equiv C-OCH_3 + OCH_3$$

m/z 239 $\qquad$ *m/z* 267 $\qquad$ *m/z* 59 $\qquad$ *m/z* 31

which, with the exception of R^+, are all readily recognizable in the spectrum.

The peak of highest abundance—the base peak (100%)—in the spectra of most methyl esters of fatty acids is found at *m/z* 74. The formation of the rearranged radical ion, the McLafferty (1963) rearrangement

involves a specific transfer of a γ-hydrogen atom in a six-membered transitional state to the carbonyl oxygen, with subsequent fission at the β-carbon atom (b).

Oxygen-containing ions (c) of the type $(CH_2)_n COOCH_3^+$ are found in the mass spectrum for all n (except $n = 1$). These ions, for example of *m/z* 87 ($n = 2$), *m/z* 101 ($n = 3$), *m/z* 115 ($n = 4$), etc., are separated by 14 atomic mass units. However, fragments corresponding to the maximum n (=M − 15) are often of extremely low abundance, and sometimes even absent.

The hydrocarbon-type ions (d) are of significant abundance only in the low-mass region. Such peaks are recognizable at *m/z* 43 (C_3) and *m/z* 57 (C_4), but they subsequently decay, becoming negligible at higher masses. The hydrocarbon fragments generally are of little diagnostic value in spectra of normal-chain methyl esters.

Details on the EI fragmentation patterns of amino acids, carbohydrates, fatty acids, and complex lipids are given in Chapters 3–5.

2.2. Chemical Ionization

By increasing the pressure in the ion source to approximately 1 Torr, for instance by introducing controlled amounts of small inorganic or organic molecules (reactant gas), the strictly unimolecular reactions characteristic of EI are supplemented with significant numbers of reactions of the ion–molecule collision type (chemical ionization, CI). The compound to be investigated, the concentration of which is only about 1/1000 that of the reactant gas, is therefore ionized by a set of reagent ions. The latter ions are formed from the reactant gas by a combination of unimolecular reactions (EI) and ion–molecule collisions. The reaction sequence postulated for methane (CH_4), a common reactant gas, may be as follows (Arsenault, 1972):

$$CH_4 + e \rightarrow CH_4^{+\cdot} + 2e$$

$$CH_4^{+\cdot} + CH_4 \rightarrow CH_5^{+} + CH_3^{\cdot}$$

$$CH_3^{+} + CH_4 \rightarrow C_2H_5^{+} + H_2$$

$$CH_3^{+} + 2CH_4 \rightarrow C_3H_7^{+} + 2H_2$$

$$CH_2^{+\cdot} + 2CH_4 \rightarrow C_3H_5^{+} + 2H_2 + H^{\cdot}$$

etc.

The mass spectrum of methane at a pressure of 1 Torr indicates CH_5^{+} and $C_2H_5^{+}$ in the respective proportions 48% and 40%, with considerably smaller amounts of ions of higher masses.

The reactant ions CH_5^{+} and $C_2H_5^{+}$ have strong acidic properties and can react with the analyte by proton transfer, hydride abstraction, and, to a minor extent, also by alkyl transfer. For the molecule BH these reactions may be expressed as follows:

$$BH + CH_5^{+} \rightarrow BH_2^{+} + CH_4 \quad \text{(protonation)}$$

$$BH + C_2H_5^{+} \rightarrow BH_2^{+} + C_2H_4 \quad \text{(protonation)}$$

$$BH + C_2H_5^{+} \rightarrow B^{+} + C_2H_6 \quad \text{(hydride abstraction)}$$

$$BH + CH_5^{+} \rightarrow BHCH_5^{+} \quad \text{(alkyl transfer)}$$

The molecule adducts formed in CI are even electron species, in contrast to the molecular ion radical produced by EI, and consequently their inherent stability is greater. Furthermore, the amount of energy transferred to the molecular ion is considerably less than the energy transfer by electrons at 70 eV. Because of these factors, the total ion current in CI is much smaller than in EI. In addition, the fragmentation of the molecule adduct usually is greatly reduced, so that the $(M + 1)^{+}$ or $(M - 1)^{+}$ ions

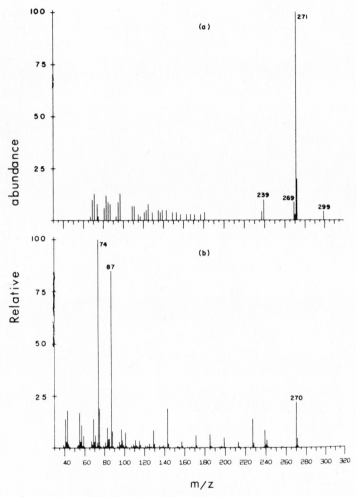

FIGURE 3. Mass spectrum of methyl palmitate. (a) Methane CI, 1 Torr, 180°C, 70 eV. (b) EI, 180°C, 70 eV. (Hertz *et al.*, 1971.)

frequently form the base peak in the spectrum. This is illustrated in Figure 3 (Hertz *et al.*, 1971), where the CI (CH₄) spectrum of methyl palmitate (a) is shown together with the corresponding EI spectrum (b). The base peak in the former spectrum is at m/z 271 $(=M+1)^+$. The fragments formed by the hydride abstraction $(=M-1)^+$ account for only about 10% of the base peak abundance, i.e., approximately the same as the proportion observed for the molecule adducts after alkyl transfer [m/z 299 ($=$M + 29)$^+$].

Numerous reactant gases have been successfully employed for CI in MS. They include methane, isobutane, ammonia, and water. The two hydrocarbons mainly transfer or abstract hydrogen, leading to (M + 1)$^+$ and/or (M − 1)$^+$ ion formation, whereas ammonia extensively leads to adducts of the (M + 18)$^+$ type. When water is used, intense (M + 1)$^+$ and (M + 19)$^+$ ions are usually observed. The results, however, depend on the chemical nature of the analyte.

CI MS is a convenient technique when there is doubt concerning the molecular weights of compounds studied with EI. In quantitative MS, moreover (Chapter 9), the possible gain in sensitivity and specificity when CI is used instead of EI for measuring adduct ions from the molecule is a factor to be considered. Disadvantages of CI include highly temperature-dependent spectra, which may result in poor reproducibility. Further, the relatively high pressure in the ion source used in CI means that contamination of the ion source tends to be more pronounced than in EI. High purity of reactant gases is an important prerequisite for reducing the risk of contamination and minimizing the background spectrum.

Several reviews on CI MS have been published, e.g., by Field (1968, 1972) and Arsenault (1972).

2.3. Field Ionization

Field ionization (FI) of a molecule occurs when the molecule interacts with a strong electric field (10^7–10^8 V/cm). Sufficiently high fields can be produced in sources where sharp blades, wires, or metal points form the anodes (Beckey 1971, 1972). This mode of ionization gives mass spectra showing high-intensity molecular ion radicals, even for substances that yield vanishingly small such ions when EI is used.

The FI mass spectrum of methyl oleate (Figure 4, Rohwedder 1971) shows molecular ions of m/z 296. Probably due to loss of CH_3OH, there are also abundant metastable ions (see Section 5.1).

In FI the samples are introduced near the ionization zone. The compound can also be applied directly to the field ion emitter, however, to become *field desorbed* (FD), provided that the emitter has a large surface area (Beckey, 1971). One advantage of FD is that no decomposition of the samples occurs when thermally unstable molecules are ionized. In FD only M$^{\pm}$ ions and/or (M + 1)$^+$ ions are formed, and consequently this technique is of particular value for determinations of molecular weight. FD requires a direct inlet system to the ion source (see Section 6.1).

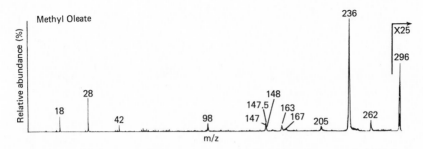

FIGURE 4. Mass spectrum (FI) of methyl oleate (Rohwedder, 1971).

2.4. Atmospheric Pressure Ionization

Atmospheric pressure ionization (API) was introduced by Horning *et al.* (1973, 1977). In this technique a carrier gas, usually N_2, is ionized by β-radiation from a ^{63}Ni source, equivalent to the principle utilized in the electron capture detector in gas chromatography (Chapter 1). Via a complex series of reactions, the carrier gas ions produce both positive and negative ions in high yield from the studied compound by charge and proton transfer. Although the ion source is maintained at atmospheric pressure, the mass spectrometer *per se* operates at low pressure (e.g., 10^{-6} Torr). A characteristic of the API mass spectrometer is its extreme sensitivity, particularly for electron-attracting compounds when recording only ions of selected m/z values (femtogram range).

2.5. ^{252}Cf Plasma Desorption

This technique (PD) was introduced in 1974 by Torgerson *et al.* It utilizes the nuclear fission fragments from ^{252}Cf decay to produce fast evaporation and ionization of nonvolatile compounds from a surface. The sample is deposited on a Ni support foil in front ot the ^{252}Cf source. When the fission fragments penetrate the Ni foil, large quantities of energy are deposited in a very small area. This process leads to formation and desorption of ions by a complex mechanism, mainly ion–molecule reactions or ion-pair formation. Both positive and negative ions are generated (MacFarlane and Torgerson, 1976). The PDMS technique so far has not been extensively used, but its ability to analyze high-molecular-weight material of biological origin is noteworthy.

2.6. Secondary Ion Mass Spectrometry

The recently introduced method of ionization (SIMS) is very useful, e.g., for analyses of nonvolatile and thermally labile compounds such as amino acids, peptides, vitamins, and nucleotides (Benninghoven and Sichtermann, 1977, 1978; Eicke *et al.*, 1980). A 2–5-keV argon primary ion beam is allowed to bombard a layer of the sample, which usually is deposited on a silver foil. The bombardment results in the emission of secondary sample ions (positive and negative) from the surface of the sample, due to the high deposit of energy from the primary ions at a small site.

The unique features of SIMS will almost certainly ensure frequent application of the method for future analytical work on biological samples. The probabilities in tissue samples are of particular interest, in view of the successful detection of muscarine in fungal tissue (Figure 5, Day *et al.*, 1980). In another experiment, these authors introduced a paper chromatogram into the SIMS spectrometer and the primary ion beam scanned along the axis of the chromatogram. The neurotransmitter acetylcholine could then be directly identified by its SIMS spectrum.

The high ion transmission of the quadrupole mass filters (see Section 4.2) makes them particularly suitable in molecular SIMS work.

2.7. Fast Atom Bombardment

Fast atom bombardment (FAB) is similar to SIMS, but uses a fast Ar atom beam instead of an Ar^+ beam to ionize the sample (Barber *et al.*, 1981a,b). This Ar atom beam is obtained by resonant charge exchange of Ar^+ ions in a collision chamber which contains Ar gas at high pressure. The Ar atom beam is then aimed at the sample deposited on a copper sample stage which is fitted onto a direct probe (Barber *et al.*, 1981a).

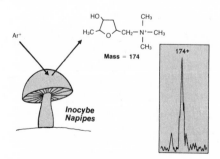

FIGURE 5. Direct sampling of a fungus by SIMS showing the presence of the quarternary alkaloid muscarine (Day *et al.*, 1980).

FAB produces relatively abundant $(M + H)^+$ ions in the positive ion spectra, and $(M - H)^-$ ions in the negative ion spectra (see Section 3.2). In addition to several structurally significant fragment ions, metastable ions (Section 3.1) can be seen. FAB thus gives not only the molecular weight, but also structural information on analyzed samples.

Impressive results have been obtained with FAB in the study of nonvolatile and thermolabile compounds of biochemical interest. These include a peptide containing 26 amino acid units (MW = 2845), vitamin B_{12} and its coenzyme, oligosaccharides, peptide antibiotics, glycopeptides, glycoside antibiotics, glycolipids, oligonucleotides, and neurotoxins (Barber, 1982).

The API, PD, SIMS, and FAB ionization methods are less widely used than the EI and CI methods at the present time. The sensitivity and scope of these newer ionization methods in analysis of high-molecular-weight biological samples nevertheless will ensure at least some of them an important part in future use of MS.

3. FURTHER ASPECTS OF ION FORMATION

3.1. Metastable Ions

The generated ions spend on average about 10^{-6} s in the ion source before they reach the transportation zone (mass filter). The ions then typically require 10^{-5} s to traverse the mass filter of magnetic instruments before reaching the detector. Ions with half-life sufficiently above these values will eventually be recorded as molecular ions, whereas those with half-life less than 10^{-6} s decompose in the ion source to form more stable units, which are subsequently accelerated (fragment ions). Metastable transitions are decompositions which occur during transportation of ions from source to detector, i.e., with half-life between 10^{-5} and 10^{-6} s. Metastable ions as a rule are easily recognized in the mass spectrum as diffuse peaks, often several mass units wide and usually centered at nonintegral mass numbers.

A galvanometer recording of the mass spectrum of isopropyl alcohol (Figure 6; McLafferty, 1973) indicates metastable peaks at m/z 16.2, 18.7, and 24.1.

The mass of a metastable ion m^x is dependent upon both the mass of the precursor (m_1) and the product ions (m_2) and obeys the equation

$$m^x = m_2^2/m_1$$

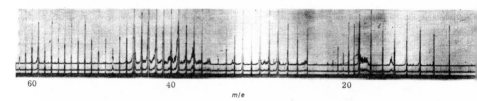

60   40   20

m/e

FIGURE 6. Galvanometer recording of the mass spectrum of isopropyl alcohol (EI) showing metastable peaks (McLafferty, 1973).

The mass spectrum often shows peaks corresponding to those precursor ions that did not decompose when traversing the mass filter and peaks from the final ions that decomposed in the ion source. The equation mentioned can be used to determine the reaction giving rise to the metastable ion. This can be done by utilizing intense characteristic peaks in the spectra as possible candidates for m_1 and m_2 and logical mass differences. Metastable ion nomographs (e.g., McLafferty, 1973) are convenient aids in finding possible values for m_1 and m_2. The uses of metastable ions in MS include elucidation of ion decomposition pathways, identification of molecular rearrangements, and determination of elemental compositions.

3.2. Negative Ions

As mentioned in Section 2.1, the bombarding electrons (EI) in the ion source may interact with the molecules to yield negatively charged ions (NI). Three fundamental types of reactions may occur in this process:
1. Nondissociative attachment:

$$AB + e \rightarrow AB^{\cdot-}$$

2. Dissociative attachment:

$$AB + e \rightarrow A + B^{\cdot-}$$
$$\rightarrow B^{\cdot-} + A$$

3. Ion-pair dissociation:

$$AB + e \rightarrow A^+ + B^- + e$$
$$\rightarrow A^- + B^+ + e$$

Ion formation according to any of these reaction types is primarily dependent on the energy of the bombarding electrons. Nondissociative and dissociative attachment predominate at energies <10 eV, while ion-pair dissociation is more frequent at higher energies.

The fundamental principles relating to the formation of negative molecular ions at low electron energies have been extensively studied by Christophorou (1976). All compounds yielding an abundance of such ions have the structural feature of conjugation, thus promoting electron capture into Π-orbitals of the molecule. Christophorou *et al.* (1973) suggested the term *electrophores* for functional groups possessing electron capture qualities in NI MS, typical exponents being nitro, cyano, carbonyl, and acid functions. Their findings indicate the type of derivatives that may promote extensive formation of molecular negative ions.

Investigations have focused also on the fragmentation process and on possibilities for predicting cleavage of negative molecular ions at the standardized EI conditions (70 eV, 10^{-7} Torr) of the ion source. Examples are the mass spectra of the p-nitrophenyl esters of normal saturated C_{12}, C_{14}, and C_{18} acids obtained by NI MS (Bowie and Stapleton, 1975). All these spectra show the p-nitrophenoxide anion $(p\text{-}O_2NC_6H_4O)^-$ as base peak and molecular anions of 30% to 70% of the base peak. In general, the NI mass spectra of such esters exhibit peaks of little diagnostic value.

Dougherty *et al.* (1972) reported on NI MS involving CI. If the pressure in the ion source is increased significantly, low-energy secondary electrons are produced via ion-pair dissociations [Eq. (3)]. These electrons may generate negative molecular ions in the presence of reactant gases. Numerous reactant gases have been used for this purpose, including methane, isobutane, methylene chloride, oxygen, and mixtures of N_2O and hydrogen. Spectra obtained with methane as reactant gas show prominent $M^{\bar{}}$ ions, whereas $(M + Cl)^-$ are characteristic for methylene chloride spectra (Tannenbaum *et al.*, 1975). Spectra produced in mixtures of N_2O and H_2, using OH^- as reacting species, have been recorded for a number of organic compounds (Smith and Field, 1977). The investigated compounds included carboxylic acids, amino acids, alcohols, thiols, ketones, esters, ethers, and amines. Most of these compound types produced abundant $(M - 1)^-$ ions.

No general statement is possible in regard to the sensitivity attainable in NICI as compared with conventional positive CI. In experiments using methane as reactant gas, aromatics containing fluorine, such as pentafluorobenzonitrile, have been reported to yield NICI spectra when the quantity of aromatic was at least two orders of magnitude less than that required for positive ion MS (Hunt *et al.*, 1976). Further research should provide information on the usefulness of NIMS for identifying and quantifying, for instance, compounds of biological origin.

4. SEPARATION OF IONS

4.1. Magnetic Instruments

When the beam of ions, with mass-to-charge ratio m/z, has been accelerated through a slit in the ion source, it is deflected by passing a magnetic field H according to $m/z = (H^2 \cdot r^2)/2V$, where H is the strength of the magnetic field, r the curvature of the pathway (instrumentally dictated), and V the voltage used to accelerate the ions. By continuous scanning of either the magnetic field or the accelerating voltage, ions of different m/z values are successively focused on the exit slit of the ion separator and subsequently measured (Figure 7). Magnetic scanning is used in most instruments because satisfactory linearity is technically easier to achieve in these systems than in accelerating voltage types, which require compensation for a discrimination against the ions of high m/z.

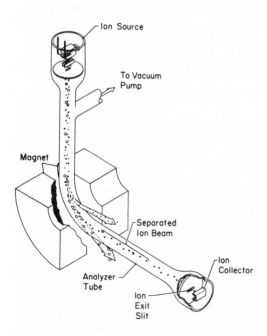

FIGURE 7. Single-focusing, magnetic mass spectrometer (McLafferty, 1973).

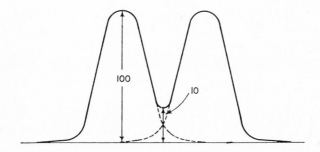

FIGURE 8. Resolved adjacent peaks according to the 10% valley definition.

The magnetic field acts only as a direction-focusing force. It disregards the distribution of initial kinetic energy in the generated ions. If this distribution is not compensated for, there will be some peak broadening at the exit, leading eventually to partial or complete overlapping of peaks corresponding to ions of similar masses.

The resolving power R, defined by $R = M/\Delta M$, where M is the mean m/z value of two adjacent ions and ΔM is the difference between the m/z values of these ions, is a measure of this overlap. Figure 8 illustrates the situation in which two ion intensities, presumed to be equal, have an overlap corresponding to 10% of the maximum intensities ("valley definition"). A practically attainable resolution around 1000, using separation to a 10% valley, is typical for magnetic instruments.

The velocity dispersion characterized by single-focusing magnetic instruments of the types described may be decreased by allowing the ion beam to traverse a radial electric field. This passage acts as a focusing device for direction and velocity before the beam enters the magnetic field (double focusing). Both types of field thus are used to bring the ions to a focal point (Figure 9). A resolving power of at least 10,000 (10% valley

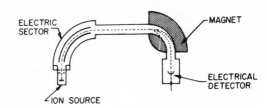

FIGURE 9. Schematic of a high-resolution mass spectrometer.

definition) is easily obtainable under scanning conditions for such an instrument. The high resolution implies possibilities of accomplishing mass determinations with an accuracy of a few ppm.

In double-focusing instruments the resolving power can be adjusted by changing the widths on either side of the mass filter unit. This implies that the sensitivity of the mass spectrometer is inversely proportional to the resolution in the observations. To achieve optimum sensitivity in the MS analysis, therefore, the selected resolution should not be higher than that necessary to solve the particular analytical problem.

4.2. Quadrupole Instruments

These nonmagnetic mass spectrometers employ as mass filter four parallel rods arranged to lie in the respective corners of an imagined square (Figure 10). Diagonally opposite rods are connected to radiofrequency (rf) and direct current (dc) voltages, one pair of rods being 180° out of phase of the other. At a given rf/dc level, only ions of a specific m/z value will avoid collision with one of the rods and will thus pass the filter along a highly complicated pathway. The complete mass spectrum is obtained by scanning the voltages from zero to their maximum values at a constant rf/dc ratio.

Several attractive features characterize the quadrupole type of mass filter. Thus the mass scale/time function is linear, which enables easy computerized handling. Rapid scanning of spectra—less than a second—and ability to maintain quality output at relatively high pressures in the ion

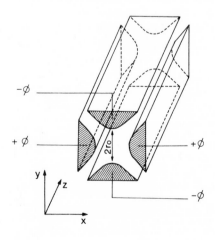

FIGURE 10. Spatial arrangement of rods in a quadrupole mass filter (reprinted with permission from Nermag S.A., Paris, France).

source are other advantages. The reason for this output maintenance is that the filter does not act on ion beams controlled by physical slits, but directs tuned masses through a large cross section to the detector.

Quadrupole mass spectrometers can work only with low resolution, resolving powers of approximately 1000 being typical. They can analyze ions of mass numbers up to about 1800.

The tolerance of relatively high pressures in the ion source makes quadrupole instruments particularly attractive in CI MS. The easy switching of the rod voltages to focus on specific m/z values makes the "quadrupoles" ideal for selected ion monitoring recordings (Chapter 9).

4.3. Time-of-Flight Instruments

The operating principle of the time-of-flight (TOF) instrument involves measurement of the time taken by an ion to travel from its source to the detector in a field-free region (Wiley, 1956; Gohlke, 1962). A burst of ions is initially produced in the source by a pulsed electron beam. Since the accelerated ions have equal kinetic energy, those of small mass will reach the detector first, followed by the heavier ions. Characteristic of TOF mass spectrometers is extreme speed of generating and recording mass spectra. The operative cycle takes about 100 μsec, thus theoretically permitting production of about 10,000 spectra per second.

4.4. Fourier Transform/Ion Cyclotron Resonance Instruments

The Fourier transform/ion cyclotron resonance (FT/ICR) instrument, which was developed simultaneously by Comisarow and Marshall, and McIver and Hunter, is a direct descendant of the cyclotron particle separators used in high-energy physics (Comisarow and Marshall, 1974; Hunter and McIver, 1977a,b; Marshall *et al.*, 1979). The ions are formed by a short electron beam pulse (EI) and trapped by strong magnetic-electrostatic fields in an analyzer cell. All the ions thereby undergo cyclotron motion, the frequency of which is mass dependent. In order to detect the ions, the radii of their cyclotron orbits are expanded by subjecting the ions to an alternating electric field. When the frequency of this field is the same as that of the cyclotrons, the ions are accelerated to a progressively enlarging radius of gyration (excitation). Fast Fourier transformation is used to convert image currents of the excited ions to specific frequency ranges. The location of each peak in the mass spectrum is characteristic of the frequency, and hence the mass, of the corresponding ion. The height of the peaks indicates the abundance of the respective ions.

Several characteristics of the FT/ICR mass spectrometer merit mention. The ion formation and the mass analysis take place in the same region. Unlike other types of mass spectrometers, therefore, the ions do not leave their source. Consequently, the source can be made gas-tight with minimum loss of ions in the vacuum system. Theoretically this should permit heightened sensitivity.

Negative ions (see Section 3.2) show the same behavior as positive ions, except that their orbits are in contrary direction. The negative ions are detected by use of a negative trap voltage. Controlled alternation of this voltage permits recording of "simultaneous" negative-ion–positive-ion spectra.

CI FT/ICR mass spectrometry is a very recent development (Ghaderi *et al.*, 1981). It is based on collisions between reactant gas ions in orbit (before excitation) and the sample ions. Despite the low pressure (approximately 10^{-8} Torr) as compared with the 0.1–1 Torr generally used in conventional CI MS, the spectra obtained with the two methods are comparable.

In many cases it is possible to use fragment ions of the sample molecule itself as reagent ions. This mode of operation, in which the need for a conventional reagent gas thus is eliminated, is called 'self-CI." Figure 11 shows a self-CI FI/ICR mass spectrum of methyl stearate employing a 200-ms reaction time between ionization and excitation (Nicolet Analytical Instruments preliminary bulletin, ASMS Conference, 1981).

With the commercially available instrument, measurements can be made in the mass range 1–1000. In principle, however, further development

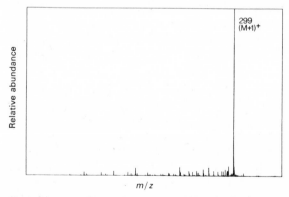

FIGURE 11. Mass spectrum of methyl stearate, self-CI, 200 ms reaction time between ionization and excitation (reprinted with permission from Nicolet Instruments, Madison, Wisconsin).

should be possible, permitting the instrument to handle a much higher molecular weight range, of the order 6000–8000. FI/ICR mass spectrometry should then be of particular interest for the study of compounds of biological origin.

5. ION DETECTION

5.1. Electron Multipliers

Most modern mass spectrometers use electron multipliers to register the ions and amplify their signals after passage through the mass filter. The electron multiplier utilizes the principle of secondary electron emission to effect rapid amplification (Figure 12). In the multiplier used for recording positive ions, the first stage consists of a conversion dynode, at which the positive ion beam is converted into an equivalent electron current. These secondary electrons are accelerated and focused on the second dynode, which ejects more electrons than those striking it (cascade effect). Electron multipliers may utilize up to 20 dynodes to yield gains in the electron current of approximately 10^6. The final dynode is connected to a conventional amplifier.

Certain disadvantages of the electron multiplier should be mentioned. One is a certain degree of mass discrimination, since the secondary electron emission at the conversion dynode depends to a minor extent on the mass, energy, and nature of the bombarding ions. This means that ions of high m/z values produce somewhat fewer secondary electrons than do ions of low m/z values. Further, the electron multipliers have a finite lifespan, the length of which varies from instrument to instrument. Some multipliers thus may lose a gain factor of 10^2 after only six months of use, while others can maintain high performance for several years.

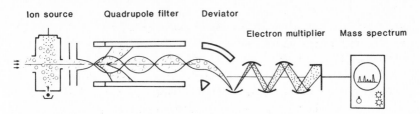

FIGURE 12. Schematic of quadrupole mass spectrometer including electron multiplier (reprinted with permission from Nermag, S.A., Paris, France).

5.2. Photographic Detection

Photographic plates are sometimes used in high-resolution MS to obtain precise mass measurements. The photoplate typically collects about 30 high-resolution mass spectra in rows, each being recorded in a few seconds. The spectra, recorded as lines on the plate, may be processed by densitometers and computers to produce highly accurate mass measurements, provided that reference compounds are incorporated in the sample. A drawback in quantitative work is that blackening of the photoemulsion cannot easily be correlated with the number of striking ions.

6. INTRODUCTION OF SAMPLE

6.1. Direct Inlet

Modern mass spectrometers usually provide several means of introducing the sample into the ion source. The simplest is the direct inlet probe technique. It utilizes a small glass or gold crucible (volume $\simeq 100$ μl) attached to a metal rod. The rod may be advanced through an airtight seal connected to a vacuum lock so that the opening of the crucible closely approaches the bombarding electron beam. The inlet probe is designed for accurate, rapid heating, by which a small quantity (μg or less) of sample can be vaporized and introduced into the beam of electrons. Precisely constructed direct inlet systems permit controlled, fractionated distillations in the source, and thus consecutive analysis of components in a mixture.

6.2. Gas Chromatography

For the interpretation of mass spectra of unknown components, purity of the analyte is imperative. Gas chromatography (GC) of volatilizable compounds usually is the optimum analytical technique to ensure purity. On-line combination of a gas chromatograph with the mass spectrometer (GC/MS), however, has fundamental complications in regard to pressure. The ion source usually operates at pressures of about 10^{-7} Torr in EI MS, whereas the pressure at the outlet of the GC column is equal to, or even higher than the atmospheric pressure. Furthermore, the sample in GC is diluted in carrier gas to a concentration of perhaps a few parts per thousand. When using packed columns in GC/MS (Chapter 1), where the carrier-gas flow is about 30 ml/min, molecular separators are necessary to concentrate the sample prior to its entry into the ion source.

Several types of molecular separators have been designed. Two of the most commonly used are the Ryhage–Stenhagen (Ryhage, 1964) type (Figure 13a) and the Watson–Biemann separator (Watson and Biemann, 1964, Figure 13b). The former utilizes two opposed separation jet assemblies in series, each in a separately pumped compartment of moderate and high vacuum, respectively. The GC effluent traverses these compartments on its way to the ion source. At least 90% of the carrier gas is estimated to be removable already in the first assembly. At a carrier-gas flow of approximately 30 ml/min, the approximate sample loss is 16% in the first compartment and 40% in the second (Ryhage, 1967).

The Watson–Biemann separator consists of a porous glass tube mantled by a second glass tube, with an intervening vacuum. The GC effluent enters through a constrictor which reduces the pressure in the porous glass tube (pore diameter $\approx 1\,\mu m$). A molecular flow through the pores preferentially of carrier gas, with the heavier molecules passing through the inner tube to the ion source, constitutes the principle for function. The efficiency (ratio of sample leaving to sample entering the separator) can reach about 30%.

A capillary column coupled on-line to the mass spectrometer presents less difficulty, since the flow-rate of the carrier gas is much lower—approximately 1 ml/min. Such columns can be introduced into the ion source without the aid of molecular separators. However, the direct coupling implies that a certain length of the capillary is kept under subatmospheric pressure, and this can negatively influence the GC resolution.

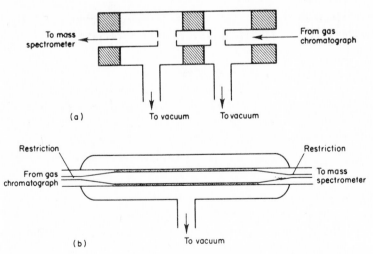

FIGURE 13. Molecular separators: (a) The Ryhage–Stenhagen separator; (b) the Watson–Biemann separator (Ryhage, 1964; Watson and Biemann, 1964).

The continuous loss of stationary phase that occurs from GC columns—bleeding—may cause complications in GC/MS. This problem can be circumvented by automatic background subtraction, using a computerized data acquisition system. Manual manipulations of this type are extremely tedious. Use of "low-load" packed or capillary columns with stationary phase of high thermal stability is advisable.

6.3. Liquid Chromatography

Much effort has been devoted to on-line coupling of liquid chromatographs (LC) to MS. As compared with GC, LC has the advantage of permitting separation of low-volatility compounds, thus reducing the need for derivatization. The main technical difficulty in designing on-line LC/MS systems is associated with the large volume of solvent that accompanies the sample. Although two differing systems have become commercially available, LC/MS is still in its developing stage. In one of these systems a heated, moving belt is used to introduce the sample into the ion source, and in the other a fraction of the LC effluent is introduced through a diaphragm, with the evaporated solvent functioning as reactant gas (CI mode). These systems are presented in more detail in Chapter 9. The moving belt interface is the more versatile of the two systems, in that it permits EI or CI MS with any reactant gas. In many instances, however, collection of LC fractions and evaporation of solvent followed by direct-inlet introduction of the sample into the ion source is a highly convenient procedure.

6.4. Reservoir

For calibration of the mass scale, a separate reservoir containing a compound such as perfluorokerosene (PFK) is frequently connected to the ion source. The pressure in the reservoir, being much higher than that in the ion source, forces the sample at a constant rate through a molecular leak in the source. PFK can be used for calibration up to mass numbers of about 800.

7. USE OF COMPUTERS

7.1. Computer Control of Low-Resolution Mass Spectrometers and Data Acquisition

Computers have become almost essential for full productivity from mass spectrometers, particularly GC/MS systems. In principle, mass spectrometers can be interfaced either to large computers also servicing other instruments in the laboratory, or to a small computer exclusive to the MS. Both models are currently employed but, partly because of the falling cost of computer hardware, MS systems with integrated minicomputers seem to be increasing, as compared with the larger systems.

A typical on-line mass spectrometer–computer system (MS/DA) includes interface electronics, disk storage of data, and a visual display unit for presentation of mass spectra and chromatograms, in addition to facilities for making permanent copies of these graphs. The main features are illustrated in Figure 14.

Computer control in MS was first applied to quadrupole instruments, since these present fewer technical problems than the magnetic instruments. The computer is programmed to control the quadrupole rods by generating the voltages required to focus each mass in sequence on the electron multiplier. The correct voltages are automatically obtained from a calibration run, using for instance PFK (cf. Section 6.4). Various functions such as background subtraction in GC/MS (for column bleed and instrumental contamination) and integration of chromatographic peaks are usually included in the computer's software facilities.

Computerized control of the voltages to focus only on a limited, selected number of ions, instead of scanning complete mass spectra, is another valuable and easily achievable function with quadrupole instruments. A consequence of this selected ion monitoring (SIM) is that the sensitivity for the ions of interest may be manifoldly increased (Chapter 9). For example, up to eight preselected ions can be used in one set, and also up to 13 consecutive sets during a single chromatographic run.

Computer-controlled scanning over a narrow mass range (e.g., 10 amu) constitutes an alternative to SIM for increasing sensitivity in the detection of selected components in mixtures. However, as in the recording of complete spectra, time is spent on numerous ions, and the sensitivity usually is inferior to that obtained with SIM.

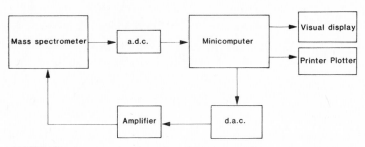

FIGURE 14. Computer configuration of a low-resolution mass spectrometer.

Technical problems with respect to the stability of the accelerating voltage alternator complicate the use of computer control for magnetic mass spectrometers. Nevertheless, computers have been used for continuous addition or subtraction of a small offset voltage in relation to the accelerating voltage, thus giving computer control also of magnetic instruments.

7.2. High-Resolution Data Acquisition

If only integral mass values are required (low resolution), continuous comparison of masses from the sample and the reference compound (PFK) is not required. Periodic calibration of the mass scale, for example daily, usually suffices. In high-resolution acquisition of data, however, scans must be made on a mixture of PFK and the analyte. The voltage output from the mass spectrometer (above a preset threshold value) is sampled at a fast rate (e.g., every 50 μs). In one method the computer then calculates the center of gravity of the peaks and lists the times and intensities of all the peaks in the mixed spectrum of analyte and PFK. Comparison with the corresponding PFK spectrum defines the mass spectrum of the analyte. The computer compares the times of the reference peaks and the sample peaks along the spectrum and converts these times to accurate masses. It is essential, however, that the computer knows the times of the first peaks in the PFK reference spectrum. The accuracy of the mass determination by this method can exceed ±10 ppm.

7.3. Repetitive Scanning

Modern mass spectrometers, of magnetic or of quadrupole type, are capable of rapid repetitive scanning of spectra. Typical values for quadrupole intruments are about 1 s for a complete cycle. The introduction of data acquisition systems having large storage capacity has made feasible the recording, in GC/MS or LC/MS, of a complete chromatogram by repetitive scanning. The technique, originally reported by Hites and Biemann (1971), is often described as "mass chromatography." The chromatogram is visualized in real time and thus represents, as a function of time, the sum of the ion intensities in the chosen scan range.

Ions of selected m/z can be extracted from the stored scan data. This facility, often known as EICP (extracted ion current profile), is particularly useful in analyses of complex mixtures. Figure 15 illustrates a mass chromatogram and EICP tracings of the methyl esters of free and bound fatty acids in cells of *Staphylococcus aureus*. It should be emphasized that EICP is merely an assortment of ion intensities, so that the achieved

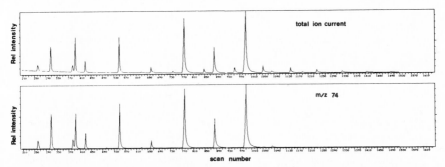

FIGURE 15. Mass chromatogram of methyl esters of free and bound fatty acids in cells of *Staphylococcus aureus*. Top tracing: total ion current. Bottom tracing: EICP at *m/z* 74, characteristic for methyl esters of saturated fatty acids (Odham and Larsson, unpublished).

sensitivity is no greater than that given by the repetitive scanning procedure. In SIM, on the other hand, the duty cycle of the mass spectrometer is maximized. The quantitative aspects of mass chromatography and EICP are dealt with in Chapter 9.

Fast scanning is particularly valuable in capillary GC/MS. With slower scans there is a risk that rapidly eluting GC peaks will be recorded as a single component.

7.4. Library Searching

The external memory of the minicomputers used in MS can accommodate data representing several thousand low-resolution mass spectra. In recent years, many research groups have assembled individual collections of such spectra. With the aim of creating a standardized mass spectrum library, a number of organizations, including the Mass Spectrometry Data Centre (UK), the Environmental Protection Agency, the National Institutes of Health, and the National Bureau of Standards (USA), agreed to cooperate in a definitive data bank. After extraction of "cleaned-up" spectra from various collections, such a bank now exists. It contains some 40,000 mass spectra. The bank is routinely offered by manufacturers of mass spectrometers, and accompanying software provides facilities for computer search in the library to match up with unknown spectra.

To save space in the external memory, the original spectra are condensed in a fashion that permits retention of their characteristic features. Various methods have been described for such abbreviation of mass spectra, and several algorithms have been formulated for the matching procedure.

Biemann and co-workers developed a system in which the spectra are abbreviated by taking the two most intense peaks in each 14 unit mass interval from m/z 9 and upwards (Hertz *et al.*, 1970; 1971). Computerized reduction of an unknown mass spectrum followed by matching leads to a printout of the compounds with the best agreement, according to a similarity index.

Other matching systems have been described, e.g., by Abramson (1973) and Clerc *et al.* (1973). The matching can be improved by adding functional group characteristics of the compound. These characteristics can be supplied, for instance, by nuclear magnetic resonance (NMR) and/or infrared (ir) spectroscopy.

REFERENCES

Abramson, F. P., 1973, American Society for Mass Spectrometry, 21st Annual Conference on Mass Spectrometry and Allied Topics, San Francisco, California, Proceedings, p. 76.

Arsenault, G. P., 1972, "Chemical Ionization Mass Spectrometry," in *Biochemical Applications of Mass Spectroscopy* (G. R. Waller, ed.), p. 817, Wiley-Interscience, New York.

Barber, M., 1982, *Mass Spectrom. Rev.* **1**. in press.

Barber, M., Bordoli, R. S., Sedgwick, R. D., and Tyler, A. N., 1981a, Fast atom bombardment of solids (FAB): A new ion source for mass spectrometry, *J. Chem. Soc. Chem. Commun.* 325.

Barber, M., Bordoli, R. S., Sedgwick, R. D., and Tetler, L. W., 1981b, Fast atom bombardment mass spectrometry of two isomeric tripeptides, *Org. Mass. Spectrom.* **16**:256.

Beckey, H. D., 1971, *Field Ionization Mass Spectrometry*, Pergamon Press, Oxford.

Beckey, H. D., 1972, "Determination of the Structures of Organic Molecules and Quantitative Analyses with the Field Ionization Mass Spectrometer," in *Biochemical Applications of Mass Spectrometry* (G. R. Waller, ed.), p. 725, Wiley-Interscience, New York.

Benninghoven, A., and Sichtermann, W. K., 1977, Secondary ion mass spectrometry. A new analytical technique for biologically important compounds, *Org. Mass. Spectrom.* **12**:595.

Benninghoven, A., and Sichtermann, W. K., 1978, Detection, identification and structural investigation of biologically important compounds by secondary ion mass spectrometry, *Anal. Chem.* **50**:1180.

Beynon, J. H., Saunders, R. A., and Williams, A. E., 1968, *The Mass Spectra of Organic Molecules*, Elsevier Publishing Company, Amsterdam.

Bowie, J. H., and Stapleton, B. J., 1975, Electron impact studies. XCVI Negative ion mass spectra of naturally occurring compounds. Nitrophenyl esters derived from long-chain acids and alcohols, *Austr. J. Chem.* **28**:1011.

Budzikiewicz, H., Djerassi, C., and Williams, D. H., 1967, *Mass Spectrometry of Organic Compounds*, Holden-Day, San Franscisco.

Christophorou, L. G., 1976, Electron attachment to molecules in dense gases ("quasi-liquids") *Chem. Rev.* **76**:409.

Christophorou, L. G., Hadjiantoniou, A., and Carter, J. G., 1973, Long-lived parent negative ions formed via nuclear-excited feshback resonances, *J. Chem. Soc. Faraday Trans. II* **69**:1704.

Clerc, J. T., Erni, F., Jost, C., Meili, T., Nageli, D., and Schwarzenbach, R., 1973, Computerunterstützte Spektreninterpretation zur Strukturaufklärung organischer Verbindungen, *Z. Anal. Chem.* **264**:192.

Comisarow, M. B., and Marshall, A. G., 1974, Fourier transform ion cyclotron resonance spectroscopy, *Chem. Phys. Lett.* **25**:282.

Day, R. J., Unger, S. E., and Cooks, R. G., 1980, Molecular secondary ion mass spectrometry, *Anal. Chem.* **52**:557A.

Dougherty, R. C., Dalton, J., and Biros, F. J., 1972, Negative chemical ionization mass spectra of polycyclic chlorinated insecticides, *Org. Mass. Spectrom.* **6**:1171.

Eicke, A., Sichtermann, W. K., Benninghoven, A., 1980, Secondary ion mass spectrometry of nucleic acid components: Pyrimidines purines, nucleosides, and nucleotides, *Org. Mass. Spectrom.* **15**:289.

Field, F. H., 1968, Chemical ionization mass spectrometry, *Acc. Chem. Res.* **1**:42.

Field, F. H., 1972, "Chemical Ionization Mass Spectrometry," in *Mass Spectrometry*, Vol. 5 (A. Maccoll, ed.), Butterworths, London.

Ghaderi, S., Kulkarni, P. S., Ledford, E. B., Wilkins, C. L., and Gross, M. L., 1981, Chemical ionization in Fourier transform mass spectrometry, *Anal. Chem.* **53**:428.

Gohlke, R. S., 1962, Time-of-flight mass spectrometry: Application to capillary column gas chromatography, *Anal. Chem.* **34**:1332.

Hertz, H. S., Evans, D. A., and Biemann, K., 1970, User-oriented computer-searchable library of mass spectrometric literature references, *Org. Mass Spectrom.* **4**:452.

Hertz, H. S., Hites, R. A., and Biemann, K., 1971, Identification of mass spectra by computer-searching a file of known spectra, *Anal. Chem.* **43**:681.

Hites, R. A., and Biemann, K., 1971, Computer evaluation of continuously scanned mass spectra of gas chromatographic effluents, *Anal. Chem.* **42**:855.

Horning, E. C., Horning, M. G., Carrol, D. I., Dzidic, I., and Stillwell, R. N., 1973, New picogram detection system based on a mass spectrometer with an external ionization source at atmospheric pressure, *Anal. Chem.* **45**:936.

Horning, E. C., Carrol, D. I., Dzidic, I., Haegele, K. D., Lin, S.-N., Oertli, C. U., and Stillwell, R. N., 1977, Development and use of analytical systems based on mass spectrometry, *Clin. Chem.* **23**:13.

Hunt, D. F., Stafford, G. C., Crow, F. W., and Russell, J. W., 1976, Pulsed positive negative ion chemical ionization mass spectrometry, *Anal. Chem.* **48**:2098.

Hunter, R. L., and McIver, R. T., Jr., 1977a, Conceptual and experimental basis for rapid scan ion cyclotron resonance spectroscopy, *J. Chem. Phys. Lett.* **49**:577.

Hunter, R. L., and McIver, R. T., Jr., 1977b, Rapid scan ion cyclotron resonance spectroscopy, *Am. Lab.* **9**:13.

Macfarlane, R. D., and Torgerson, D. F., 1976, Californium-252 plasma desorption mass spectrometry, *Science* **191**:920.

McLafferty, F. W., 1963, "Decompositions and Rearrangements of Organic Ions," in *Mass Spectrometry of Organic Ions* (F. W. McLafferty, ed.), p. 309, Academic Press, New York.

McLafferty, F. W., 1973, *Interpretation of Mass Spectra*, W. A. Benjamin, Inc., Reading Massachusetts.

Marshall, A. G., Comisarow, M. B., and Panisod, G., 1979, Relaxation and spectral line shape in Fourier transform ion resonance spectroscopy, *J. Chem. Phys.* **71**:4434.

Nicolet Fourier transform mass spectrometer. A preliminary bulletin prepared for the 1981 ASMS Conference, Nicolet Analytical Instruments, Madison, Wisconsin.

Rohwedder, W. K., 1971, Field ionization mass spectrometry of long chain fatty methyl esters, *Lipids* **6**:906.

Ryhage, R., 1964, Use of a mass spectrometer as a detector and analyzer for effluents emerging from high-temperature gas liquid chromatography columns, *Anal. Chem.* **36**:759.

Ryhage, R., 1967, Efficiency of molecule separators used in gas chromatograph–mass spectrometer applications, *Arkiv Kemi* **26**:305.

Ryhage, R., and Stenhagen, E., 1960, Mass spectrometry in lipid research, *J. Lipid. Res.* **1**:361.

Smith, A. L. C., and Field, F. H., 1977, Gaseous anion chemistry. Formation and reactions of OH⁻, Reactions of anions with N₂O˙, OH⁻ negative chemical ionization, *J. Am. Chem. Soc.* **99**:6471.

Tannenbaum, H. P., Roberts, J. D., and Dougherty, R. C., 1975, Negative chemical ionization mass spectrometry—Chloride attachment spectra. *Anal. Chem.* **47**:49.

Torgerson, D. F., Skowrouski, R. P., and Macfarlane, R. D., 1974, New approach to the mass spectrometry of nonvolatile compounds, *Biochem. Biophys. Res. Commun.* **60**:616.

Waller, G. R., ed., 1972, *Biochemical Applications of Mass Spectrometry*, Wiley-Interscience, New York.

Waller, G. R., and Dermer, O. C., eds., 1980, *Biochemical applications of Mass Spectrometry,"* First supplementary volume. Wiley-Interscience, New York.

Watson, J. T., and Biemann, K., 1964, High-resolution mass spectra of compounds emerging from a gas chromatograph. *Anal. Chem.* **36**:1135.

Wiley, W. C., 1956, Bendix time-of-flight mass spectrometer. *Science* **124**:817.

PART II

ANALYSIS OF
MICROBIAL CONSTITUENTS

3

FATTY ACIDS AND COMPLEX LIPIDS

Cécile Asselineau and Jean Asselineau, Centre de
Recherche de Biochimie et Génétique Cellulaires
du CNRS, 118 route de Narbonne, 31062
Toulouse Cedex, France

Fatty acids are the basic components of the lipids of all microorganisms, except the group of Archaebacteria. Frequently unusual fatty acids are detected in bacterial lipids. Due to the easy volatilization of their methyl esters, fatty acids are conveniently analyzed by gas chromatography GC, but precise assignment of structure to a fatty acid detected as a peak in GC investigations is not always an easy task, and mass spectrometry MS has an important role to play. Much work has been performed on the analysis of fatty acids by GC/MS, mainly in the field of bacterial lipids, so the part devoted to fatty acids in this review is much larger than that concerning unvolatile complex lipids.

1. FATTY ACIDS

GC and MS, working separately or on tandem, have provided a major contribution to structure determination of the fatty acids obtained after saponification of the lipids of microorganisms. At the present time, many fatty acids are known, so that, when standard samples are available, the use of GC can allow the identification of every component of a mixture of fatty acids, if the resolving power of the apparatus under the conditions chosen for the analysis is high enough. However, mass spectrometry is a helpful and sometimes indispensable method for choosing between hypothetical structures compatible with the results of the GC analysis. MS is indispensable in the case of unusual fatty acids with new structural features, and allows their structure determination by using a minimum amount of sample. The sequential use of GC and MS is often due to material reasons, such as the lack of suitable apparatus.

At first we shall consider the kind of information that can be expected on the structures of fatty acids, from the use of GC and MS, working independently or in line.

1.1. Preparation of the Samples

Many works have been performed by direct transesterification of fatty acids from the complex lipids in the cells with methanol, followed by GC/MS analysis of the resulting fatty acid methyl esters.

Transesterification in alkaline medium (sodium methanolate) can produce artefacts by epimerization when the mixture to be analyzed contains α-substituted fatty acids. Transesterification catalyzed by boron trihalides more or less destroys cyclopropane fatty acids; unsaturated fatty acids can also be partially modified. Moreover, the recovery of some fatty acids, particularly hydroxy acids, is less than with classical saponification (Moss and Dees, 1975). For these reasons, saponification has to be preferred. It must be noted that some very long chain fatty acids can be partly extracted with ether because of the dissociation of their potassium salts. The presence of nonacidic lipid components in the mixture to be analyzed by GC can be an advantage (Asselineau et al., 1979).

Methylation of the free fatty acids can be performed by using an ether solution of diazomethane. It must be noted that α,β-unsaturated fatty esters can add diazomethane on their double bond to give pyrrazolines (Eistert, 1948).

The methylation step can be performed by addition of a few milliliters of commercial boron-trichloride-methanol reagent (Marshall et al., 1970). This procedure gives artefacts as mentioned above. Practical details on the preparation of samples for GC/MS can be found in Chapter 1.

If the amount of sample is large enough, preliminary chromatographic fractioning on a microcolumn or by thin-layer chromatography can be useful to separate polar and non-polar fatty esters. In every case, the mixture of methyl esters is dissolved in about 0.1 ml of ether or hexane, and analyzed as soon as possible. When necessary the sample solutions are kept in a freezer (about −10°C) to avoid any alteration of the unsaturated components.

1.2. Structural Information Given by GC Analysis

If the amount of sample is so low that it cannot be further fractionated, gas chromatography can be used to get information on the functional groups present in the various components of the mixture. Comparative studies by means of a pair of chromatographic columns, one filled with a polar phase (such as DEGS), and the other one with a nonpolar phase (such as SE 30) have to be performed. Such studies can be deepened by examining the chromatographic behavior of suitably prepared derivatives (Jamieson, 1970).

Usually the relative amounts of the mixture components (evaluated by the peak areas) do not exhibit important changes with the nature of the stationary phases, so the relative peak sizes may be used to correlate the various peaks found on the pair of chromatograms. The relative retention times of the components of the mixture are rather similar on both kinds of columns when these components have the same functional groups (for example saturated fatty acids, branched chain, or normal chain fatty acids).

Straight chain fatty esters are easily separated on polar or nonpolar phases from their methyl-branched isomers while separation of isomeric methyl-branched chain fatty esters require high separating powers, obtained by use of capillary columns. Retention times depend on the position on the chain of the methyl branch: isomeric esters having their methyl branch located in the middle of the chain have similar retention times and cannot be separated (Kuksis, 1978). (Such separations have been performed in the case of hydrocarbons; see page 79.) Methyl esters of 2-methyl-, 16-methyl-, 17-methyl- and 10-methyl-octadecanoic acids can be separated, whereas 9- and 10-methyl-octadecanoates cannot. Branched chain esters have a shorter retention time than that of the isomeric straight chain esters. The more numerous the methyl branches, the shorter the retention time. For example the methyl ester of a tetramethylhexadecanoate (C_{20}) is eluted between methyl n-heptadecanoate and n-octadecanoate on polar or nonpolar phases (Kuksis, 1978).

When an unknown mixture of fatty acid esters is chromatographed in parallel on nonpolar and polar columns, the two chromatograms obtained are similar if the polarities of all the components of the mixture are identical (for example, mixture of saturated fatty acid esters, mixture of hydroxyacid esters). If only some components of the mixture carry polar groups, they can be detected on comparison of the two chromatograms by the shift of their relative retention times. One double bond slightly decreases the retention time of an unsaturated fatty ester with respect to that of the analogous saturated ester on a nonpolar column, whereas its retention time is increased and becomes larger than that of the analogous saturated ester on a polar column. An α,β-*trans*-unsaturated ester (artifact often observed when β-hydroxy esters are present) exhibits a retention time higher than that of the analogous saturated ester, even on a nonpolar column. Confirmation of the presence of unsaturated esters in a mixture can be easily obtained by looking for the changes occurring in the peaks of the chromatogram, either after catalytic microhydrogenation, or after treatment by a bromine solution (James, 1960; Stein *et al.*, 1967). Separation of isomeric unsaturated fatty esters according to the position of the double bond (except for α,β-unsaturation) requires chromatographic conditions giving high separating powers such as those obtained by capillary gas chromatography (Panos, 1965). An accumulative effect is observed when several double bonds are present.

With respect to a saturated fatty ester used as a standard, hydroxy fatty esters show an important increase of retention time from a nonpolar to a polar column. The fact that the retention time is altered after treatment of the hydroxy ester by reagents specific for the alcohol group confirms the presence of a hydroxy ester.

1.3. Structural Information Given by Mass Spectrometry

Methyl esters are the derivatives of fatty acids most often used for mass spectrometry. Mass spectra of normal long-chain saturated methyl esters are dominated by peaks due to ions containing oxygen, corresponding to $(CH_2)_n$—$COOCH_3^+$ (m/z 87, 101, 115, etc.). In the high-mass region, the molecular peak, usually well defined, is accompanied by a peak at M-31, arising from the loss of a methoxyl group and formation of an acylium ion:

$$CH_3-(CH_2)_n-C(OCH_3)=\overset{\cdot}{O}{}^+ \rightarrow CH_3-(CH_2)_n-C\equiv O^+$$
$$\text{M} \qquad\qquad\qquad\qquad \text{M-31}$$

A peak at M-43 is formed by a more complex intramolecular rearrangement followed by the loss of a propyl radical ($C_3H_7 = 43$).

The base peak of these methyl esters is observed at m/z 74; it is due to ions formed on 2–3 cleavage with simultaneous migration of one hydrogen atom from the lost fragment. This cleavage is known as the McLafferty rearrangement. It is observed in the case of all fatty acid methyl esters having one hydrogen atom on carbon 4. If a branch R occurs at position 2, this peak is observed at m/z 73 + R: in the case of a methyl branch, the peak is at m/z 88 (73 + 15) [formula (**1**)].

When one or several branches occur on the chain, preferential cleavages of the molecular ion next to one of the tertiary carbon atoms are observed; the positive charge is kept by the fragment-ion with the highest relative stability. A competition between such fragmentation and the McLafferty rearrangement may occur, so that the peak at m/z 74 may not be the base peak of the spectrum. Then the occurrence of a branch on the chain gives rise to a higher number of peaks, compared with the spectrum of a normal chain ester containing a series of peaks differing by 14 mass units with decreasing intensities from the peak at m/z 74 to the high mass region. The location of the branch on the carbon chain can be deduced from the presence of high-intensity peaks due to ions arising from cleavages next to the tertiary carbon atom.

The presence of unsaturation is directly shown by the molecular peak value. But the mass spectrum gives no information on the location of a double bond, except for a double bond in the α,β-position to the methoxy-carbonyl group (Ryhage *et al.*, 1961). Electron impact provokes migration of the double bond along the carbon chain, and the isomerization is faster than fragmentation in the α-position to the double bond. The mass spectra of isomeric unsaturated fatty esters are quite similar. When a branch occurs on the chain, the double bond preferentially remains on the tertiary carbon atom and a mass spectrum is obtained which is characteristic of the position of the branch, but not of the double bond. Mass spectra of unsaturated fatty esters differ from those of saturated esters by the fact that the peak at m/z 74 is not the base peak any longer and that the peak at M-31 is accompanied by a peak at M-32 (sometimes of higher intensity) arising from a loss of methanol.

Therefore, location of a double bond on the carbon chain requires its chemical modification in nonisomerizing conditions, by addition of groups able to induce specific fragmentations. A detailed review of this problem has recently been published (Minninkin, 1978).

The simplest and most reliable method of double bond location is based on the successive transformation of the monoenoic ester into a dihydroxy ester and a bis-trimethylsilyloxy derivative. The hydroxylation of the double bond can be performed as follows: about 5 mg of fatty acid methyl esters are dissolved in 2 ml of dioxane-pyridine (8:1 v/v) in a centrifuge tube. A small crystal of OsO_4 is added and the mixture is left in the dark at room temperature for 2 h, with occasional shaking. Gaseous H_2S is then bubbled through the orange-red solution until saturation; the excess H_2S is displaced by bubbling nitrogen for 1 h. The black precipitate of osmium salts is removed by centrifugation. The supernatant is collected in a second centrifuge tube. The solvents are evaporated *in vacuo*. To the residue, dissolved in 8 drops of dry pyridine, are successively added 4 drops

SCHEME 1

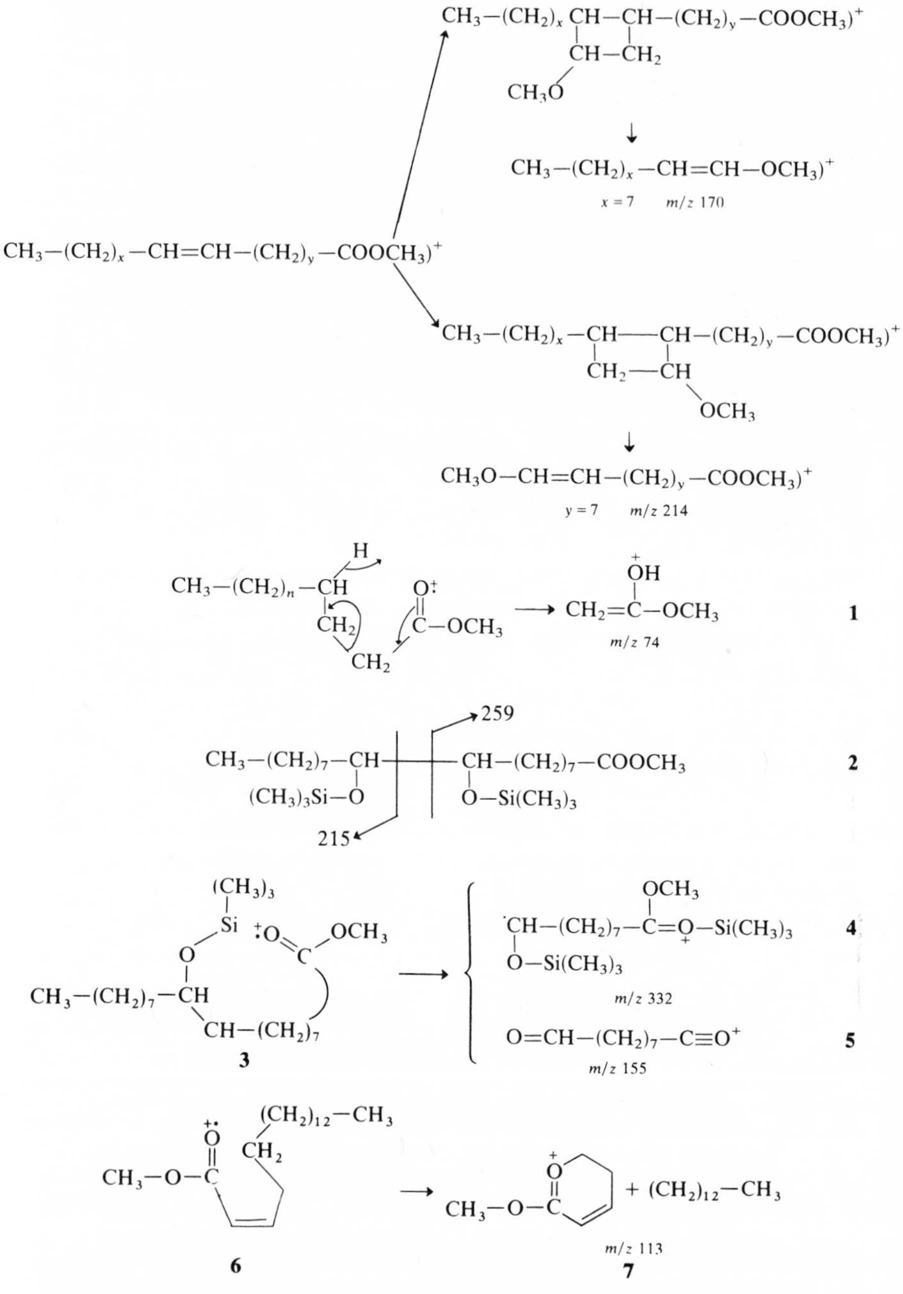

$$CH_3-(CH_2)_x \underset{\underset{CH_3\overset{|}{O}}{\overset{|}{CH}-\overset{|}{CH_2}}}{\overset{|}{CH}-\overset{|}{CH}}-(CH_2)_y-COOCH_3)^+$$

↓

$$CH_3-(CH_2)_x-CH=CH-OCH_3)^+$$

$x=7 \qquad m/z\ 170$

$$CH_3-(CH_2)_x-CH=CH-(CH_2)_y-COOCH_3)^+$$

$$CH_3-(CH_2)_x-\underset{\underset{OCH_3}{\overset{|}{CH_2}-\overset{|}{CH}}}{\overset{|}{CH}}-CH-(CH_2)_y-COOCH_3)^+$$

↓

$$CH_3O-CH=CH-(CH_2)_y-COOCH_3)^+$$

$y=7 \qquad m/z\ 214$

$$CH_3-(CH_2)_n-\underset{\underset{CH_2}{\overset{|}{CH_2}}}{\overset{H}{\overset{\curvearrowright}{CH}}} \quad \overset{O^+}{\underset{\curvearrowleft}{\overset{\|}{C}}}-OCH_3 \longrightarrow CH_2=\overset{\overset{OH}{|}}{C}-OCH_3 \qquad \mathbf{1}$$

$m/z\ 74$

$$\overset{259}{\nearrow}$$

$$\underset{\underset{215}{\nwarrow}}{\overset{}{CH_3-(CH_2)_7-CH}}\,\,\Big|\,\,\underset{O-Si(CH_3)_3}{\overset{}{CH}}-(CH_2)_7-COOCH_3 \qquad \mathbf{2}$$

(CH_3)_3Si-O

$$\begin{Bmatrix} \overset{\cdot}{CH}-(CH_2)_7-\overset{\overset{OCH_3}{|}}{C}=\underset{+}{O}-Si(CH_3)_3 & \mathbf{4} \\ O-Si(CH_3)_3 & \\ m/z\ 332 & \\ O=CH-(CH_2)_7-C\equiv O^+ & \mathbf{5} \\ m/z\ 155 & \end{Bmatrix}$$

$$\longrightarrow$$

$$\mathbf{3}$$

$$CH_3-O-\overset{\overset{\cdot\cdot}{O}}{\underset{}{C}}\underset{CH-(CH_2)_7}{} \quad \overset{(CH_3)_3}{\underset{}{Si}} \quad :O=\overset{}{\underset{}{C}}-OCH_3$$

$$CH_3-O-\overset{\overset{\cdot\cdot}{O}}{\underset{}{C}}\underset{\overset{}{CH_2}}{\overset{(CH_2)_{12}-CH_3}{}} \longrightarrow CH_3-O-\overset{\overset{+}{O}}{\underset{}{C}} + (CH_2)_{12}-CH_3$$

$m/z\ 113$

$\mathbf{6}$ $\qquad\qquad\qquad\qquad \mathbf{7}$

of hexamethyldisilazane and 2 drops of chloromethylsilane. After a few minutes at room temperature, the mixture is centrifuged; the supernatant contains the bis-trimethylsilyloxy derivatives.

The mass spectrum of the TMS derivatives exhibits intense peaks due to cleavage between the trimethylsilyl ether functions. For example, in the case of methyl Δ^9-octadecenoate, peaks at m/z 215 and 259 are formed. Parasitic peaks at m/z 332 and 155 are also observed, due to rearrangement by cyclization of the molecule [(2)–(5)].

Trimethylsilyloxy derivatives of unsaturated fatty esters have been analyzed by gas chromatography–chemical ionization mass spectrometry, isobutane being the reagent gas (Murata *et al.*, 1978). In the case of methyl oleate, the base peak is at m/z 385, corresponding to elimination of one trimethylsilanol; the peaks at m/z 215 and 259 allow the location of the double bond at position 9,10. Small differences of peak intensities might be used to determine the double bond geometry (*cis* or *trans*).

The mass spectra of methyl *cis*- and *trans*-2-octadecenoate show a characteristic peak at m/z 113 (with a high intensity in the case of the *cis* isomer) (Ryhage *et al.*, 1961), which might be due to a cyclization reaction [(6)–(7)].

Another approach has been used for double bond location, based on the replacement of the ester group by a group that is ionized in preference to the double bond. Tertiary amides, in particular pyrrolidides, can be used for this purpose. In this case, fragmentations in the α position to the double bond can be observed. The use of pyrrolidides has been developed by Andersson and Holman (1974; Andersson, 1978).

In the mass spectrum of N-octadec-9-enoyl-pyrrolidine, the base peak is due to the usual McLafferty rearrangement ion at m/z 113. The molecular ion is at m/z 335. Between these two peaks are a set of cluster peaks, each cluster being centered on a main peak. The main peaks of these clusters (at m/z 320, 306, 292, etc.) are regularly spaced by 14 mass units, except in the vicinity of the double bond where the interval is 12 mass units (observed between m/z 196 and 208 in the present example). These ions correspond to the fragments containing the 8th and 9th carbon atoms of the acyl chain, respectively. When such a couple of two peaks (differing by only 12 mass units) is found in the mass spectrum, the second value (9 in this example) gives the first carbon atom bearing the double bond: the double bond is thus located in position 9, 10 (see Figure 1).

These results gave rise to the following statement:

> if an interval of 12 atomic mass units instead of the regular 14, is observed between the most intense peaks of the clusters of fragments containing n and $n - 1$ carbon atoms of the acid moiety, a double bond occurs between carbons n and $n + 1$ in the molecule (Andersson, 1978).

Pyrrolidides are prepared as follows: 10 µl of methyl esters are dissolved in 1 ml of freshly distilled pyrrolidine and 0.1 ml of acetic acid. The solution is heated in a sealed tube for half an hour. The mixture is diluted with methylene chloride and the solution washed with dilute hydrochloric acid and with water. After drying on sodium sulfate, the solution is evaporated *in vacuo* (Andersson and Holman, 1974).

Location of a double bond in olefinic compounds can be directly performed, without using chemically modified derivatives, by chemical ionization mass spectrometry. The use of the reagent gas system 75% N_2/20% CS_2/5% vinyl methyl ether gives spectra containing easily identified ions characteristic of the location of the double bond, as well as abundant molecular ions. It is assumed that there is formation of intermediary methoxycyclobutyl ions in the spectrometer; they would give rise to the ions shown at the top of Scheme 1 (Chal and Harrison, 1981).

The location of a cyclopropane ring cannot be directly deduced from the mass spectrum of a fatty acid ester, but requires the use of amide derivatives such as pyrrolidides (see page 74). Because of the instability of the ring under electron impact, the mass spectrum of a cyclopropane fatty

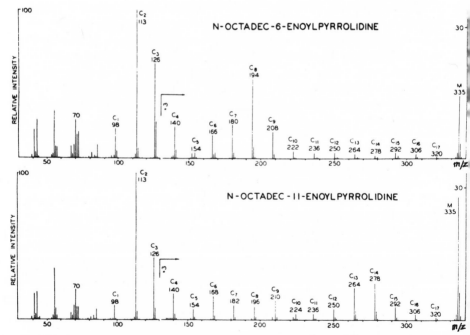

FIGURE 1. Mass spectra of the pyrrolidide derivatives of Δ^6- and Δ^{11}-octadecenoic acids (reproduced from *Lipids*, by permission of the American Oil Chemists' Society).

acid methyl ester is similar to that of the corresponding monoenoic ester with the same molecular weight. Gas chromatography allows a better distinction than MS between cyclopropane fatty esters and isomeric unsaturated fatty esters. A cyclopropane ring can even be located by capillary gas chromatography (Panos and Henrickson, 1968). The presence of a cis-cyclopropane ring in a fatty ester can be easily detected by NMR spectroscopy by looking for bands at 9.4 and 10.3 τ in the spectrum.

The mass spectrometry of the pyrrolidides has been studied, but the interpretation of the spectra does not seem to be as easy as in the case of enoic esters. Three main solutions have been proposed to locate a cyclopropane ring on the carbon skeleton of a fatty ester.

Catalytic hydrogenation of cyclopropane compounds (8) leads to a mixture of a straight chain ester (9) (10%–20%) and two isomeric methyl-branched esters [(10) and (11)] (80%–90%). GC/MS analysis of the mixture of methyl-branched esters, after the splitting of the molecule on both sides of the methyl branch in the mass spectrometer, allows the cyclopropane to be located (McCloskey and Law, 1967). For example methyl 11,12-methylene-octadecanoate is characterized, after hydrogenolysis, by peaks at m/z 185, 186, 187; 199, 200, 201, and 213 and 227 [(10) and (11)].

The hydrogenolysis reaction is performed on about 100 μg of ester in 0.2 ml of acetic acid, by using 2–3 mg of Adam's catalyst (PtO$_2$). The mixture is shaken in a suitable apparatus under a hydrogen pressure of 20 psi, at room temperature, for 20 h. After dilution of the mixture with ether, the reduced catalyst is filtered off and the solvents evaporated in vacuo. The product is ready to be analyzed by GC/MS.

Another approach has been used by Minnikin and Polgar (1967a, b, Minnikin, 1972): to yield three couples of positional isomeric methoxy-esters (Scheme 2) and, as by-products, a complex mixture of unsaturated esters. The mass spectrum of the mixture exhibits six intense peaks arising from cleavages adjacent to the oxygenated functions [(13), (14); (15), (16); (17), (18)].

The third approach relies on the fact that chromium trioxide oxidation of a cyclopropane derivative transforms one methylene group adjacent to the cyclopropane ring into a keto group (Promé and Asselineau, 1966). The oxidation is performed by using a solution containing 4% of CrO$_3$ in a mixture of acetic acid–water (9 : 1, v/v), at 20°C for 90 h. The two isomeric α-keto-cyclopropane esters (arising from the oxidation of a cyclopropane fatty acid methyl ester) can usually be separated by thin-layer chromatography and studied by mass spectrometry (Promé, 1968). It should be noted that the α-keto-cyclopropane derivatives undergo partial decomposition by gas chromatography. Figure 2 shows the application of this method

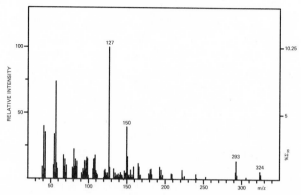

FIGURE 2. Mass spectrum of an α-keto-cyclopropane fatty acid methyl ester, prepared by chromic acid oxidation of methyl 9,10-methyleneoctadecanoate (Promé, 1968) (reproduced by permission of the *Bulletin de la Société Chimique de France*).

to the structure determination of methyl 9,10–methylene-octadecanoate. Intense peaks corresponding to the ketone fragment formed by splitting the molecule between the keto group and the cyclopropane ring are observed in the mass spectrum. The mixture of the two isomeric α-keto derivatives can be directly used for mass spectrometry.

Oxygenated functions in the chain of a fatty ester induce molecule cleavages adjacent to the carbon atom bearing this function. Fragmentations of the molecule are higher in the case of a methoxyl group than in that of a hydroxyl group, and higher in the case of a keto group than in that of an ether group. The abundance of the specific ions arising from such cleavages can vary considerably according to the position of the oxygenated group on the chain. Rearrangements with hydrogen transfers can be observed in the case of hydroxy esters. Keto groups, like the methoxycarbonyl groups, induce McLafferty rearrangements.

1.4. GC/MS Analysis of Microbial Fatty Acids

As many GC/MS analyses of microbial fatty acids (mainly bacterial) have been performed, only characteristic examples will be quoted here.

1.4.1. Iso and Anteiso Fatty Acids

Most often bacterial branched chain fatty acids are *iso* and *anteiso* fatty acids. The demonstration of such a structure by mass spectrometry is a difficult problem (Ryhage and Stenhagen, 1960b; Campbell and Naworal, 1969a; McCloskey, 1970; Kaneda, 1977). Mass spectra of

SCHEME 2

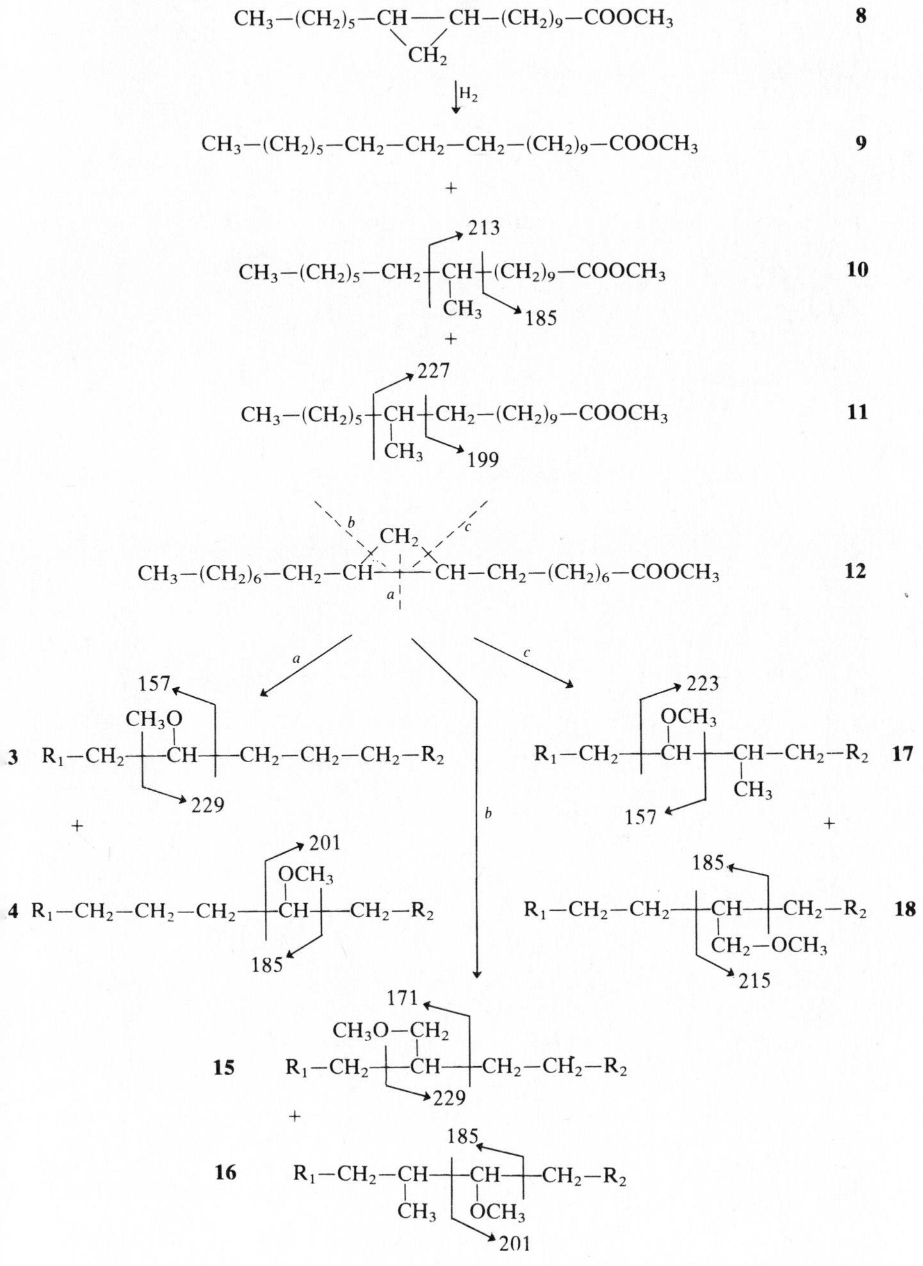

$CH_3-(CH_2)_5-CH\!\!-\!\!-\!\!CH-(CH_2)_9-COOCH_3$ **8**
                  CH_2

$\downarrow H_2$

$CH_3-(CH_2)_5-CH_2-CH_2-CH_2-(CH_2)_9-COOCH_3$ **9**

$+$

            213

$CH_3-(CH_2)_5-CH_2\!\!\vdash\!\!CH\!\!\vdash\!\!(CH_2)_9-COOCH_3$ **10**
                    CH_3 185

$+$

       227

$CH_3-(CH_2)_5\!\!\vdash\!\!CH\!\!\vdash\!\!CH_2-(CH_2)_9-COOCH_3$ **11**
             CH_3 199

           b CH_2 *c*

$CH_3-(CH_2)_6-CH_2-CH\!\!-\!\!-\!\!CH-CH_2-(CH_2)_6-COOCH_3$ **12**
                     a

     a   *c*

157   223
 CH_3O OCH_3

3 $R_1-CH_2\!\!\vdash\!\!CH\!\!\vdash\!\!CH_2-CH_2-CH_2-R_2$ $R_1-CH_2\!\!\vdash\!\!CH\!\!\vdash\!\!CH-CH_2-R_2$ **17**
             229   CH_3
                                    157

 $+$ $+$

             201   185
             OCH_3

4 $R_1-CH_2-CH_2-CH_2\!\!\vdash\!\!CH\!\!\vdash\!\!CH_2-R_2$ $R_1-CH_2-CH_2\!\!\vdash\!\!CH\!\!\vdash\!\!CH_2-R_2$ **18**
      185   CH_2-OCH_3
                                    215

             b

          171
        CH_3O-CH_2

15 $R_1-CH_2\!\!\vdash\!\!CH\!\!\vdash\!\!CH_2-CH_2-R_2$
            229

 $+$

       185
16 $R_1-CH_2-CH\!\!\vdash\!\!CH\!\!\vdash\!\!CH_2-R_2$
          CH_3 OCH_3
                 201

anteiso methyl esters exhibit a peak at M-57 [elimination of $CH(CH_3)-CH_2-CH_3$], but a similar peak is observed in the spectrum of ω-5-methyl-branched esters (due to the elimination of $CH_2-CH_2-CH_2-CH_3$), and at M-29 (elimination of CH_2-CH_3). The *anteiso* structure is suggested by the unusual relative intensities of peaks M-29 > M-31 (see for instance Heefner and Claus, 1978).

Spectra of *iso* esters exhibit a peak at M-43 [elimination of $CH(CH_3)-CH_3$], but such a peak is found in the spectrum of any kind of fatty acid methyl ester, sometimes with a higher intensity (Ryhage and Stenhagen, 1960a,b). A peak of very low intensity at M-65 [due to M-$(CH_3 + CH_3OH + H_2O)$] seems to be more significant (McCloskey, 1970). Cambell and Naworal (1969a) compare the relative intensities of peaks at M-29, M-31, and M-43 (measured at 17 eV): normal chain fatty esters give the respective values (16 to 28)/(32 to 60)/100; *iso* esters (10 to 20)/(16 to 36)/100; *anteiso* esters (32 to 58)/(16 to 38)/100.

Karlsson *et al.* (1973) proposed the transformation of the methyl esters to be analyzed into the methyl ethers of the corresponding fatty alcohols (by reduction with $LiAlH_4$, followed by methylation with CH_3I and Ag_2O): intense peaks at m/z 56 [$CH_3-CH(CH_3)-CH_2 - 1H$] and m/z 70 [$CH_3-CH_2-CH(CH_3)-CH_2 - 1H$] are observed in the mass spectra, as well as peaks at M-(32 + 28) and M-(32 + 56) for *iso* esters, and M-(32 + 29) and M-(32 + 57) for *anteiso* esters (better seen on mass spectra recorded at low voltage: 22.5 eV). The use of pyrrolidide derivatives of fatty acids may also be beneficial (Andersson, 1978).

The most reliable information is obtained from the elution migration rate of the *iso* and *anteiso* methyl esters on a capillary column gas chromatograph.

1.4.2. α-Methyl-Branched Fatty Acids

The pioneer work of Ryhage and Stenhagen (1960a) has shown that the mass spectra of methyl 2-methylalkanoates exhibit an intense peak at m/z 88, due to $CH_3-CH=C(OH)OCH_3$ [instead of m/z 74: $CH_2=C(OH)OCH_3$], through a McLafferty rearrangement. When the methyl branch is in position 3, the usual peak at m/z 74 constitutes the base peak of the spectrum, and a fragment at m/z 101 [$CH(CH_3)-CH_2-COOCH_3$] is observed, with a lower intensity. Thus, characterization of a branch (methyl or a longer branch) in the α position to the methoxycarbonyl group is quite easy.

When a second methyl branch is present in position 4 (2,4-dimethyl-alkanoates), the peak at m/z 88 is still a strong one, and two important peaks at m/z 101 and 129 are observed, arising from cleavage of the bonds on both sides of the tertiary carbon atom 4: these latter peaks are characteristic of the presence of a methyl branch in position 4. A third methyl branch at position 6 gives rise to a supplementary set of peaks at m/z 143 and 171 (cleavage of $C_5—C_6$ and $C_6—C_7$ bonds). The intensities of these couples of peaks decrease when supplementary methyl branches are introduced further along the chain: this is the case for C_{32}-mycocerosic acid, a metabolite of *Mycobacterium tuberculosis*, which is 2D,4D,6D,8D-tetramethyloctoacosanoic acid (Asselineau *et al.*, 1957, 1959; Odham *et al.*, 1970) (see Fig. 3). A more complex similar system of methyl branches, occurring in phthioceranic acids, has been studied by mass spectrometry by Goren *et al.* (1971).

By using these characteristic peaks, Julak *et al.* (1980) identified 2-methyltetradecanoic acid in the fatty acids of *M. gordonae* and 2,4-dimethyltetradecanoic acid in those of *M. kansasii*. A single peak was observed in the gas chromatogram of the latter fatty ester; by comparison with the chromatographic behavior of synthetic diastereoisomeric products, the presence of a single diastereoisomer was demonstrated.

The α,β-unsaturated fatty acids having three methyl branches at positions 2,4,6 are found in the lipids of virulent strains of *M. tuberculosis* (Asselineau, 1966). The main representatives of this group of acids are C_{25}- and C_{27}-phthienoic acids. Mass spectra of their methyl esters exhibit high-intensity peaks at m/z 88 and 101, showing that a double bond shift occurs before cleavage of the $C_2—C_3$ and $C_3—C_4$ bonds. Moreover the methyl esters of these acids give rise to a highly characteristic peak at m/z 169 [$CH(CH_3)—CH_2—CH(CH_3)—CH=C(CH_3)—COOCH_3$]. Comparison of the mass spectra of methyl *cis*- and *trans*-DL-erythro-2,4,6-trimethyl-Δ^2-tetracosenoate shows a slight difference of peak intensities, mainly at m/z 322 (M-100) and 335 (M-87) (Ahlquist *et al.*, 1958; Ryhage *et al.*, 1961).

1.4.3. Methyl Branch in the Middle of the Chain

Several bacterial species, belonging to the Actinomycetes, synthesize fatty acids having one methyl branch in the middle of the aliphatic chain. The best-known representative of this group is tuberculostearic acid (10D-methyloctadecanoic acid) (Asselineau, 1966).

The mass spectrum of methyl tuberculostearate shows characteristic ion fragments arising from cleavage of the carbon–carbon bonds on both sides of the methyl-bearing methine group: a strong peak at m/z 199 [$CH(CH_3)—(CH_2)_8—COOCH_3$] and a smaller one at m/z 171 [$(CH_2)_8—COOCH_3$]. This latter peak is accompanied by peaks at m/z

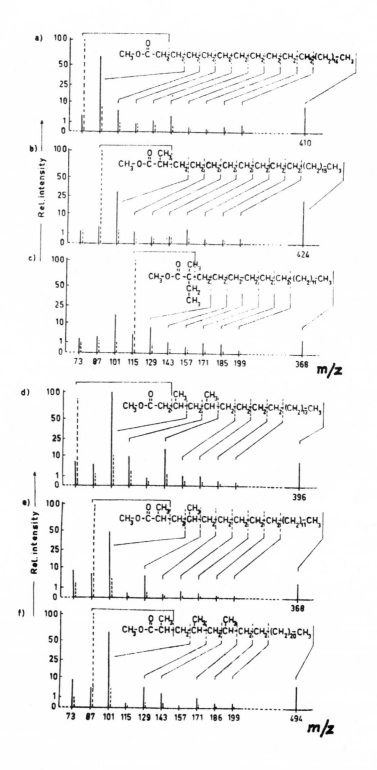

SCHEME 3

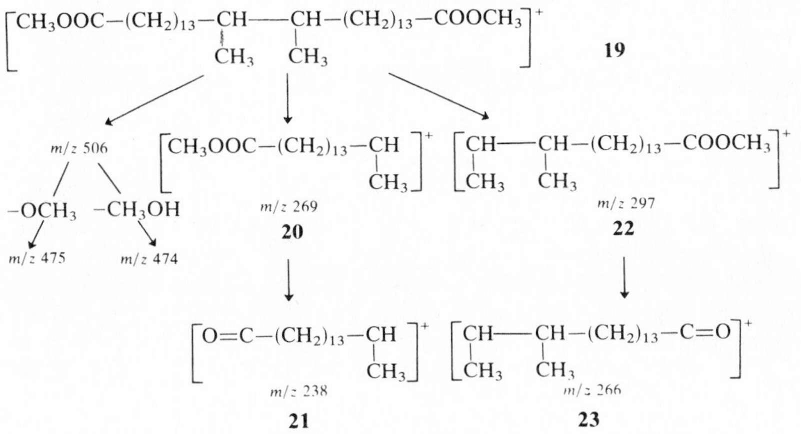

$$\left[CH_3OOC-(CH_2)_{13}-\underset{\underset{CH_3}{|}}{CH}-\underset{\underset{CH_3}{|}}{CH}-(CH_2)_{13}-COOCH_3 \right]^+ \quad 19$$

$$m/z\ 506 \quad \left[CH_3OOC-(CH_2)_{13}-\underset{\underset{CH_3}{|}}{CH} \right]^+ \quad \left[\underset{\underset{CH_3}{|}}{CH}-\underset{\underset{CH_3}{|}}{CH}-(CH_2)_{13}-COOCH_3 \right]^+$$

$$-OCH_3 \quad -CH_3OH \qquad\qquad m/z\ 269 \qquad\qquad\qquad\qquad m/z\ 297$$

$$m/z\ 475 \qquad m/z\ 474 \qquad\qquad\quad \textbf{20} \qquad\qquad\qquad\qquad\qquad\qquad \textbf{22}$$

$$\left[O=C-(CH_2)_{13}-\underset{\underset{CH_3}{|}}{CH} \right]^+ \qquad \left[\underset{\underset{CH_3}{|}}{CH}-\underset{\underset{CH_3}{|}}{CH}-(CH_2)_{13}-C=O \right]^+$$

$$m/z\ 238 \qquad\qquad\qquad\qquad\qquad\qquad m/z\ 266$$

$$\textbf{21} \qquad\qquad\qquad\qquad\qquad\qquad\qquad \textbf{23}$$

172 and 173 arising from rearrangements of one or two hydrogen atoms, respectively (Ryhage and Stenhagen, 1960a; Ryhage, 1962).

Methyl 10-methylheptadecanoate and 10-methyloctadecanoate, obtained from the lipids of *Microbispora parva*, have been similarly characterized by the peaks at m/z 199, 173, 172, and 171 in their mass spectra. The same set of peaks were obtained in both cases, as the methyl branch is located in the same position with respect to the methoxycarbonyl group (Ballio and Barcellona, 1971).

Tuberculostearic acid has been identified in cells of several species of Mycobacteria and Nocardiae by GC/MS using selected ion monitoring. Monitoring at m/z 312 (molecular ion) gives a single peak in the chromatograms. It is possible to detect tuberculostearate by this technique in as little as 2 ng of lyophilized cells (containing about 20 pg of tuberculostearate) (Larsson *et al.*, 1979).

Lyophilized samples of autoclaved sputum (2–4 ml) of patients with pulmonary tuberculosis were extracted with 1.5 ml of chloroform–methanol (2 : 1 v/v); 1 ml of 3% methanolic HCl was added to the dry extract, and the mixture was heated at 80°C for 20 h. The supernatants were evaporated

FIGURE 3. Mass spectra of methyl 2-methylhexacosanoate (b), 2-methyl-2-ethyl-eicosanoate (c), 2,4-dimethyl-tricosanoate (d), 2,4-dimethyl-heneicosanoate (e) and 2,4,6-trimethyl-nonacosanoate (f), in comparison with the mass spectrum of methyl hexacosanoate (a). Unbroken vertical lines represent normal fragments, and dotted lines rearranged fragments in which the m/z is equal to m/z of the normal fragments plus one unit (Asselineau *et al.*, 1957) (reproduced by permission of *Acta Chemica Scandinavica*).

to dryness and extracted with 100 μl of n-hexane. After thin-layer chromatography (n-hexane-diethyl ether 9:1 v/v, as solvent), the esters, recovered by scraping the ester band and extracting with ethyl acetate, were injected into the gas chromatograph. Ion monitoring at m/z 312 allowed the detection of tuberculostearate in 5 samples out of 6 (Odham et al., 1979).

A branched dicarboxylic acid, $C_{34}H_{66}O_4$, is a major fatty acid of the lipids of *Butyrivibrio* sp. grown on palmitic acid. The name "diabolic acid" is proposed for this diacid because mass spectra of its dimethyl ester show peaks arising from the loss of methanol rather than of the methoxy group, from the molecular ion. The main peaks of the mass spectrum of the dimethyl ester are explained in scheme 3 [formulaes (19)–(23)].

1.4.4. Unsaturated Fatty Acids

Double bonds on fatty acid methyl ester chains have been located by using the various methods described above. For example, the pyrrolidide method was used to identify 11-methyloctadec-11-enoic acid in the fatty acids isolated from *Rhizobium* strains (Gerson et al., 1975), and mono- and poly-unsaturated C_{16} and C_{18} acids in the lipids of photosynthetic dinoflagellates (Joseph, 1975). It can be noted that the pyrrolidide method can be conveniently applied to unsaturated fatty acids of medium chain length, but the results obtained are more difficult to interpret in the case of very long-chain fatty acids. Takayama et al. (1978), in the course of an analysis of the unsaturated fatty acids of *M. tuberculosis* (strain H37 Ra), used the pyrrolidide method to identify C_{15}–C_{21} monounsaturated esters, but preferred to study the ozonolysis products of C_{22}–C_{34} unsaturated esters. Most often unsaturated fatty esters are oxidized by the permanganate–periodate method (von Rudloff, 1950), and the resulting mono- and diesters are analyzed by GC and/or MS: see for instance studies on the fatty acids of blue-green algae (Sato et al., 1979) and *M. phlei* (Asselineau et al., 1970). Dihydroxylation by OsO_4 and formation of the bis-TMS derivative have been used for the location by mass spectrometry of the double bond of tetradec-7-enoic acid, a fatty acid isolated from an acyl phosphatidylglycerol of *E. coli* (Kobayashi et al., 1980).

Occurrence of a double bond in *iso* and *anteiso* acids may help to locate the methyl branch. For instance a mixture of *iso* and *anteiso* monoenoic acids were isolated from *Desulfovibrio desulfuricans* and transformed into methyl esters. The mass spectra of the *iso* esters exhibited peaks at M-55, M-87, and M-105, and those of *anteiso* esters, the same set of peaks with 14 mass units more: M-69, M-101, and M-119. These peaks are formed by elimination of C_4H_7 (*iso* esters) and C_5H_9 (*anteiso* esters), CH_3OH and H_2O. It was suggested that the double bond migrated towards the end of the chain and that allylic cleavage took place. This mechanism [shown in formulas (24)–(26)] was studied on methyl 15-deutero-*iso*- Δ^8- and -Δ^9-heptadecenoates: the tertiary hydrogen atom seems to play a key role in the process (Boon *et al.*, 1977a,b).

R = CH_3 or CH_3-CH_2 $\qquad$ $m + n + 2 = p$

$$CH_3-(CH_2)_{19}-CH\underset{CH_2}{\overset{}{\diagdown\diagup}}CH-(CH_2)_{14}\underset{16}{}-CH\underset{CH_2}{\overset{}{\diagdown\diagup}}CH-(CH_2)\underset{\substack{11 \\ 13}}{}{}_9COOCH_3$$

27

Usually bacterial lipids do not contain polyunsaturated acids (except the blue-green algae). However, small amounts of diunsaturated acids have recently been detected by using GC/MS. For example a hexadecadienoic acid has been characterized in the lipids of *Leptospira interrogans* (Degré *et al.*, 1980), and a dodecadienoic acid containing two conjugated double bonds (λ_{max} 232 nm, hexane) has been observed in the lipids of *Pseudomonas mildenbergii* (Roussel and Asselineau, 1980). In some special cases larger amounts of polyunsaturated acids have been detected. $C_{18:1}$, $C_{18:2}$, $C_{18:3}$, and $C_{18:4}$, as well as $C_{20:4}$ and $C_{20:5}$, have been observed in the GC–MS analysis of *Flexibacter polymorphus* (Johns and Perry, 1977). The content of $C_{20:5}$ is 18% of the total fatty acids. However, it is puzzling that, only on the basis of the retention time of the ester $C_{20:5}$ on two kinds of capillary column (SE 30 and BDS), the authors can deduce the complete location of 5 double bonds in the chain (ester ω-3, with methylene-interrupted double bond system).

1.4.5. Cyclopropane Fatty Acids

Cyclopropane fatty acids are frequently encountered in the lipids of bacteria. The structure of a C_{19} cyclopropane fatty acid isolated from *Thiobacillus thiooxidans* was determined by Knoche and Shively (1969) by studying the mass spectrum of its hydrogenolysis products. Lactobacillic acid, isolated from the lipids of *Brucella melintensis*, was identified by mass spectrometry of the chromium trioxide products (Thiele *et al.*, 1969).

The BF_3-methanol method and the CrO_3-oxidation method were used in parallel to locate cyclopropane rings in "meromycolic acids," very long-chain fatty acids arising from the pyrolysis of mycolic acids (from Mycobacteria) (see page 76). Similar results, leading to structures such as (27), were obtained by both methods (Minnikin and Polgar, 1967; Asselineau *et al.*, 1969).

Mass spectrometry of pyrrolidide derivatives of cyclopropane fatty acids have also been used to locate the ring:

> If an interval of 12 a.m.u. instead of the regular 14, is observed between the most intense peaks of clusters of fragments containing n-2 and n-1 and n, or n and $n + 1$ carbon atoms of the acid moiety, a cyclopropane ring occurs between carbon n and $n + 1$ in the molecule (Andersson 1978).

Pyrrolidide derivatives were used in parallel with the α-keto-cyclopropane derivatives to make a comparative study of a bis-cyclopropane meromycolic acid sample by mass spectrometry: identical results were obtained (Gensler and Marshall, 1977).

It has been claimed that cyclopropane fatty aldehydes give mass spectra allowing the direct location of the cyclopropane ring (Lamonica and Etemadi, 1967). By using synthetic model compounds, Puzo and Promé (1973) have observed that the mass spectra of cyclopropane aldehydes can show, when the relative locations of the cyclopropane ring and the aldehyde group are favorable, two intense peaks. But the presence of these peaks, and their intensities, depend on the number of methylene groups between the ring and the carbonyl group, making their use in establishing the structure of unknown cyclopropane aldehyde compounds uncertain.

1.4.6. ω-Cyclohexyl Fatty Acids

ω-Cyclohexyl-undecanoic and -tridecanoic acids were found in 10 strains of acido-thermophilic bacteria (isolated from Japanese hot springs). They were the main fatty acids found in their lipids. Mass spectra of these acids (measured on the methyl esters) exhibited a high intensity peak at m/z 83 (cyclohexyl ion) and a smaller one at M-38. Other peaks at m/z 69, 55 (base peak), and 41 arose from the splitting of the cyclohexyl ring: they have been used for biosynthetic studies (Oshima and Ariga, 1975).

1.4.7. Hydroxy Fatty Acids

Hydroxy acids, found in the lipids of microorganisms, most often have their hydroxyl group in position 2 or 3. The mass spectra of 2-hydroxy acid methyl esters exhibit a characteristic peak at M-59 (elimination of COOCH₃) and those of 3-hydroxy acid methyl esters contain an intense peak at m/z 103 [HOCH—CH=C(OH)OCH₃], making their identification easy. GC/MS analysis of a mixture of fatty acid methyl esters with monitoring at m/z 73 and M-59 gives the locations of 2-hydroxy fatty acid derivatives in the gas chromatograms (Laine *et al.*, 1974).

For example, mass spectrometry allowed the identification of 2-hydroxy-13-methyl- and 2-hydroxy-12-methyl-tetradecanoic acids in the lipids of several species of *Streptomyces* (Lanéelle *et al.*, 1968), of 2-hydroxy-14-methylpentadecanoic acid in *Arthrobacter simplex* (Yano *et al.*, 1969), of 2-hydroxydodecanoic acid in the lipid A of the lipopolysaccharide of *Pseudomonas ovalis* (Kawahara *et al.*, 1979) and of 2-hydroxy-octadecanoic acid in *Spirillum lipoferum* (Moss and Lambert, 1976).

The 3-hydroxy-13-methylpentadecanoic acid and 3-hydroxy-15-methylhexadecanoic acid were identified as components of siolipin (produced by *Str. sioyaensis*) by mass spectrometry (Kawanami and Otsuka, 1969). 3-hydroxy-$C_{12:0}$, -$C_{14:0}$, and -$C_{16:0}$ acids were identified in the fatty acids of several species of *Pseudomonas* by GC/MS (Moss and Dees, 1975), and 3-hydroxydodecanoic acid was characterized as the major fatty acid of the bound lipids of a moderately halophilic bacterium (Ohno *et al.*, 1976).

Eight species of the mold *Mucor*, grown under oxygen limiting conditions produce 7-hydroxy-decanoic, -dodecanoic, and -tetradecanoic acids, characterized by gas chromatography and mass spectrometry. The main peak of the mass spectra corresponds to the ion [O=C=CH—(CH₂)₅—CHOH⁺], at m/z 127 (Tahara *et al.*, 1980).

GC/MS was used to determine lipid A and endotoxin in serum. The technique was based on the detection of β-hydroxymyristic, a specific component of lipid A, which was released by hydrolysis, esterified, separated from nonhydroxylated esters by chromatography on silica gel, and transformed into its TMS derivative. Selected ion monitoring at mass unit 315.4 (M-15) was used for detection. Estimation was obtained by comparison with a known amount of the TMS pentadeutero derivative of methyl β-hydroxymyristate added to the sample (with monitoring at mass unit 320). As little as 200 fmol of β-hydroxymyristic acid could be detected (Maitra *et al.*, 1978).

The identification of 12-methoxy-11-methyl- and 11-methoxy-12-methyl-octadecanoic acids in the fatty acids of a species of *Rhizobium* by using gas chromatography and mass spectrometry may also be mentioned (Gerson *et al.*, 1975). The splitting of the corresponding esters under electron impact is quite similar to that of α-methyl-branched methoxy esters obtained by reacting BF₃-methanol with cyclopropane fatty acid esters (see Scheme 2, page 67).

A special case of 3-hydroxy fatty acids is presented by mycolic acids, β-hydroxy long chain fatty acids having a long aliphatic chain in the α-position. These acids are produced by bacteria belonging to the CMN group (Corynebacteria, Mycobacteria, Nocardiae) (Lederer et al., 1975; Barksdale, 1977). Their structures become progressively more complex, and the molecular weights higher, as one goes from the corynomycolic acids from Corynebacteria, to the nocardomycolic acids from Nocardiae, and then to the mycolic acids from Mycobacteria. The common property of all these mycolic acids is that they are split into a meromycolic aldehyde and a fatty acid, either by pyrolysis or by electron impact:

$$R-CHOH-CH(R')-COOCH_3 \rightarrow R-CHO + R'-CH_2-COOCH_3$$

The TMS derivatives of methyl corynomycolates from *Corynebacterium hofmanii* have been separated by gas chromatography according to their carbon number, but it was found more convenient to fractionate their methyl esters according to their degree of unsaturation by thin-layer chromatography on silver nitrate impregnated silica gel before performing mass spectrometry (Welby-Gieusse et al., 1970). By using GC/MS, the methyl corynomycolate fraction from *C. ulcerans* (as TMS derivatives) has been resolved into 12 molecular species differing by the number of carbon atoms (even series from C_{20} to C_{32}) and the degree of unsaturation (0, 1, or 2 double bonds) (Yano and Saito, 1972). More recently this method has been applied to the analysis of corynomycolic acids isolated from skin corynebacteria (Corina and Sesardic, 1980).

The molecular species of nocardomycolic acids were studied in 16 strains of Nocardia and related taxa, by using GC/MS of the TMS derivatives of the methyl esters. The molecular weights and the number of double bonds were determined by using the ions M^+, $[M-15]^+$, and $[M-90]^+$. The size of the meromycolic moiety was established by looking for the ion $[R-CH-O-Si(CH_3)_3]^+$ (arising from cleavage of the C_2-C_3 bond), and the size of the α-branch, by looking for the ion $[(CH_3)_3Si-O-CH-CH(R')-COOCH_3]^+$ (arising from cleavage of the C_3-C_4 bond). True Nocardiae, such as *N. asteroides*, contain $C_{44}-C_{58}$-nocardomycolic acids with from 1 to 4 double bonds; $C_{56}-C_{68}$-nocardomycolic acids with from 1 to 5 double bonds have been observed in the lipids of *N. polychromogenes* and *Gordona bronchialis* (*Rhodococcus bronchialis*) (Yano et al., 1978).

By using a gas chromatograph equipped with a glass column filled with 2% OV-1 on Chromosorb W working at 300–340°, the mycolic acids of *M. phlei* were studied by GC/MS (see a gas chromatogram in Figure 4). Seven monocarboxylic mycolic acids, ranging from C_{72} to C_{80} were detected,

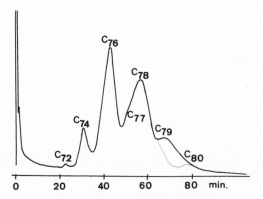

FIGURE 4. Gas chromatogram of the TMS derivatives of methyl α-mycolates isolated from *Myco-bacterium phlei*, measured iso-thermically at 345°C (2% OV-1 column) (Toriyama *et al.*, 1978) (reproduced by permission of *FEBS Letters*).

all of them having 2 double bonds (or two cyclopropane rings). The same kind of ions (as those observed in the case of nocardomycolate derivatives) were used to determine the size of the meromycolic moiety and of the α-branch. Dicarboxylic mycolic acids were a mixture of acids from C_{56} to C_{64} with only one unsaturation. The increase in the growth temperature of the bacteria (23°C to 51°C) led to a slight increase in the length of the meromycolic moiety (2 carbon atoms more), without any change in the length of the α-branch (Toriyama *et al.*, 1978).

This work has been extended to the study of a mycolic acid fraction containing the lowest components, isolated from *M. smegmatis*, *M. vaccae*, and *M. lepraemurium* (Yano *et al.*, 1978a). Similar studies have been performed on the mycolic acids isolated from several strains of *Nocardia* (Yano *et al.*, 1978b), from *N. rubra* (Tomiyasu *et al.*, 1981), and more recently, on the mycolic acids (C_{22}–C_{39}) isolated from *Bacterionema matruchotii* (Wada *et al.*, 1981).

1.4.8. Characteristic Fatty Acid Components of Bacterial Species

Tables 1 and 2 show the distribution of a few characteristic fatty acid components in some bacterial species arbitrarily chosen among the most studied ones. These analyses play an important role in bacterial taxonomy, as shown in Chapter 8. This is illustrated by the GC/MS analysis of the fatty acids of *Legionella pneumophila*, rapidly performed after the isolation of this new pathogenic bacterial species: its characteristic fatty acid composition showed that it could not be included in any known bacterial genus. *L. pneumophila* is characterized by the unusual preponderance of *iso*-hexadecanoic acid and the simultaneous occurrence of *iso* and *anteiso* acids and cyclopropane fatty acids (Fisher-Hoch *et al.*, 1979).

TABLE 1. Distribution of a Few Characteristic Fatty Acid Components in Some Gram Negative Bacteria[a]

Bacteria	Iso and anteiso fatty acids	Cyclopropane fatty acids	Saturated fatty acids $\geq C_{20}$	3D-hydroxy fatty acids	Vaccenic acid	Reference
Acetobacter xylinum				14		Dekker *et al.* (1977)
Acinetobacter				*12,14*		Nishimura *et al.* (1979)
Actinobacillus (actinomycetemcomitans)				14		Jantzen *et al.* (1980)
Agrobacterium		19		16	+	Kaneshiro and Marr (1962)
Alcaligenes faecalis		17		14		Moss and Dees (1975)
Azotobacter		17,19		10,12		Su *et al.* (1979)
Bacteroides	*anteiso* 15			*iso* 17		Shah and Collins (1980)
Bordetella				14		Thiele and Schwinn (1973)
Brucella		19				Thiele *et al.* (1969)
Chromatium vinosum					+	Hurlbert *et al.* (1976)
Chromobacterium violaceum	*anteiso* 15			10, *12,14*		Hase *et al.* (1977)
Desulfovibrio	*iso* 15			16, *iso* 15, *iso* 17	+	Boon *et al.* (1977)
Escherichia		17,19		14	+	Cronan *et al.* (1974)
Francisella tularensis			26	16,18		Jantzen *et al.* (1979)
Neisseria				*12,14*		Jantzen *et al.* (1978)
Proteus		17,19		14,16		Gmeiner and Martin (1976)
Pseudomonas		19[b]		*10,12*	+	Moss and Dees (1976)
Rhizobium[c]		19		14	+	Gerson *et al.* (1975)
Salmonella typhi		17,19		14	+	Jantzen *et al.* (1980)
Serratia marcescens		17,19		*10,12*	+	Bishop and Still (1963)
Thiobacillus		17,19				Agate and Vishniac (1973)
Vibrio		19[b]		10,12,14		Boe and Gjerde (1980)

[a] For every individual group of bacteria one important work is quoted in order to have a clue to the literature pertaining to this group.
[b] Only in a few species.
[c] This bacterial species also contains 11-methyl-Δ^{11}-octadecenoic acid.

2. ALIPHATIC HYDROCARBONS

A few bacterial species have a relatively high content of long-chain hydrocarbons.

The hydrocarbon fraction of *Sarcina lutea* and *S. flava* was studied by GC/MS. This fraction is a complex mixture of normal and branched chain hydrocarbons in the range C_{22}–C_{30} (main members C_{27}, C_{28}, C_{29}). Most of them have methyl branches located either in *iso* or *anteiso* configurations and contain one double bond. The elution order of the isomeric hydrocarbons from the gas chromatograph (stationary phase: OV-17 or Apiezon L) is as follows: *iso–iso, anteiso–iso, anteiso–anteiso, iso*–normal and *anteiso*–normal, normal–normal. The position of the double bond (in or near the centre of the chain) was determined by osmium tetraoxide dihydroxylation followed by mass spectrometric analysis of the bis-TMS derivatives (Tornabene and Markey, 1971).

Hydrocarbons are also found in blue-green algae (0.05% to 0.12% of the dry weight) as a homologous series of medium chain alkenes ranging from C_{15} to C_{18} (predominant homologue: C_{17}). In *Nostoc muscorum* and *Anacystis nidulans*, 7- and 8-methylheptadecane were detected, that were separated by capillary gas chromatography. Their structures were established by mass spectrometry (McCarthy *et al.*, 1968; Hun *et al.*, 1968). Related methyl-branched hydrocarbons were found in other species of blue-green algae.

Normal chain hydrocarbons were detected by GC–MS in *Anacystis montane*, another species of blue-green algae, and in *Botryococcus braunii* (golden-brown alga). The major members were C_{17} and C_{29} in the former species and C_{29} in the latter. They were mixtures of mono-, di-, and trienes (except for $C_{17:0}$) (Gelpi *et al.*, 1968). Very long-chain *n*-alkenes were isolated from the microscopic alga *Emiliana huxleyi* and identified by GC/MS. The alkenes were mainly dienes, trienes, and tetraenes from C_{31} to C_{37} (Volkman *et al.*, 1980b).

3. TERPENIC DERIVATIVES AND STEROLS

Small amounts of terpenic derivatives are detected in many strains of bacteria, mainly as metabolic intermediates (for example squalene and phytofluene) or as activated forms of transport of metabolites (e.g., polyprenylphosphates and pyrophosphates). Mass spectrometry was used for their identification (see, for instance, Sasak and Chojnacki, 1977; Amdur *et al.*, 1978). But in some cases terpenic compounds can play an important role as structural components of the bacterial cell: this review is limited to these cases.

TABLE 2. Distribution of a Few Characteristic Fatty Acid Components in Some Gram Positive and Actinomycetales Bacteria

	Iso and anteiso fatty acids	Cyclopropane fatty acids	Tuberculo-stearic acid type	Saturated fatty (acids ≥C_{20})	Mycolic acid type	Main monoenoic acid	2-OH fatty acid	References
Gram positive bacteria:								
Bacillus	anteiso 15							Kaneda (1977)
Clostridium		13,15,15:1		20		Δ^{11}		Chan et al. (1971)
Lactobacillus	Tr	19		b		Δ^{11}		Veerkamp (1971)
Listeria monocytogenes	anteiso 15							Carroll et al. (1968)
Micrococcus luteus	anteiso 15							Jantzen et al. (1974)
Propionibacterium	anteiso 15							Moss et al. (1979)
Sarcina	anteiso 15							Jantzen et al. (1979)
Staphylococcus	anteiso 15			20				Jantzen et al. (1979)
Streptococcus		19						Drucker (1974)
Actinomycetales and related genera:								
Arthrobacter	{anteiso 15				−			Bowie et al. (1972)
					+		+	Keddie and Cure (1977)

Corynebacterium						
Aerobic species				+	Δ^9	Asselineau (1966)
Anaerobic species	*anteiso* 15			−		Keddie and Cure (1977)
Nocardia	[c]	+		+	Δ^9	Bordet and Michel (1963)
Mycobacterium		+	24,26	+	Δ^9	Asselineau (1966)
Rhodococcus[d]		+	24	+		Goodfellow *et al.* (1978)
Actinoplanes	[e]					Ballio and Barcellona (1968)
Micromonospora	*iso* 15	[c]				Ballio and Barcellona (1968)
Streptomyces	*anteiso* 15	$a + iC_{15}$				Ballio and Barcellona (1968)

[a] See also footnote of Table 1.
[b] Up to C_{24} in *L. heterohiochii* (Uchida, 1974).
[c] Found only in some species.
[d] Formerly *Gordona*.
[e] The nature of the major member depends on the strain.

3.1. Aliphatic Terpene Alcohols

The main lipids of extremely halophilic bacteria, such as *Halobacterium cutirubrum*, are devoid of ester-linked fatty acids, but contain phytanol (dihydrophytol) linked by ether bond to hydroxyl 2 and 3 of the *sn*-1-glycerophosphate moiety (Kates, 1978). Treatment of these lipids with hydroiodic acid gives an alkyl iodide which can be converted (through the corresponding acetate) into a terpene alcohol identified as phytanol by mass spectrometry (Kates *et al.*, 1965). Later, it was proved that the bacterial phytanol was 3R,7R,11R,15 (or 3D,7D,11D,15)-tetramethyl-hexadecanol. It can be noted that phytanol (C_{20}) is eluted between the methyl esters of C_{17} and C_{18} acids, because of the four methyl branches (Kates, 1978).

The lipids of the thermo-acidophile bacteria contain ether-lipids made of biphytanol, a diol resulting from the tail-to-tail linkage of two phytanol molecules. Biphytanol has been isolated by means of acidic hydrolysis and studied by GC/MS of the bis-TMS derivative. For example, *Thermoplasma acidophilum* contains a mixture of biphytanol derivatives differing by their degree of cyclization. Figure 5 shows the mode of cleavage of a noncyclized molecule under electron impact, and Figure 6, the mode of cleavage of a molecule containing two pentacyclic rings. Four pentacyclic rings can also occur in the molecule (Yang and Haug, 1979). Similar results have been obtained on the lipids of the *Caldariella* group of bacteria (De Rosa *et al.*, 1980).

In methanogenic bacteria, both kinds of ether-linked lipids are found, one with phytanyl chains such as diethers of glycerol, and the other with biphytanyl chains such as tetraethers of glycerol (Tornabene and Langworthy, 1978).

All these bacterial species, able to grow in extreme environments (high temperature, low pH, high salinity, or a combination of these factors), have a lipid composition characterized by the absence of acyl ester groups and the presence of ether-linked terpene alcohols. It has been suggested that these bacteria, which probably share a common evolutionary episode, represent ancestral life-forms, designated as *Archaebacteria* (Fox *et al.*, 1977).

3.2. Sterols and Polycyclic Terpene Alcohols

3.2.1. Bacteria

Most bacteria are devoid of significant amounts of sterols. Sometimes very low amounts of plant sterols (derivatives of ergosterol and stigmasterol)

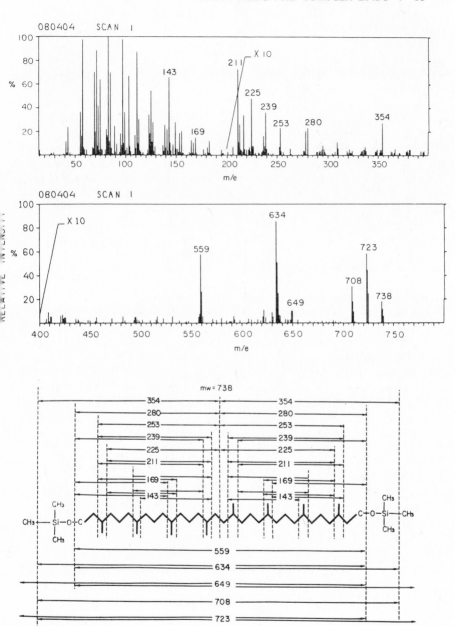

FIGURE 5. Mass spectrum of the bis-TMS derivative of biphytanol (top), isolated from *Thermoplasma acidophilum*, and assignments of the ion fragments (bottom) (Yang and Haug 1979) (reproduced by permission of *Biochimica Biophysica Acta*).

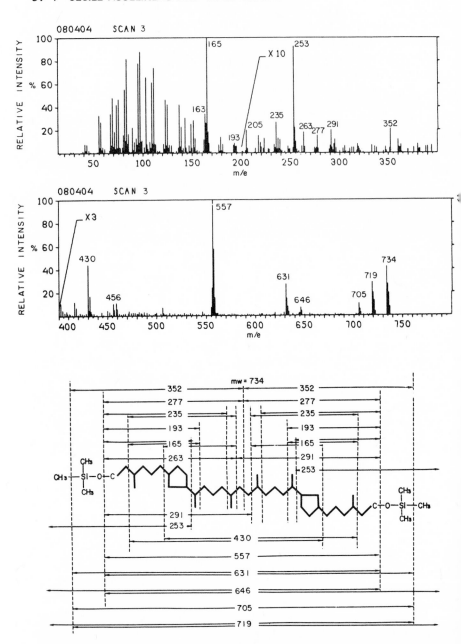

FIGURE 6. Mass spectrum of the bis-TMS derivative of a dicyclic-C_{40}-diol isolated from *Thermoplasma acidophilum* (top), and assignments of the ion fragments (bottom) (Yang and Haug 1979) (reproduced by permission of *Biochimica Biophysica Acta*).

(0.01% to 0.003% dry weight) can be detected and identified by mass spectrometry (Schubert *et al.*, 1968). However, Bird *et al.* (1971) isolated 0.22% (dry weight) of a "sterol fraction" from the lipids of *Methylococcus capsulatus* grown on methane. GC/MS analysis of this fraction showed that it was a mixture of 4α-methyl- and 4,4-dimethyl- 5α-cholest-8(9 or 14)-en-3β-ol and 4α-methyl- and 4,4-dimethyl- 5α-cholest-8(9 or 14), 24-dien-3β-ol. Later, Bouvier *et al.* (1976) demonstrated that the nuclear double bond was in position 8,14.

Sterols were found in blue-green algae. The sterol fraction isolated from *Phormidium luridum* var. *olivaceae* was studied by gas chromatography and mass spectrometry: the molecular ions were at m/z 414, 412, and 410, corresponding to C_{29} steroids with one, two, or three double bonds. Peaks at m/z 231 and 229, due to the tetracyclic carbon skeleton of the steroids, showed that the additional carbon atoms (with respect to the C_{27} sterols) were located in the side chain, and the possibility of 4,4-dimethyl-steroids was thus rejected. It was suggested that these sterols could be 24-ethyl-Δ^7- cholestenol, 24-ethyl-$\Delta^{7,22}$-cholestadienol, and 24-ethyl-$\Delta^{5,7,22}$-cholestatrienol (De Souza and Ness, 1968).

The isolation of hopane derivatives (pentacyclic triterpenoids) from the organic matter of sedimentary rocks led to the discovery and identification by mass spectrometry of bacteriophanetetrol from *Acetobacter xylinum* (Rohmer and Ourisson, 1976a). Simultaneously the same compound was identified as a component of the glycolipid fraction of *Bacillus acidocaldarius* (*Caldariella* group of bacteria) by Langworthy and Mayberry (1976). The mode of cleavage of the peracetyl derivative of bacteriophanetetrol under electron impact is shown in formula (**28**).

28

In the lipids of *B. acidocaldarius*, bacteriophanetetrol occurs in the free form and as the *N*-acyl D-glucosaminide of the primary alcohol group of the pentanetetrol chain (Langworthy *et al.*, 1976).

Bacteriophanetetrol and/or related derivatives have been found in other species of *Acetobacter*, in *Acetomonas oxydans*, *Methylococcus capsulatus* (which also contains 4α-methyl- and 4,4-dimethyl-steroids), *Methylosinus trichosporium*, and in the blue-green algae *Nostoc muscorum* and *Anabaena variabilis* (Rohmer and Ourisson, 1976b).

3.2.2. Nonbacterial Microorganisms

Many papers deal with the identification of sterols in molds and lower fungi. Among the most recent papers, the study of the sterols of the spores of *Glomus caledonius* by GC/MS can be quoted (Beilby and Kidby, 1980): cholesterol, 24-methyl-cholesterol, 24-methylene-cholesterol, and 24-ethyl-cholesterol were identified.

Using computerized capillary GC/MS, nine sterols have been characterized in the lipids of the marine diatom *Biddulphia sinensis* (Volkman *et al.*, 1980). The presence of 24-methylene-cholesterol in the lipids of the nonphotosynthetic diatom *Nitzchia alba* has been demonstrated by studies involving GC/MS (Kates *et al.*, 1978).

The mode of cleavage in electron impact mass spectrometry of tetrahymanol, a pentacyclic triterpenic alcohol present in the lipids of the protozoan *Tetrahymena pyriformis*, has been discussed by Caspi (1980).

4. SPHINGOSINE DERIVATIVES

The occurrence of sphingosine bases is very rare in bacterial lipids and is practically limited to the genera *Bacteroides* and *Flavobacterium*.

B. melaninogenicus contains ceramide phospholipids. The sphingosine moiety was studied by GC/MS of the *N*-acetyl-*O*-trimethylsilyl ether derivatives: C_{17}-, C_{18}-, and C_{19}-dihydrosphingosine derivatives were characterized. The major components were *iso*-branched (except for the C_{18} member, which had a partly *iso*-branched and partly normal chain). The molecular ion was usually nondetected, but a prominent peak was found at M-15. The main ions observed in the mass spectrum of these derivatives are shown in formula (**29**) (White *et al.*, 1969; see also Vaver and Ushkov, 1980).

$$CH_3-CH-(CH_2)_{11}-CH | CH | CH_2-O-TMS$$

29

In *B. fragilis*, sphingosine was found partly as free ceramides (*N*-acyl sphingosine), which were directly studied by GC/MS, as TMS derivatives.

Gas chromatography was performed by using a column of 2% OV-17 on Chromosorb W, 0.3×50 cm, at 230°C. A mixture of C_{17}-, C_{18}-, and C_{19}-dihydrosphingosines (branched and normal chain) was observed (Miyagawa et al., 1979).

Dihydrosphingosine was also characterized by GC/MS as the component of a glycoside of glucuronic acid and ceramide, in the lipids of *Flavobacterium devorans* (*Pseudomonas paucimobilis*) (Yabuuchi et al., 1979).

In yeasts, a derivative of dihydrosphingosine with an additional hydroxyl group (phytosphingosine) is often found. For example, C_{18}- and C_{20}-phytosphingosines were found in the free form in *Candida intermedia* (Kimura, 1976). They were identified by mass spectrometry of the tetra(trimethylsilyl) derivative. The trimethylsilylation was performed as follows: about 1 mg of the sample was dissolved in 0.2 ml CH_3CN, and 0.2 ml of *N,O*-bis-(trimethyl-silyl)-acetamide were added. The tube was carefully capped and heated at 75°C in an oil bath for 30 min. After cooling, the mixture was directly used for injection into the GC/MS apparatus (equipped with a 0.3×20 cm glass column packed with 3% OV-1 on Chromosorb W 60–80 mesh.) (Kimura, 1976).

A mixture of C_{16}- and C_{18}-dihydrosphingosine and phytosphingosine (major component: C_{18}-phytosphingosine) was characterized by GC/MS in the lipids of *Hansenula ciferrii* (Karlsson, 1966; Kulmacz and Schroepfer, 1978). GC/MS was used to analyze the components of a cerebroside fraction isolated from *Aspergillus oryzae*: the major base was an *iso*-C_{19}-sphingadienine (Fujino and Onishi, 1977).

MS and GC were used to determine the structure of the unusual sphingosine derivative $CH_3-(CH_2)_8-C(CH_3)=CH-(CH_2)_2-CH=CH-CH(OH)-CH(NH_2)-CH_2OH$, isolated as *N*-2'-hydroxy-3'-*trans*-octadecenoyl 1β-D-glucopyranoside from the phytopathogenic fungus *Fusicoccum amygdali* (Ballio et al., 1979).

In the protozoan *Tetrahymena pyriformis*, branched chain C_{17}- and C_{19}-sphingosines were characterized by gas chromatography of their TMS derivatives (Carter and Gaver, 1967). C_{18}-dihydrosphingosine (sphinganine) and 17-methyl-C_{18}-sphinganine were identified by GC/MS as components of a lipopeptidophosphoglycan isolated from *Trypanosoma cruzi* (De Lederkremer et al., 1978).

5. COMPLEX LIPIDS

Complex lipids are of course not volatile enough, even after derivatization, to be separated by gas chromatography prior to mass spectrometric studies. GC/MS can be applied only to degradation products of complex lipids.

5.1. Phospholipids

Fatty acid components can be analyzed by GC/MS as described above. Fatty aldehydes, obtained by acid hydrolysis of alkenyl ether phospholipids (plasmalogens) can be transformed into dimethylacetals, quite suitable for GC/MS analysis. For example, such an analysis was performed on the aldehydes obtained by hydrolysis of the plasmalogens isolated from *Treponema hyodysenteriae* (Matthews *et al.*, 1980).

The molecular species occurring in a mixture of phosphatidyl-ethanolamine, phosphatidylglycerol, and cardiolipids can be studied by using the monoacetyldiglycerides obtained after acetolysis. The glycerides prepared from the phospholipids of a thermophilic bacterium were analyzed by GC/MS: gas chromatography was performed in helium (flow rate 26 ml/min) on a glass column (3m × 3 mm) filled with 1% OV-1 on Chromosorb W, at 280°C. Excess of helium was eliminated by means of a separator working at 300°C and the monoacetyl-diglycerides were fragmented by chemical ionization with NH₃ (0.8 Torr) (Kagawa and Ariga, 1977).

Deacylation and Smith degradation of a phosphoglycolipid isolated from *Streptococcus faecalis* gave glycolphosphorylglycerol. The trimethyl-silyl derivative of this compound was identified by gas chromatography on comparison with known products. The following relative retention times of the TMS derivatives were observed: glycolphosphorylglycol 1.00; glycol-phosphorylglycerol 1.89; glycerolphosphorylglycerol 2.97 (column 210 cm × 4 mm; 2% OV-17 on Chromosorb WAW; temperature 165–195°C, with a programmation of 2°C/min.). The identification was checked by EI mass spectrometry of the TMS derivative of the intact product and of the partially deuterated compound (obtained through the use of NaB^2H_4 in the course of the Smith degradation) (Fischer and Landgraff, 1975). *Hexakis*-(trimethylsilyl)-glycolphosphorylglycerophosphoglycerol (containing two deuterium atoms) was analyzed by EI mass spectrometry. The highest peak was at m/z 789 (M—CH₃) and high intensity peaks were observed at m/z 714 (M − TMSOH), 713 (M − $TMSO^2H$), and 700 (M − $TMSO—CH^2H$) [see formula (30)] (Laine and Fischer, 1978; see also Vaver and Ushakov, 1980).

30

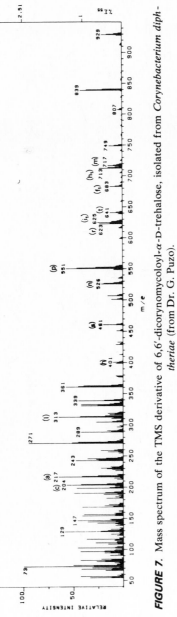

FIGURE 7. Mass spectrum of the TMS derivative of 6,6'-dicorynomycoloyl-α-D-trehalose, isolated from *Corynebacterium diphtheriae* (from Dr. G. Puzo).

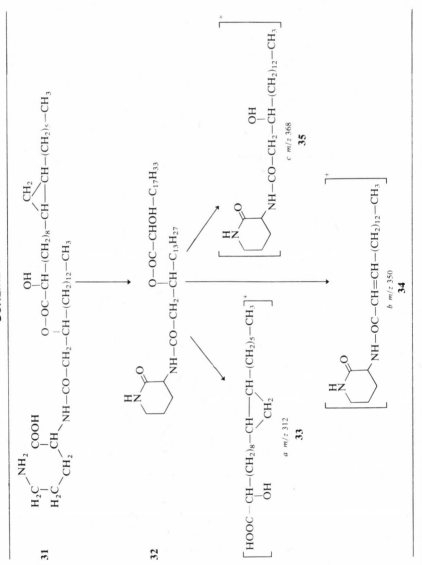

SCHEME 4

5.2. Glycolipids

In spite of the fact that monoacyl-hexoses can be separated by gas chromatography according to the location of the acyl group on the sugar ring (Martin and Asselineau, 1969), neither whole glycolipids nor sugar-containing degradation products appear to have been analyzed by gas chromatography. The high molecular weight of these compounds (mainly after derivatization) precludes the use of gas chromatography. The carbohydrate moieties of glycolipids and lipopolysaccharides, after suitable derivatization, can be studied by GC/MS, as shown in Chapter 4.

Most of the structural studies of glycolipids isolated from microorganisms rely on mass spectrometry. The following examples can be given for structure determination of a trigalactosyl-diacylglycerol from Cyanobacteria (Zepke et al., 1978), a trihexosylsulphate-diphytanylglyceroldiether from *Halobacterium salinarium* (Falk et al., 1980), and 6,6'-diacyl-α-D-trehalose (Cord Factor fraction from *Corynebacterium diphtheriae*) (see Figure 7) (Puzo et al., 1978).

5.3. Lipoamino Acids and Peptidolipids

Amino acids covalently linked to fatty acids are often found in bacterial lipids (Asselineau, 1966). The most simple forms are N-acyl amino acids (lipoamino acids). "Ornithine-containing lipids" are probably the most widely distributed lipoamino acids, as they have been isolated from Mycobacteria, *Streptomyces*, *Bacillus*, and mainly from gram negative bacteria (*Brucella* sp., *Desulfovibrio gigas*, *Gluconobacter cerinus*, *Pseudomonas* sp., *Rhodospirillum rubrum*, *Thiobacillus thiooxidans*, etc.).

It has been observed that the ornithine-containing lipid from *Th. thiooxidans*, in an underivatized form, can be introduced into the mass spectrometer source by gentle pyrolysis from the direct inlet probe. A molecule of water is eliminated by cyclization of the ornithine moiety [(**31**)), (**32**)], leading to a 2-piperidone derivative, sufficiently volatile to give mass spectra. The high-mass region exhibits peaks corresponding to the fragmentation into ions a (**33**), b (**34**), and c (**35**), as shown in scheme 4 (Hilker et al., 1978). Spectra obtained by thermolysis of this lipid in a chemical ionization source (working with methane or isobutane) exhibit a base peak at m/z 351 (corresponding to the protonated ion b) (Hilker et al., 1979).

The protonated molecular ions of ornithine-containing lipids, isolated from *Erwinia aroideae*, were obtained by chemical ionization (NH_3) mass spectrometry of the derivatives prepared by treatment with dimethyl-formamide dimethylacetal (**36**) (Madhavan et al., 1981).

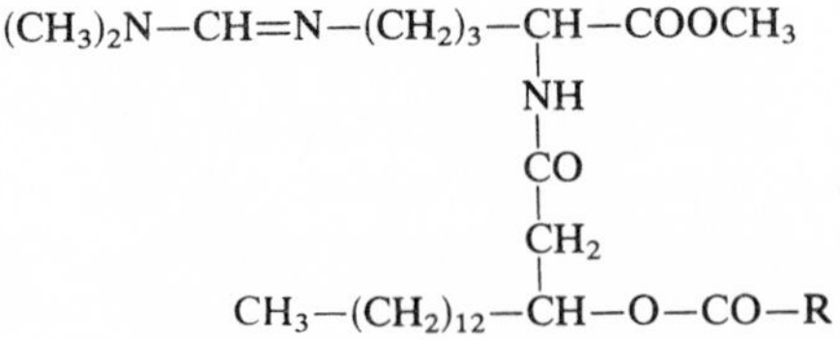

$$(CH_3)_2N-CH=N-(CH_2)_3-\underset{\underset{\underset{\underset{CH_3-(CH_2)_{12}-CH-O-CO-R}{CH_2}}{CO}}{NH}}{CH}-COOCH_3$$

36

$$R = C_{15}H_{31} \text{ or } C_{17}H_{33}$$

Mass spectrometry (mainly by electron impact) was intensively used for the structure determination of bacterial peptidolipids. The pioneer work of Das and Lederer allowed the structure elucidation of peptidolipin NA (lipoheptapeptide) from *Nocardia asteroides* (Barber *et al.*, 1965b), fortuitine (liponapeptide) from *Mycobacterium fortuitum* (Barber *et al.*, 1965a) and peptidolipid JA (lipopentapeptide) from *M. paratuberculosis* (Lanéelle *et al.*, 1965). In every case, these compounds were not volatile enough, even after derivatization, to be separated by gas chromatography.

GC/MS was used in the course of the structure determination of iturine A, an antibiotic peptidolipid produced by *Bacillus subtilis*. Hydrolysis of this peptidolipid released an amino fatty acid which was transformed into the *N*-acetyl methyl ester: GC/MS (with a column of SE 30) showed that it was a mixture of the derivatives of 3-amino-12-methyltridecanoic acid and 3-amino-12-methyltetradecanoic acid. The most intense peak in the mass spectra of both compounds was at m/z 144, corresponding to the ion $CH_3CO-N\overset{+}{H}=CH-CH_2-COOCH_3$. These amino fatty acids were also separated as *N*-trifluoroacetyl *n*-butyl esters on a column containing 0.65% polyethyleneglycol adipate on Chromosorb W (Peypoux *et al.*, 1973). The mass spectrum of the permethylated iturine A (M^+ 1294 and 1280) is in agreement with formula (**37**) for iturine A (Peypoux *et al.*, 1978).

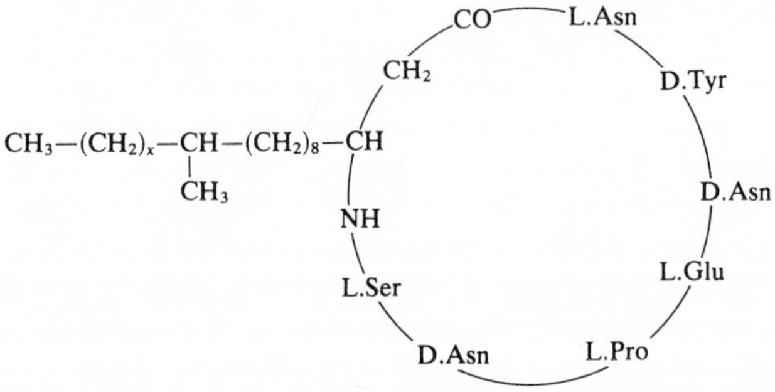

$$CH_3-(CH_2)_x-\underset{CH_3}{CH}-(CH_2)_8-CH$$

37

CO—L.Asn
CH$_2$
D.Tyr
D.Asn
NH
L.Glu
L.Ser
L.Pro
D.Asn

$x = 1$ or 2

The important role played by mass spectrometry in the structure determination of viscosine from *Pseudomonas viscosa* (Hiramoto *et al.*,

$CH_3-(CH_2)_{24}-CHOH-CH_2-CO-NH-CH-CO-NH-CH-CO-NH-CH-CO-NH-CH-CH_2$

CH_2 — C_6H_5

CH — O CH_3

CH_3

CH_3

O

CH_3 O CH_3

CH_3O OCH_3

CH_3

O O

H
2 CH_3-CO

38

1970), surfactine from *Bacillus subtilis* (Kakinuma *et al.*, 1969), globomycine from *Streptomyces halstedii* (Nakajima *et al.*, 1978) and lipopeptine A from *Str. violaceochromogenes* (Nishii *et al.*, 1980) can also be quoted.

However, even though mass spectrometry is a mighty tool for structure elucidation, requiring only a few micrograms of sample, it does not completely avoid the necessity for more classical methods of investigation. This is well illustrated by the study of a glycoside of a peptidolipid, mycoside C, produced by several species of Mycobacteria (MAIS group: *M. avium*, *M. intracellulare*, *M. scrofulaceum*). A first study of mycoside C, mainly relying on mass spectrometry, could not go further than a partial structure (Vilkas *et al.*, 1966). The isolation of the last unidentified component and its subsequent identification as alaninol (by gas chromatography of its *N*-trifluoroacetyl and *N,O*-ditrifluoroacetyl derivatives) led to the complete structure (**38**), which is supported by its mass spectrum (Lanéelle and Asselineau, 1968).

6. CONCLUSION

GC/MS is more and more often used to study the fatty acid composition of bacterial species. The mass spectrometric identification of the chromatographic peaks avoids erroneous labeling and moreover helps to detect new fatty acid species, which may play the role of markers for taxonomic purposes. The introduction of elaborate techniques now allows the positions of double bond or cyclopropane rings in a fatty acid chain to be determined on less than one milligram. The identification of very minor components in a complex mixture may help elucidate metabolic pathways, which can receive support by the use of precursors labeled with stable isotopes such as ^2H, ^{13}C, ^{15}N, or ^{18}O. In this respect, mass spectrometry has the advantage of giving information on the location of the isotope(s) in the molecule by working on very minute amounts of sample.

One may wonder why such powerful techniques have so far been put to rather little use in the field of fungi and other eukaryotic microorganisms. This can probably be explained by the fact that eukaryotic lipids usually contain fatty acids having no peculiar features in their carbon skeleton, and which are rich in polyunsaturation. No simple identification of poly-unsaturated systems by mass spectrometry is yet available.

The use of GC/MS to study complex lipids is most often hampered by their lack of volatility and only degradation products can usually be studied. A combination of high performance liquid chromatography and mass spectrometry will soon afford a solution to such structural problems. The complementary use of GC/MS and LC/MS techniques will become indispensable in lipid investigations, mainly when limited amounts of sample are available, as is often the case for lipids from microorganisms.

REFERENCES

Agate, A. D., and Vishniac, W., 1973, Characterization of *Thiobacillus* species by gas–liquid chromatography of cellular fatty acids, *Arch. Microbiol.* **89**:257.

Ahlquist, L., Asselineau, C., Asselineau, J., Ställberg-Stenhagen, S., and Stenhagen, E., 1958, Synthesis of the *cis* and *trans* isomers of DL-*erythro*-2,4,6-trimethyl-$\Delta^{2:3}$-tetracosanoic acid, *Ark. Kemi* **13**:543.

Amdur, B. H., Szabo, E. I., and Socransky, S. S., 1978, Presence of squalene in gram positive bacteria, *J. Bacteriol.* **135**:161.

Andersson, B. A., 1978, Mass spectrometry of fatty acid pyrrolidides, *Prog. Chem. Fats Other Lipids*, **16**:279.

Andersson, B. A., and Holman, R. T., 1974, Pyrrolidides for mass spectrometric determination of the position of the double bond in monounsaturated fatty acids, *Lipids* **9**:185.

Asselineau, C., Asselineau, J., Ryhage, R., Ställberg-Stenhagen, S., and Stenhagen, E., 1959, Synthesis of (−) methyl 2D, 4D, 6D-trimethylnonacosanoate, and identification of C_{32}-mycocerosic acid as a 2,4,6,8-tetramethyloctacosanoic acid, *Acta Chem. Scand.* **13**:822.

Asselineau, C., Baess, I., Kolman, A., Lapchine, L., Puzo, G., and Wickmann, K., 1979, Comparative studies of the strains PA and PN of *Mycobacterium phlei* leading to their reclassification: examination of lipids and DNA, biochemical tests and phage typing, *Ann. Microbiol. (Inst. Pasteur)* **130B**:385.

Asselineau, C., Lacave, C., Montrozier, H., and Promé, J. C., 1970, Relations structurales entre les acides mycoliques insaturés et les acides inférieurs insaturés synthétisés par *Mycobacterium phlei*, *Eur. J. Biochem.* **14**:406.

Asselineau, C., Montrozier, H., and Promé, J. C., 1969, Structure des acides α-mycoliques isolés de la souche Canetti de *Mycobacterium tuberculosis*, *Bull. Soc. Chim. Fr.* 592.

Asselineau, J., ed., 1966, *The Bacterial Lipids*, Hermann, Paris and Holden-Day, San Francisco.

Asselineau, J., Ryhage, R., and Stenhagen, E., 1957, Mass spectrometric studies on long chain methyl esters. A determination of the molecular weight and structure of mycocerosic acid, *Acta Chem. Scand.* **11**:196.

Ballio, A., and Barcellona, S., 1968, Relations chimiques et immunologiques chez les Actinomycétales. I. Les acides gras de 43 souches d'Actinomycètes aèrobies, *Ann. Inst. Pasteur* **114**:121.

Ballio, A., and Barcellona, S., 1971, Identification of 10–methyl-branched fatty acids in *Microbispora parva* by combined gas chromatography–mass spectrometry, *Gazz. Chim. Ital.* **101**:635.

Ballio, A., Casinovi, C. G., Framondino, M., Marino, G., Nota, G., and Santurbano, B., 1979, A new cerebroside from *Fusicoccum amygdali* DEL., *Biochim. Biophys. Acta* **573**:51.

Barber, M., Jollès, P., Vilkas, E., and Lederer, E., 1965a, Determination of amino acid sequences in oligopeptides by mass spectrometry. I. The structure of fortuitine, an acyl-nonapeptide methyl ester, *Biochem. Biophys. Res. Commun.* **18**:469.

Barber, M., Wolstenholme, W. A., Guinand, M., Michel, G., Das, B. C., and Lederer, E., 1965b, Determination of amino acid sequences in oligopeptides by mass spectrometry. II. The structure of peptidolipin NA, *Tetrahedron Lett.* 1331.

Beilby, J. P., and Kidby, D. K., 1980, Sterol composition of ungerminated and germinated spores of the vesicular arbuscular mycorrhizal fungus *Glomus caledonus*, *Lipids* **15**:375.

Bird, C. W., Lynch, J. M., Pirt, F. J., Reid, W. W., Brooks, C. J. W., and Middletlitch, 1971, Steroids and squalene in *Methylococcus capsulatus* grown on methane, *Nature (London)* **230**:473.

Bishop, D. G., and Still, J. L., 1963, Fatty acid metabolism in *Serratia marcescens*. III. The constituent fatty acids of the cell, *J. Lipid Res.* **4**:81.

Boe, B. and Gjerde, J., 1980, Fatty acid patterns in the classification of some representatives of the *Enterobacteriaceae* and *Vibrionaceae*, *J. Gen. Microbiol.* **116**:41.

Boon, J. J., De Leeuw, J. W., van den Hoek, G. J., and Vosjan, J. H., 1977a, Significance and taxonomic value of *iso* and *anteiso* monoenoic fatty acids and branched β-hydroxy acids in *Desulfovibrio desulfuricans*, *J. Bacteriol.* **129**:1183.

Boon, J. J., van de Graaf, B., Schuyl, P. J. W., De Lange, F., and De Leeuw, J. W., 1977b, The mass spectrometry of *iso* and *anteiso* monoenoic fatty acids, *Lipids* **12**:717.

Bordet, C., and Michel, G., 1963, Etude des acides gras isolés de plusieurs espèces de Nocardia, *Biochim. Biophys. Acta* **70**:613.

Bouvier, P., Rohmer, M., Benveniste, P., and Ourisson, G., 1976, $\Delta^{8(14)}$-Steroids in the bacterium *Methylococcus capsulatus*, *Biochem. J.* **159**:267.

Bowie, I. S., Grigor, M. R., Dunckley, G. G., Loutit, M. W., and Loutit, J. S., 1972, The DNA base composition and fatty acid constitution of some gram positive pleomorphic soil bacteria, *Soil Biol. Biochem.* **4**:397.

Campbell, I. M., and Naworal, J., 1969a, Mass spectral discrimination between monoenoic and cyclopropanoid, and between normal, *iso* and *anteiso* fatty acid methyl esters, *J. Lipid Res.* **10**:589.

Campbell, I. M., and Naworal, J., 1969b, Composition of the saturated and monounsaturated fatty acids of *Mycobacterium phlei*, *J. Lipid Res.* **10**:593.

Carroll, K. K., Cutts, J. H., and Murray, E. G. D., 1968, The lipids of *Listeria monocytogenes*, *Can. J. Biochem.* **46**:899.

Carter, H. E., and Gaver, R. C., 1967, Branched chain sphingosines from *Tetrahymena pyriformis*, *Biochem. Biophys. Res. Commun.* **29**:886.

Caspi, E., 1980, Biosynthesis of tetrahymanol by *Tetrahymena pyriformis*: Mechanistic and evolutionary implications, *Accounts Chem. Res.* **13**:97.

Chal, R., and Harrison, A. G., 1981, Location of double bonds by chemical ionization mass spectrometry, *Anal. Chem.* **53**:34.

Chan, M., Himes, R. H., and Akagi, J. M., 1971, Fatty acid composition of thermophilic, mesophilic and psychrophilic Clostridia, *J. Bacteriol.* **106**:876.

Corina, D. L., and Sesardic, D., 1980, Profile analysis of total mycolic acids from skin corynebacteria and from named *Corynebacterium* strains by gas–liquid chromatography and GC/MS, *J. Gen. Microbiol.* **116**:61.

Cronan, J. E., Nunn, W. D., and Batchelor, J. G., 1974, Studies on the biosynthesis of cyclopropane fatty acids in *Escherichia coli*, *Biochim. Biophys. Acta* **348**:63.

Degré, R., Higgins, R., Carbonneau, M., Bilodeau, M., and Hebert, J., 1980, Fatty acid composition of three strains of *Leptospira interrogans* serotype *icterohemorrhagiae* from highly virulent to avirulent, *FEMS-Microbiol. Lett.* **8**:275.

Dekker, R. F., Tietschel, E. T., and Sandermann, H., 1977, Isolation of α-glucan and lipopolysaccharide fractions from *Acetobacter xylinum*, *Arch. Microbiol.* **115**:353.

De Lederkremer, R. M., Casal, O. L., Tanaka, C. T., and Colli, W., 1978, Ceramide and inositol content of the lipopeptidophosphoglycan from *Trypanosoma cruzi*, *Biochem. Biophys. Res. Commun.* **85**,1268.

De Rosa, M., De Rosa, S., Gambacorta, A., and Bu'Lock, J. D., 1980, Structure of calditol, a new branched chain nonitol, and of the derived tetraether lipids in thermoacidophile archaebacteria of the *Caldariella* group, *Phytochemistry* **19**:249.

De Souza, N. J., and Nes, W. R., 1968, Sterols: Isolation from a blue-green alga, *Science* **162**:363.

Drucker, D. B., 1974, Aerobic streptococcal fatty acid fingerprints, *Microbios* **11A**:15.

Eistert, B., 1948, "Syntheses with Diazomethane," in *Newer Methods of Preparative Organic Chemistry*, Wiley-Interscience, New York, pp. 513–570.

Fak, K. E., Karlsson, K. A., and Samuelsson, B. E., 1980, Structural analysis by mass spectrometry and NMR spectroscopy of the glycolipid sulfate from *Halobacterium salinarium* and a note on its possible function, *Chem. Phys. Lipids* **27**:9.

Fischer, W., and Landgraf, H. R., 1975, Glycerophosphoryl phosphatidyl kojibiosyl diacylglycerol, a novel phosphoglucolipid from *Streptococcus faecalis*, *Biochim. Biophys. Acta* **380**:227.

Fisher-Hoch, S., Hudson, M. J., and Thompson, M. H., 1979, Identification of a clinical isolate as *Legionella pneumophila* by gas chromatography and mass spectrometry of cellular fatty acids, *The Lancet* No. **8138**:323.

Fox, G. E., Magrum, L. J., Balch, W. E., Wolfe, R. S., and Woese, C. R., 1977, Classification of methanogenic bacteria by 16S ribosomal RNA characterization, *Proc. Natl. Acad. Sci. U.S.A.*, **74**:4537.

Fujino, Y., and Onishi, M., 1977, Structure of cerebroside in *Aspergillus oryzae*, *Biochim. Biophys. Acta* **486**:161.

Gailly, C., Sandra, P., Verzele, M., and Cocito, C., 1982, Analysis of mycolic acids of a group of Corynebacteria by capillary gas chromatography and mass spectrometry, *Eur. J. Biochem.*, **125**:83.

Gelpi, E., Oró, J., Schneider, H. J., and Bennett, E. O., 1968, Olefins of high molecular weight in two microscopic algae, *Science* **161**:700.

Gensler, W. J., and Marshall, J. P., 1977, Structure of mycobacterial bis-cyclopropane mycolates by mass spectrometry, *Chem. Phys. Lipids*, **19**:128.

Gerson, T., Patel, J. J., and Nixon, L. N., 1975, Some unusual fatty acids of *Rhizobium*, *Lipids* **10**:134.

Gmeiner, J., and Martin, H. H., 1976, Phospholipid and lipopolysaccharide in *Proteus mirabilis* and its stable protoplast L-form, *Eur. J. Biochem.* **67**:487.

Goodfellow, M., Orlean, P. A. B., Collins, M. D., Alshamaony, L., and Minnikin, D. E., 1978, Chemical and numerical taxonomy of strains received as *Gordona aurantiaca, J. Gen. Microbiol.* **109**:57.

Goren, M. B., Brokl, O., Das, B. C., and Lederer, E., 1971, Sulfolipid I of *Mycobacterium tuberculosis*, strain H37 Rv. Nature of the acyl substituents, *Biochemistry* **10**:72.

Han, J., McCarthy, E. D., Calvin, M., and Benn, M. H., 1968, Hydrocarbon constituents of the blue-green algae *Nostoc muscorum, Anacystis nidulans, Phormidium luridum* and *Chlorogloea fritschii, J. Chem. Soc. (C)* 2785.

Harvey, D. J., 1982, Picolinyl esters as derivatives for the structural determination of long chain branched and unsaturated fatty acids, *Biomed. Mass Spectrom.*, **9**:33.

Hase, S., and Rietschel, E. T., 1977, The chemical structure of the lipid A component of lipopolysaccharides from *Chromobacterium violaceum* NCTC 9694, *Eur. J. Biochem.* **75**:23.

Heefner, D. L., and Claus, G., 1978, Lipid and fatty acid composition of *Gluconobacter oxydans* before and after cytoplasmic membrane formation, *J. Bacteriol.* **134**:38.

Hilker, D. R., Gross, M. L., Knoche, H. W., and Shively, J. M., 1978, The interpretation of the mass spectrum of an ornithine-containing lipid from *Thiobacillus thiooxidans, Biomed. Mass Spectrom.* **5**:64.

Hilker, D. R., Knoche, H. W., and Gross, M. L., 1979, Thermolysis chemical ionization of a complex polar lipid, *Biomed. Mass Spectrom.* **6**:356.

Hiramoto, M., Okada, K., and Nagai, S., 1970, The revised structure of viscosin, a peptide antibiotic, *Tetrahedron Lett.* 1087.

Hunter, S. W., Fujiwara, T., and Brennan, P., 1982, Structure and antigenicity of the major specific glycolipid antigen of *Mycobacterium leprae, J. Biol. Chem.*, **257**:15072.

Hurlbert, R. E., Weckesser, J., Mayer, H., and Fromme, I., 1976, Isolation and characterization of the lipopolysaccharide of *Chromatium vinosum, Eur. J. Biochem.* **68**:365.

James, A. T., 1960, "Qualitative and Quantitative Determination of the Fatty Acids by Gas–Liquid Chromatography", in *Methods of Biochemical Analysis*, Wiley Interscience, New York, Vol. 8, pp. 1–59.

Jamieson, G. R., 1970, "Structure Determination of Fatty Esters by Gas-Liquid Chromatography," in *Topics in Lipid Chemistry*, Logos Press Ltd., London, Vol. 1, pp. 107–159.

Jantzen, E., Berdal, B. P., and Omland, T., 1979, Cellular fatty acid composition of *Francisella tularensis, J. Clin. Microbiol.* **10**:928.

Jantzen, E., Berdal, B. P., and Omland, T., 1980, Cellular fatty acid composition of *Haemophilus* species, *Pasteurella multocida, Actinobacillus actinomycetemcomitans* and *Haemophilus vaginalis, Acta Path. Microbiol. Scand., Sect. B,* **88**:89.

Jantzen, E., Bergan, T., and Bøvre, K., 1974, Gas chromatography of bacterial whole cell methanolysates. VI. Fatty acid composition of strains within *Micrococcaceae, Acta Path. Microbiol. Scand., Sect. B,* **82**:785.

Jantzen, E., Bryn, K., Hagen, N., Bergan, T., and Bøvre, K., 1978, Fatty acids and monosaccharides of *Neisseriaceae* in relation to established taxonomy, *NIPH Ann.* **1**:59.

Jantzen, E., and Lassen, J., 1980, Characterization of *Yersinia* species by analysis of whole cell fatty acids, *Intern. J. System. Bacteriol.* **30**:421.

Johns, R. B., and Perry, G. J., 1977, Lipids of the marine bacterium *Flexibacter polymorphus, Arch. Microbiol.* **114**:267.

Joseph, J. D., 1975, Identification of 3,9,12,15-octadecapentaenoic acid in laboratory-cultured photosynthetic dinoflagellates, *Lipids* **10**:395.

Júlak, J., Tureček, F., and Mikova, Z., 1980, Identification of characteristic branched chain fatty acids of *Mycobacterium kansasii* and *M. gordonae* by gas chromatography–mass spectrometry, *J. Chromatog.* **190**:183.

Kagawa, Y., and Ariga, T., 1977, Determination of molecular species of phospholipids from thermophilic bacterium PS3 by mass chromatography, *J. Biochem. (Tokyo)* **81**:1161.

Kakinuma, A., Sugino, H., Isono, M., Tamura, G., and Arima, K., 1969, Determination of fatty acid in surfactin and elucidation of the total structure of surfactin, *Agr. Biol. Chem.* **33**:973.

Kaneda, T., 1977, Fatty acids of the genus *Bacillus*: An example of branched chain chain preference, *Bacteriol. Rev.* **41**:391.

Kaneshiro, T., and Marr, A. G., 1962, Phospholipids of *Azotobacter agilis*, *Agrobacterium tumefaciens* and *Escherichia coli*, *J. Lipid Res.* **3**:184.

Karlsson, K. A., 1966, The chemical structure of phytosphingosine of human origin and a note on the lipid composition of the yeast *Hansenula ciferii*, *Acta Chem. Scand.* **20**:2884.

Karlsson, K. A., Samuelsson, B. E., and Steen, G. O., 1973, Improved identification of monomethyl paraffin chain branching (close to the methyl end) of long chain compounds by gas chromatography and mass spectrometry, *Chem. Phys. Lipids* **11**:17.

Kates, M., 1978, The phytanyl ether-linked polar lipids and isoprenoid neutral lipids of extremely halophilic bacteria, *Prog. Chem. Fats Other Lipids* **15**:301.

Kates, M., Tremblay, P., Anderson, R., and Volcani, B. E., 1978, Identification of the free and conjugated sterol in a nonphotosynthetic diatom, *Nitzchia alba*, as 24-methylene-cholesterol, *Lipids* **13**:34.

Kates, M., Yengoyan, L. S., and Sastry, P. S., 1965, A diether analog of phosphatidyl-glycerophosphate in *Halobacterium cutirubrum*, *Biochim. Biophys. Acta* **98**:252.

Kawahara, K., Uchida, K., and Aida, K., 1979, Direct hydroxylation in the biosynthesis of hydroxy fatty acids in Lipid A of *Pseudomonas ovalis*, *Biochim. Biophys. Acta* **572**:1.

Kawanami, J., and Otsuka, H., 1969, Lipids of *Streptomyces sioyaensis*. VI. On the β-hydroxy fatty acids in siolipin, *Chem. Phys. Lipids* **3**:135.

Keddie, R. M., and Cure, G. L., 1977, The composition of the cell wall and distribution of free mycolic acids in named strains of Coryneform bacteria and in isolates from various natural sources, *J. Appl. Bacteriol.* **42**:229.

Kimura, A., 1976, Presence of free bases of C_{18}- and C_{20}-phytosphingosine in a yeast, *Candida intermedia*, *Agr. Biol. Chem.* **40**:239.

Klein, R. A., Hazlewood, G. P., Kemp, P., and Dawson, R. M. C., 1979, A new series of long chain dicarboxylic acids with vicinal dimethyl branching found as major components of the lipids of *Butyrivibrio* spp., *Biochem. J.* **183**:691.

Knoche, H. W., and Shively, J. M., 1969, The identification of *cis*- 11,12-methylene-2-hydroxyoctadecanoic acid from *Thiobacillus thiooxidans*, *J. Biol. Chem.* **244**:4773.

Kobayashi, T., Nishijima, M., Tamori, Y., Nojima, S., Seyama, Y., and Yamakawa, T., 1980, Acyl phosphatidylglycerol of *Escherichia coli*, *Biochim. Biophys. Acta*, **620**:356.

Kuksis, A., 1978, "Separation and Determination of Structure of Fatty Acids," in *Handbook of Lipid Research. 1. Fatty Acids and Glycerides*, (A. Kuksis, ed.), Plenum Press, New York, pp. 1–76.

Kulmaczand, R. J., and Schroepfer, G. J., 1978, Dramatic alteration of sphingolipid bases of *Hansenula ciferii* by exogenous fatty acids, *Biochem. Biophys. Res. Commun.* **82**:371.

Laine, R. A., and Fischer, W., 1978, On the relationship between glycerophospholipids and lipoteichoic acids of gram positive bacteria. III. Di(glycerophospho)-acylkojibiosyl-diacylglycerol and related compounds from *Streptococcus lactis* NCDO 712, *Biochim. Biophys. Acta* **529**:250.

Laine, R. A., Young, N. D., Gerber, J. N., and Sweeley, C. C., 1974, Identification of 2-hydroxy fatty acids in complex mixtures of fatty acid methyl esters by mass chromatography, *Biomed. Mass Spectrom.* **1**:10.

Lamonica, G., and Etemadi, H., 1967, Sur la coupure spécifique, en spectrométrie de masse, des aldéhydes linéaires comportant des cycles propaniques, *Bull. Soc. Chim. Fr.* 4275.

Lanéelle, G., and Asselineau, J., 1968, Structure d'un glycoside de peptidolipide isolé d'une mycobactérie, *Eur. J. Biochem.* **5**:487.

Lanéelle, M. A., Asselineau, J., and Castlenuovo, G., 1968, Relations chimiques et immunologiques chez les Actinomycétales. IV. Composition chimique des lipides de quatre souches de *Streptomyces* et d'une souche de *N. (Str.) gardneri, Ann. Inst. Pasteur* **114**:305.

Lanéelle, G., Asselineau, J., Wolstenholme, W. A., and Lederer, E., 1965, Détermination de séquences d'acides aminés dans des oligopeptides par la spectrométrie de masse. III. Structure d'un peptidolipide de *Mycobacterium johnei, Bull. Soc. Chim. Fr.* 2133.

Langworthy, T. A., and Mayberry, W. R., 1976, A 1,2,3,4-tetrahydroxypentane-substituted pentacyclic triterpene from *Bacillus acidocaldarious, Biochim. Biophys. Acta* **431**:570.

Langworthy, T. A., Mayberry, W. R., and Smith, P. F., 1976, A sulfonolipid and novel glucosaminyl glycolipids from the extreme acidothermophile *Bacillus acidocaldarius, Biochim. Biophys. Acta* **431**:550.

Larsson, L., Märdh, P. A., and Odham, G., 1979, Detection of tuberculostearic acid in mycobacteria and nocardiae by gas chromatography and mass spectrometry using selected ion monitoring, *J. Chromatog., Biomed. Appl.* **163**:221.

Lederer, E., Adam, A., Ciorbaru, R., Petit, J. F., and Wietzerbin, J., 1975, Cell walls of mycobacteria and related organisms, *Mol. Cell. Biochem.* **7**:87.

Leonhardt, B. A., De Vilbiss, E. D., and Klun, J. A., 1983, Gas chromatographic mass spectrometric indication of double bond position in mono-unsaturated primary acetates and alcohols without derivatization, *Org. Mass Spectrom.*, **18**:9.

McCarthy, E. D., Han, J., and Calvin, M., 1968, Hydrogen atom transfer in mass spectrometric fragmentation patterns of saturated aliphatic hydrocarbons, *Anal. Chem.* **40**:1475.

McCarthy, J. A., 1970, "Mass Spectrometry of Fatty Acid Derivatives", in *Topics in Lipid Chemistry*, (F. D. Gunstone, ed.), Logos Press Ltd., London, Vol. 1, pp. 369–440.

McCloskey, J. A., and Law, J. H., 1967, Ring location in cyclopropane fatty acid esters by a mass spectrometric method, *Lipids* **2**:225.

Madhavan, V. N., Done, J., and Vine, J., 1981, Characterization of two ornithine-containing lipids from *Erwinia aroideae, Chem. Phys. Lipids* **28**:79.

Maitra, S. K., Schotz, M. C., Yoshikawa, T. T., and Guze, L. B., 1978, Determination of lipid A and endotoxin in serum by mass spectroscopy, *Proc. Natl. Acad. Sci. U.S.A.*, **75**:3993.

Marshall, J. L., Erickson, K. C., and Folsom, T. K., 1970, The esterification of carboxylic acids using a boron trifluoride-etherate-alcohol reagent, *Tetrahedron Lett.* 4011.

Martin, G., and Asselineau, J., 1969, Chromatographie en phase gazeuse de dérivés palmitoylés de sucres simples, *J. Chromatog.* **39**:322.

Matthews, H. M., Yang, Y. K., and Jenkin, H. M., 1980, Alk-1-enyl ether phospholipids (plasmalogens) and glycolipids of *Treponema hyodysenteriae, Biochim. Biophys. Acta* **618**:273.

Minnikin, D. E., 1972, Ring location in cyclopropane fatty acid esters by boron trifluoride-catalyzed methoxylation followed by mass spectroscopy, *Lipids* **7**:398.

Minnikin, D. E., 1978, Location of double bonds and cyclopropane rings in fatty acids by mass spectrometry, *Chem. Phys. Lipids* **21**:313.

Minninkin, D. E., and Polgar, N., 1967a, Structural studies on the mycolic acids, *Chem. Commun.* 312.

Minnikin, D. E., and Polgar, N., 1967b, The methoxymycolic and ketomycolic acids from human tubercle bacilli, *Chem. Commun.* 1172.

Miyagawa, E., Azuma, R., Suto, T., and Yano, I., 1979, Occurrence of free ceramides in *Bacteroides fragilis* NCTC, *J. Biochem. (Tokyo)* **86**:311.

Moss, C. W., and Dees, S. B., 1975, Identification of microorganisms by gas chromatographic-mass spectrometric analysis of cellular fatty acids, *J. Chromatog.* **12**:595.

Moss, C. W., and Dees, S. B., 1976, Cellular fatty acids and metabolic products of *Pseudomonas* species obtained from clinical specimens, *J. Clin. Microbiol.* **4**:492.

Moss, C. W., Dowell, V. R., Farshtchi, D., Raines, L. J., and Cherry, W. B., 1969, Cultural characteristics and fatty acid composition of Propionibacteria, *J. Bacteriol.* **97**:561.

Murata, T., Ariga, T., and Araki, E., 1978, Determination of double bond positions of unsaturated fatty acids by a chemical ionization mass spectrometry computer system, *J. Lipid Res.* **19**:172.

Nakajima, N., Imukai, M., Haneishi, T., Terahara, A., Arai, M., Kinoshita, T., and Tamura, C., 1978, Structural determination of globomycin, *J. Antibiot. (Tokyo)* **31**:426.

Nishii, M., Kihara, T., Isono, K., Higashijima, T., and Miyazawa, T., 1980, The structure of lipopeptin A, *Tetrahedron Lett.* **21**:4627.

Nishimura, Y., Yamamoto, H., and Iizuka, H., 1979, Taxonomical studies of *Acinetobacter* species. Cellular fatty acid composition, *Z. Allg. Mikrobiol.* **19**:307.

Odham, G., Larsson, L., and Mårdh, P. A., 1979, Demonstration of tuberculostearic acid in sputum from patients with pulmonary tuberculosis by selected ion monitoring, *J. Clin. Invest.* **63**:813.

Odham, G., Stenhagen, E., and Waern, K., 1970, Stereospecific total synthesis of mycocerosic acids, *Ark. Kemi* **31**:533.

Ohno, Y., Yano, I., Hiramatsu, T., and Masui, M., 1976, Lipids and fatty acids of a moderately halophilic bacterium, *Biochim. Biophys. Acta* **424**:337.

Oshima, M., and Ariga, T., 1975, ω-Cyclohexyl fatty acids in acidophilic thermophilic bacteria, *J. Biol. Chem.* **250**:6963.

Panos, C., 1965, Separation and identification of positional isomers of bacterial long chain monoethenoid fatty acids by Golay column chromatography, *J. Gas Chromatog.* **3**:278.

Panos, C., and Henrickson, C. V., 1968, Resolution of positional isomers of bacterial long chain cyclopropane ring containing fatty acids by capillary column chromatography, *J. Gas. Chromatog.* **6**:551.

Peypoux, F., Guinand, M., Michel, G., Delcambe, L., Das, B. C., Varenne, P., and Lederer, E., 1973, Isolement de l'acide 3-amino-12-méthyltétradécanoïque and 3-amino-12-méthyltridécanoïque à partir de l'iturine, antibiotique de *Bacillus subtilis*, *Tetrahedron* **29**:3455.

Peypoux, F., Guinand, M., Michel, G., Delcambe, L., Das, B. C., and Lederer, E., 1978, Structure of iturin A, a peptidolipid antibiotic from *Bacillus subtilis*, *Biochemistry* **17**:3992.

Promé, J. C., 1968, Localisation d'un cycle propanique dans une substance aliphatique par examen du spectre de masse des cétones obtenues par oxydation, *Bull. Soc. Chim. Fr.* 655.

Promé, J. C., and Asselineau, C., 1966, Sur l'oxydation chromique de dialcoyl-1,2 cyclopropanes, *Bull. Soc. Chim. Fr.* 2114.

Puzo, G., and Promé, J. C., 1973, Fragmentation des aldéhydes cyclopropaniques en spectrométrie de masse. Intervention d'interactions bifonctionnelles, *Tetrahedron* **29**:3619.

Puzo, G., Tissié, G., Lacave, C., Aurelle, H., and Promé, J. C., 1978, Structural determination of "Cord Factor" from a *Corynebacterium diphtheriae* strain by a combination of mass spectral ionization methods, *Biomed. Mass Spectrom.* **5**:699.

Rohmer, M., and Ourisson, G., 1976a, Structure des bacteriophanetetrols d'*Acetobacter xylinum*, *Tetrahedron Lett.* 3633.

Rohmer, M., and Ourisson, G., 1976b, Dérivés du bactériophane: Variations structurales et répartition, *Tetrahedron Lett.* 3637.

Roussel, J., and Asselineau, J., 1980, Fatty acid composition of the lipids of *Pseudomonas mildenbergii*. Presence of a fatty acid containing two conjugated double bonds, *Biochim. Biophys. Acta* **619**:689.

Ryhage, R., 1962, Mass spectrometric analysis of the methylstearic acid from *Mycobacterium phlei*, *J. Biol. Chem.* **237**:670.

Ryhage, R., Ställberg-Stenhagen, S., and Stenhagen, E., 1961, Mass spectrometric studies. VII. Methyl esters of α,β-unsaturated long chain acids. On the structure of C_{27}-phthienoic acid, *Ark. Kemi* **18**:179.

Ryhage, R., and Stenhagen, E., 1960a, Mass spectrometric studies. IV. Esters of monomethyl-substituted long chain carboxylic acids, *Ark. Kemi* **15**:291.

Ryhage, R., and Stenhagen, E., 1960b, Mass spectrometry in lipid research, *J. Lipid Res.* **1**:361.

Sasak, W., and Chojnacki, T., 1977, The identification of lipid acceptor and the biosynthesis of lipid-linked glucose in *Bacillus stearothermophilus*, *Arch. Biochem. Biophys.* **181**:402.

Sato, N., Murata, N., Miura, Y., and Ueta, N., 1979, Effect of growth temperature and fatty acid composition in the blue-green algae, *Anaboena variabilis* and *Anacystis nidulans*, *Biochim. Biophys. Acta* **572**:19.

Schubert, K., Rose, G., Watchel, H., Hörhold, C., and Ikekawa, N., 1968, Zum Vorkommen von Sterinen in Bakterien, *Eur. J. Biochem.* **5**:246.

Shah, H. N., and Collins, M. D., 1980, Fatty acid and isoprenoid quinone composition in the classification of *Bacteroides melaninogenicus* and related taxa, *J. Appl. Bacteriol.* **48**:75.

Stein, R. A., Slawson, V., and Mead, J. F., 1967, Gas–liquid Chromatography of Fatty Acids and Derivatives", in *Lipid Chromatographic Analysis* (G. V. Marinetti, ed.), Marcel Dekker, New York, Vol. 1, pp. 361–400.

Su, C. J., Reusch, R., and Sadoff, H. L., 1979, Fatty acids in phospholipids of cells, cysts and germinating cysts of *Azotobacter vinelandii*, *J. Bacteriol.* **137**:1434.

Sugatani, J., Kino, M., Saito, K., Matsuo, T., Matsuda, H., and Katakuse, I., 1982, Analysis of molecular species of phospholipids by field desorption mass spectrometry, *Biomed. Mass Spectrom.*, **9**:293.

Tahara, S., Hosokawa, K., and Mizutani, J., 1980, Occurrence of 7-hydroxyalkanoic acids in *Mucor* species, *Agr. Biol. Chem.* **44**:193.

Takayama, K., Qureshi, N., and Schnoes, H. K., 1978, Isolation and characterization of the monounsaturated long chain fatty acids of *Mycobacterium tuberculosis*, *Lipids* **13**:575.

Thiele, O. W., Lacave, C., and Asselineau, J., 1969, On the fatty acids of *Brucella abortus* and *Brucella melitensis*, *Eur. J. Biochem.* **7**:393.

Thiele, O. W., and Schwinn, G., 1973, The free lipids of *Brucella melitensis* and *Bordetella pertussis*, *Eur. J. Biochem.* **34**:333.

Tomiyasu, I., Toriyama, S., Yano, I., and Masui, M., 1981, Changes in molecular species composition of nocardomycolic acids in *Nocardia rubra* by the growth temperature, *Chem. Phys. Lipids* **28**:41.

Toriyama, S., Yano, I., Masui, M., Kusunose, M., and Kusunose, E., 1978, Separation of C_{50-60} and C_{70-80} mycolic acids and molecular species and their changes by growth temperatures in *Mycobacterium phlei*, *FEBS Lett.* **95**:111.

Tornabene, T. G., and Langworthy, T. A., 1978, Diphytanyl and Dibiphytanyl glycerol ether lipids of methanogeneic Archaebacteria, *Science* **203**:51.

Tornabene, T. G., and Markey, S. P., 1971, Characterization of branched monounsaturated hydrocarbons of *Sarcina lutea* and *Sarcina flava*, *Lipids* **6**:190.

Uchida, K., 1974, Occurrence of saturated and monounsaturated fatty acids in the unusually long chains (C_{20}–C_{30}) in *Lactobacillus heterohiochii*, an alcohophilic bacterium, *Biochim. Biophys. Acta* **348**:86.

Vaver, V. A., and Ushakov, A. N., 1980, High temperature gas–liquid chromatography in lipid analysis, *Methods Biochem. Anal.*, **26**:328.

Veerkamp, J. H., 1971, Fatty acid composition of *Bifidobacterium* and *Lactobacillus* strains, *J. Bacteriol.* **108**:861.

Vilkas, E., Rojas, A., Das, B. C., Wolstenholme, W. A., and Lederer, E., 1966, Détermination de séquences d'acides aminés dans les oligopeptides par la spectrométrie de masse. VI. Structure du mycoside C_b, peptidoglycolipide de *Mycobacterium butyricum*, *Tetrahedron* **22**:2809.

White, D. C., Tucker, A. N., and Sweeley, C. C., 1969, Characterization of the *iso*-branched sphinganines from the ceramide phospholipids of *Bacteroides melaninogenicus*, *Biochim. Biophys. Acta* **187**:527.

Volkman, J. K., Eglinton, G. and Corner, E. D. S., 1980, Sterols and fatty acids of the marine diatom *Biddulphia sinensis*, *Phytochemistry* **19**:1809.

Volkman, J. K., Eglinton, G., Corner, E. D. S., and Forsberg, T. E. V., 1980, Long chain alkenes and alkenones in the marine coccolithophoride *Emiliania huxleyi*, *Phytochemistry* **19**:2619.

Von Rudloff, E., 1956, Periodate-permanganate oxidations. V. Oxidation of lipids in media containing organic solvents, *Can. J. Chem.* **34**:1413.

Wada, H., Okada, H., Suginaka, H., Tomiyasu, I., and Yano, I., 1981, Gas chromatographic and mass spectromtric analysis of molecular species of bacterionemamycolic acids from *Bacterionema matruchotii*, *FEMS Microbiol. Lett.* **11**:187.

Welby-Gieusse, M., Lanéelle, M. A., and Asselineau, J., 1970, Structure des acides corynomy-coliques de *Corynebacterium hofmanii* et leur implication biogénétique, *Eur. J. Biochem.* **13**:164.

Yabuuchi, E., Tanimura, E. Ohyama, A., Yano, I., and Yamamoto, A., 1979, *Flavobacterium devorans* ATCC 10829: A strain of *Pseudomonas paucimobilis*, *J. Gen. Appl. Microbiol.* **25**:95.

Yang, L. L., and Haug, A., 1979, Structure of membrane lipids and physicobiochemical properties of the plasma membrane of *Thermoplasma acidophilum* adapted to growth at 37°C, *Biochim. Biophys. Acta* **573**:308.

Yano, I., Furukawa, Y. and Kusunose, M., 1969, Occurrence of α-hydroxy fatty acids in Actinomycetales, *FEBS Lett.* **4**:96.

Yano, I., and Saito, K., 1972, Gas chromatographic and mass spectrometric analysis of molecular species of corynomycolic acids from *Corynebacterium ulcerans*, *FEBS Lett.* **23**:352.

Yano, I., Toriyama, S., Masui, M., Kusunose, M., Kusunose, E., and Akimori, N., 1978a, Gas chromatographic and mass spectrometric analysis of C_{50-60} mono- and dicarboxy mycolic acids in Mycobacteria, *Proc. 3d Meeting Japan. Soc. Medical Mass Spectrom.* **3**:169.

Yano, I., Kageyama, K., Ohno, Y., Masui, M., Kusunose, E., Kusunose, M., and Akimori, N., 1978b, Separation and analysis of molecular species of mycolic acids in Nocardia and related taxa by gas chromatography–mass spectrometry, *Biomed. Mass. Spectrom.* **5**:14.

Zepke, H. D., Heinz, E., Radunz, A., Linscheid, M., and Pesch, R., 1978, Combination and positional distribution of fatty acids in lipids of blue-green algae, *Arch. Microbiol.* **119**:157.

4

CARBOHYDRATES

Danielle Patouraux-Promé and Jean-Claude Promé,
Centre de Recherche de Biochimie et Génétique
Cellulaires du CNRS, Université Paul Sabatier,
118 route de Narbonne, 31062 Toulouse
Cedex, France

1. INTRODUCTION

Gas chromatography–mass spectrometry (GC/MS) has become a very important technique in the determination of carbohydrates structures and important and detailed reviews have been published (see for example Chizhov and Kochetkov, 1966, Lönngren and Svensson, 1974; Radford and De Jongh, 1972.

Most mono- and oligosaccharides are thermally unstable substances which possess high polarity and low volatility. The direct analysis of monosaccharides is often possible by the classical electron-impact or chemical ionization methods using a direct inlet probe. The molecules of lower volatility, such as oligosaccharides, need the desorption techniques to become ionized with a minimal extent of thermal degradation. By field desorption, for example, a free hexasaccharide was ionized and its cationized molecular ion was observed (Puzo and Promé, 1977). The direct analysis of some carbohydrates is also possible by means of the "in-beam" methods, in which the organic solid layer is directly exposed to a chemical ionization plasma. However, ions issued from thermal degradation products are generally observed in the spectra, together with ions derived from intact sugar molecules. The analysis of underivatized sugar molecules will not be presented in this chapter devoted to GC/MS of carbohydrates.

To enhance the volatility of sugars, and to develop good gas–liquid chromatographic properties, these polar molecules must be converted into derivatives. In this way, the strong intermolecular hydrogen or ionic bonds developed by free sugars are transformed into weak hydrophobic bonds by chemical modifications before analysis. The formation of these derivatives, and their chromatographic properties, have been extensively reviewed (Dutton, 1973; 1974). For GC/MS analysis, the structural information which can be expected from a mass spectrometric study governs the choice of the derivative. Moreover, although the electron-impact mode has been the most widely used ionization method, complementary data can be obtained by chemical ionization. This soft ionization technique can give reliable information about the molecular weight and stereochemistry provided that a suitable derivative is used.

Free sugars exist generally as mixtures of anomeric and cyclic forms which are in equilibrium in solution. Upon derivatization, each form is able to generate a stable and volatile product which can give a characteristic GC peak. Thus, one particular sugar could give rise to multiple GC peaks, which could be characterized by a mass spectrum. As will be further presented, the EI mass spectra are only slightly influenced by the stereochemistry of the carbon atoms, and the anomeric forms generally possess quite identical mass spectra. On the contrary, the spectra are greatly modified by the ring size of the derivative, and furanoid or pyranoid forms are easily distinguished. The multiplicity of GC peaks produced by a single sugar, together with the corresponding structural features and ring sized identified by mass spectrometry, could be advantageous for the identification of components in a very simple mixture. However, the chromatogram can become complicated with possible overlapping of peaks. For example, fructose gives three GC peaks after trimethylsilylation (Semenza *et al.*, 1967) and 3-deoxy-D-erythro-hexosulose produces not less than 11 peaks (El Dash and Hodge, 1971). To avoid this difficulty, derivatives of acyclic forms of sugars can be synthesized. One or two GC peaks are generally obtained from a single sugar. Moreover, the interpretation of the corresponding mass spectrum becomes much easier because the main abundant ions arise from single fragmentations. However, not any reliable difference with respect to the stereochemistry of the carbon atoms could be observed in the EI spectrum. Under CI conditions only, and by a careful control of the experimental conditions, some stereoisomers become distinguishable by their mass spectra.

A good way to determine precisely the stereochemical characteristics of the sugars seems to be to synthesize cyclic derivatives of the cyclic forms of sugars, by means of bifunctional reagents. The stereochemical control of the derivatization reactions allows the formation of structurally different isomers from stereoisomers which can be easily differentiated by mass spectrometry. This method seems to be applicable only to simple mixtures since multiple GC peaks are often obtained from a single component.

The spectra of derivatives of simple oligosaccharides has been investigated. Although the mass spectra become more complicated, the fragmentation pattern follows the same general rules as depicted for monosaccharides. However, some specific ions can be used to differentiate the mode of linkage between the different sugars.

An immense amount of work has been done on the mass spectrometry of carbohydrate derivatives. The aim of this chapter is to present some basic rules for the interpretation of the mass spectra of the most useful derivatives used for GC analysis and to evaluate the possibilities and limitations of the technique. In any case, this presentation should not be considered an exhaustive review.

2. DERIVATIVES FROM MONOFUNCTIONAL REAGENTS OF THE CYCLIC FORMS OF MONOSACCHARIDES

2.1. Formation of Derivatives and GC Separations

Methyl ethers were the first volatile derivatives of carbohydrates used for gas–liquid chromatography. The time required for the preparation has mitigated against their general use, particularly after the trimethylsilyl ether derivatives were introduced. However, the methylation procedures are widely used for structural investigation of polysaccharides and the per-methylated polysaccharides give, upon hydrolysis, a mixture of partially methylated carbohydrates. The determination of their structures, by GC/ MS, is a very important problem which may be resolved by several and different ways. The methylation procedures have been reviewed extensively (Dutton, 1974; Björndal *et al.*, 1970; Lindberg, 1972; Lindberg and Lönngren, 1978; Aspinall, 1976). Although the mass spectrometric identification of the partially methylated sugars was done generally on derivatives of their acyclic forms, some workers have attempted to determine their structure by examination of the mass spectra of the cyclic forms of sugars. Numerous reports were presented about GC separations of derivatives of partially methylated glycosides. The separation of partially methylated sugars by the formation of peralkylated derivatives was described in several papers. For example, Saier and Ballou (1968) have detected two different O-methyl glucose moieties in a polysaccharide from *Mycobacterium phlei*. For their identification by GC, they have derivatized these sugars by peralkylation. They found, however, that the perethyl ethers of the methyl 6-O-methyl-D-glucosides and of the 3-O-methyl isomers were not separable by GC, but satisfactory results were obtained using a propylation of the hydroxyl groups. The interpretation of the spectra of the peralkylated derivatives will be presented hereafter and the presence of other groups which derivatize the remaining free hydroxyl groups does not change drastically the fragmentation patterns.

The most widely used derivatization procedure of sugars is their conversion into trimethylsilyl ethers. This transformation is most commonly achieved by a fast reaction with the pyridine–hexamethyldisilazane–trimethylchlorosilane reagent, as described by Sweely *et al.*, (1963). In a typical procedure, a dry sample containing ca. 10 mg of carbohydrate was mixed with 1 ml of anhydrous pyridine, 0.2 ml hexamethyldisilazane (HMDS), and 0.1 ml trimethylchlorosilane (TMCS). The reaction is generally complete within some minutes, but when some sugars react more slowly, the reaction could be conducted at 75–85°C or allowed to continue for 24 h.

This procedure could be applied to microgram amounts (Stewart *et al.*, 1968) and even incompletely dried samples, provided sufficient TMS reagent was used to react with all the water in the sample. Various reagents other than HMDS–TMCS–pyridine have been recommended for trimethylsilylation of sugars. Trimethylsilyl imidazole (TSIM) was found to be a powerful trimethylsilylating reagent. It is relatively tolerant to moisture and reacts rapidly with most samples (Pierce, 1968).

The above-mentioned reagents did not react or reacted very slowly with free amino groups. The full trimethylsilylation of amino sugars needed the addition of other reagents, such as bis(trimethylsilyl) acetamide (BSA). Kärkkäinen and Vihko (1969) used HMDS–TMCS–pyridine–BSA (1:0.5:10:1) (30 min at room temperature) for the preparation of TMS derivatives of amino hexoses. This method produced mono N-TMS, per O-TMS derivatives. The use of BSA–TMCS–CH_3CN (100:1:400) at 60°C during 5 min gave bis-N-TMS per O-TMS derivatives (Hurst, 1973).

Many stationary phases were used for GC separation of TMS derivatives. Most of the workers used nonpolar or slightly polar stationary phases such as SE 30, OV 1, SE 52, OV 17, Carbowax 6000 or 20 M. For example Gray and Ballou (1971) distinguished the 2,3,4,6-tetra-O-methyl D-mannose, the 2,3,6 and the 3,4,6-tri-O-methyl D-mannose and the 2,3, bis-O-methyl D-mannose by GC separation of their TMS ethers on Carbowax 20 M. These partially methylated sugars arose from a degradation study of a mannan from *Mycobacterium phlei.*

Acetates and trifluoroacetates were also used for GC analysis of sugars or partially methylated sugars. These derivatives are readily prepared, but some decomposition problems are observed at high column temperatures, and the interpretation of their mass spectra is somewhat more complicated. A typical preparation is the following: 1 mg is mixed with 30 µl dry pyridine and 30 µl acetic anhydride. Steam bath 1 h in a sealed tube. A comparative study of the GC separation of acetates, trimethylsilyl ethers and trifluoroacetates of several hexosides in the pyranose or furanose forms have been reported (Yoshida *et al.*, 1969).

Generally, the best system for the separation of the different components of a complex mixture of partially methylated sugars depends on their degree of methylation. The formation of different derivatives and the use of several kinds of stationary phases are often necessary to improve a complete resolution. For example, the separation of 2,3,4,6-tetra-*O*-methyl-D-glucose, -D-mannose, and -D-galactose on one liquid phase is nearly imposible when these compounds are obtained as methyl glycosides. However, the acetates of these free sugars could be separated.

2.2. Electron Impact Mass Spectra of Trimethylsilyl Ethers, Methyl Ethers, and Acetate Derivatives of Monosaccharides

The EI mass spectra of trimethylsilyl derivatives of carbohydrates have been studied in detail, using deuterium labeling and high-resolution measurements (De Jongh *et al.*, 1969) and the fragmentation pathway was found to follow the general rules edited by Kochetkov and Chizhov (1966) for the decomposition of methyl ethers. The degradation pattern of the peracetylated glycosides resembles the preceding ones (Biemann *et al.*, 1963), but the fragments may further decompose by a multiplicity of ways, since the acetyl groups can be eliminated as acetic acid (60 mass units), acetoxy radical (59 mass units), ketene (42 mass units), or acetic anhydride (102 mass units).

The general decomposition rules will be depicted simultaneously for these three kinds of derivatives. The mass-over-charge values of the fragments will depend on the nature of the derivative. Their relative intensities would not be expected to be similar, since their relative speeds of formation and decomposition are dependent upon the nature of the protective group. Finally, some fragments are specific to the nature of the substituent.

Only slight differences resulting from the relative stereochemistry of the hydroxyl groups are observed. The spectra are much more dependent upon the ring size of the cyclic form of the sugar. Thus, the mass spectral study permits the determination of structural peculiarities of the sugar (i.e.,

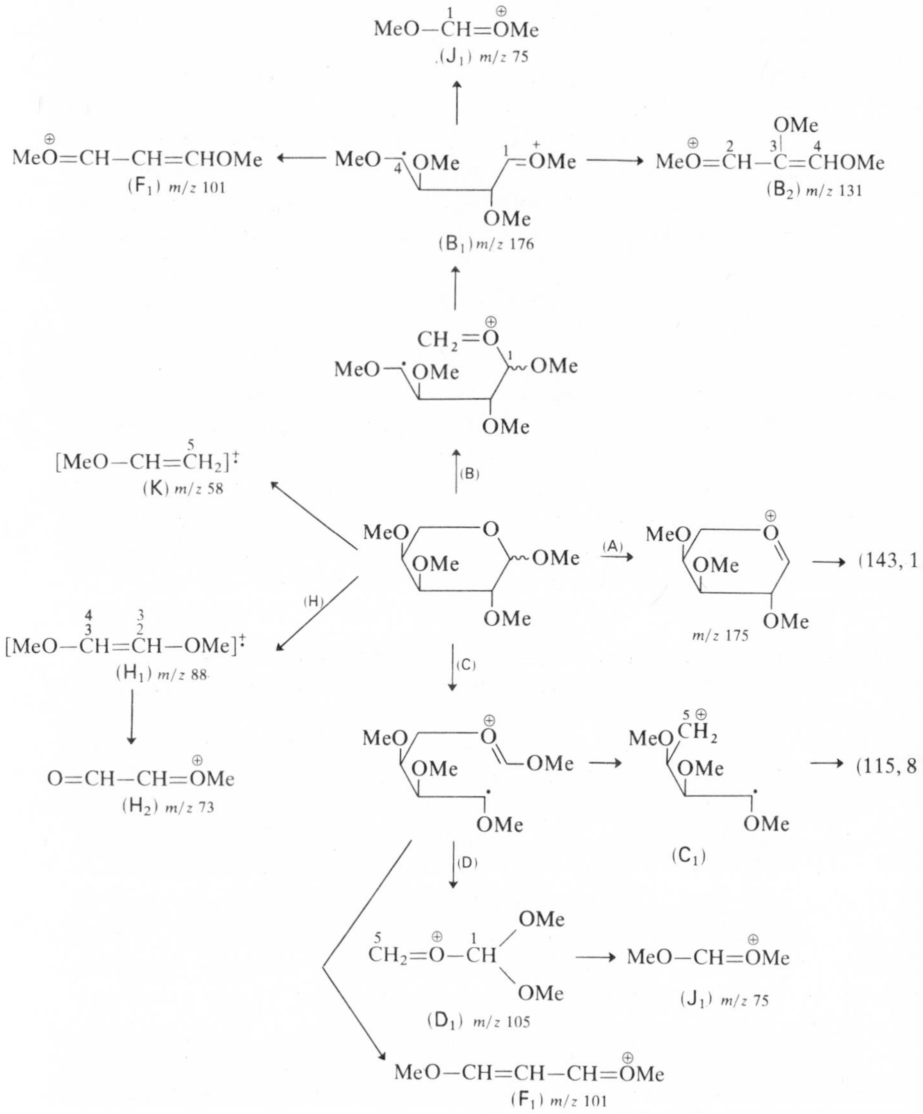

FIGURE 1. Decomposition pathway of a methyl 2,3,4-tri-O-methyl pentopyranoside under electron impact.

molecular weight, number of hydroxyl groups and their relative positions, and ring size), but generally cannot give information allowing the differentiation of stereoisomers. The determination of the gas chromatographic parameters of the sample and their comparison with those of authentic samples is necessary for a complete structural identification.

Kochetkov and Chizhov (1966) have classified the decomposition pattern of the permethyl derivatives of monosaccharides into different ways, giving several ion series, designed by capital letters: A, B, C, etc.

This pattern is shown in Figure 1 using a pentopyranose as a typical example. For hexopyranoses, the fragments containing C5 are shifted by the mass of the C5 substituent towards greater m/z values.

The A-series is produced by the loss of the substituent from C1 giving an oxonium ion with subsequent elimination of the substituents [CH_3OH for methyl ethers, $(CH_3)_3$ SiOH for TMS ethers, acetic acid, ketene or acetic anhydride for acetate esters].

The ions of the B-series are induced by a cleavage of the C4, C5 bond, followed by the elimination of the C5 carbon. Further decompositions produce ions containing three carbon atoms of the sugar ring (B_2 and F_1) and an ion which retains C1 (ion J_1).

1   2

Analogous fragments to the ion J_1 which appears at m/z 75 for permethylated derivatives (as depicted), are found at m/z 191 for pertrimethylsilyl ethers of carbohydrates, such as 1 and 2, and at m/z 133 for pertrimethylsilyl ethers of methylglycosides (ions 3 and 4).

$$Me_3Si-O-\overset{1}{C}H=\overset{+}{O}-SiMe_3 \qquad CH_3-O-\overset{1}{C}H=\overset{+}{O}-SiMe_3$$

m/z 191 m/z 133

3   4

The ion J_1 could also be produced by a C1-C2 cleavage according to (C) followed by a C4–C5 rupture.

The ion B_2, which retains C_2, C_3, and C_4 and three substituent residues mainly, is found at m/z 131 for permethylated derivatives. Its analog is found at m/z 305 for trimethyl silyl ethers (ion 5)

$$Me_3Si-\overset{+}{O}=CH-\underset{\underset{OSiMe_3}{|}}{C}=CH-OSiMe_3 \qquad m/z\ 305$$

5

The loss of one substituent from B_2 gives ion F_1 (m/z 101) in permethyl derivatives. A recent study (De Jong et al., 1980b) shows that this intense ion possesses essentially the depicted allylic structure and not a cyclic one.

About 70% of this ion retains the C2, C3, C4 part of the ring, together with the methoxyl groups which originated from substituents on C2 and C3. The remaining isobaric ions (30%) are formed in another way, and retain the C1, C2, C3 moiety together with the methoxyl groups on C1 and C3. An analogous intense ion is found in the TMS-derivatives at m/z 217 (ion **6**)

$$Me_3SiO-CH=CH-CH=O^+SiMe_3 \qquad m/z\ 217$$

6

The H-series furnishes an intense ion H_1 which retains either C2, C3, or C3–C4 moieties of the sugar ring. The loss of a substituent gives the ion H_2. The H_1 ions contain an odd electron number, and thus are easily discernible by their even mass number. They appear at m/z 88 in permethyl ethers (as depicted in Figure 1). Their analogs are found at m/z 204 in trimethylsilyl ethers. Their intensity (ion **7**) is drastically reduced in the mass spectra of furanoid forms of sugars, and the relative intensity ratio of m/z 204 and 205 is a good diagnostic value for the assignment of ring size in trimethylsilyl ethers of aldohexoses (MacLeod *et al.*, 1976). The latter ion arises from a direct cleavage of the C4–C5 bond and retains C5 and C6 (Figure 2).

$$[Me_3SiO-\overset{\overset{2}{3}}{C}H=\overset{\overset{3}{4}}{C}H-OSiMe_3]^{\ddagger}$$
$$m/z\ 204$$

7

The elimination of the side chain is also found in the permethyl derivatives and produces ions of the E-series which retain all the carbon atoms of the sugar ring. In the furanoid form of aldohexoses, a two-carbon fragment is expelled (Figure 3).

Fragments containing 4-carbon atoms of the sugar backbone are also found in the mass spectra. In permethyl derivatives, the primary odd-electron ions B_1 retain, respectively, C1 through C4. An even-electron ion,

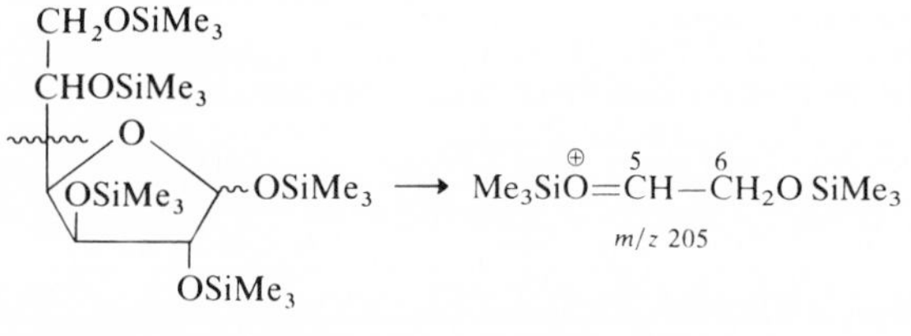

FIGURE 2. Formation of m/z 205 from a pertrimethylsilyl ether of an hexofuranose.

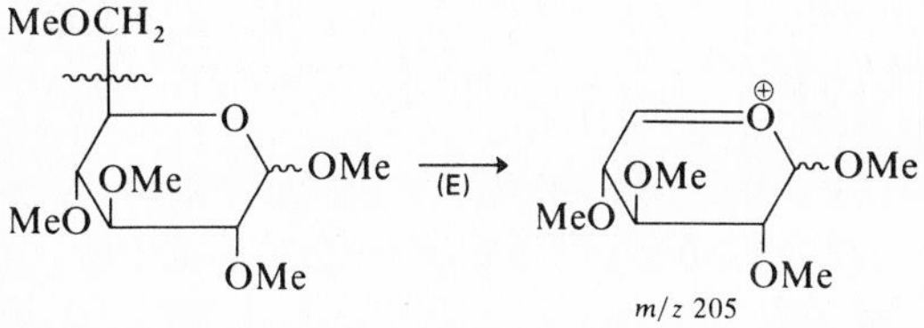

FIGURE 3. Side chain cleavage of permethyl ethers of hexopyranoses.

particularly intense in the furanoid form of the TMS derivatives of hexoses, is found at m/z 319. This ion retains C2 through C5 mainly and contains three trimethylsilyl groups. To our knowledge, its structure is not yet established.

The presence of an acetamido group on the sugar ring modifies the spectrum to some extent. For example, in the TMS-derivatives of methyl-2 acetamido-2-deoxy-, or 3-acetamido-3-deoxy-glucopyranosides, a very intense ion is observed at m/z 173 (**8**). This ion is the analog of the ion H_1.

$$[Me_3SiO-CH=CH-NHCOCH_3]^{+\cdot}$$

m/z 173

8

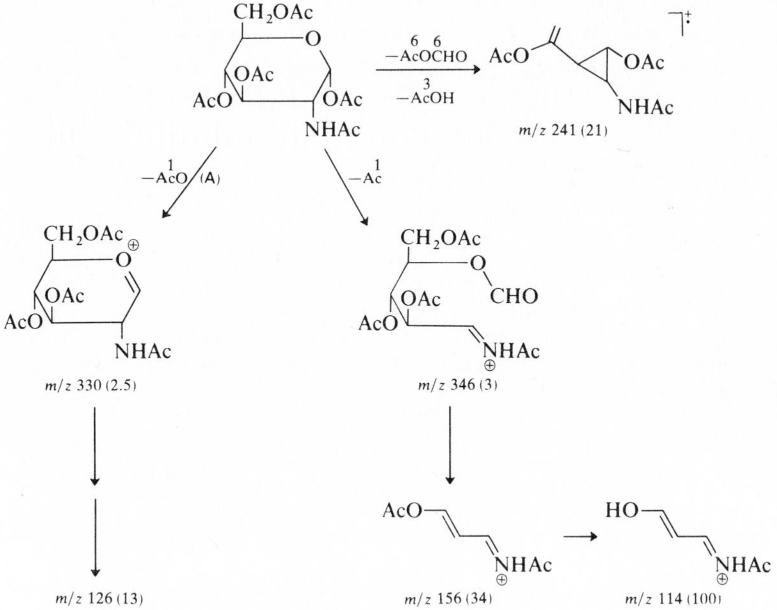

FIGURE 4. Fragmentation pathway of *N*-acetylglycosamine peracetate.

This ion is shifted to m/z 131 or m/z 203 for TMS-derivatives of free amino sugars possessing, respectively, either a free amino group or a monosilylated one (Kärkkäinen and Vihko, 1969).

The ring size could also be determined by the examination of the intensities of some ions, as for neutral monosaccharides (Coduti and Bush, 1977).

The decomposition pathway of the peracetylated 2-acetamido-2-deoxy glycopyranose was studied in detail (Dougherty *et al.*, 1973) and some differences were observed with the pattern of neutral sugars (Figure 4). Cleavages of the C5–C6 bond induce the opening of the sugar ring and the formation of ion m/z 241 for which a cyclopropanoid structure was proposed. The F_1 ion is found at m/z 156 and gives the base peaks at m/z 114 by loss of a ketene molecule. The A-series form a set of ions of low intensity.

The spectra of derivatives of carbohydrates also contain ions that are characteristic of the functional groups which are introduced for derivatization.

For example, the TMS-derivatives easily lose a methyl group from the molecular ion giving a $(M - 15)^+$ which further decomposes by successive losses of trimethylsilanol molecules. The TMS-groups also give rise to an intense trimethylsilyl cation m/z 73 (**9**) and to an ion at m/z 147 formed by cleavage of two vicinal TMS groups. The intensity of this ion could lead to an indication of the relative stereochemistry of the hydroxyl groups (Diekman *et al.*, 1968) (Figure 5).

The acylated glycosides contain an intense acetyl ion CH_3CO^+ at m/z 43, but also form diacetoxonium (m/z 103) and triacetoxonium (m/z 145) ions

$$Me_3Si^+ \qquad CH_3-\overset{\overset{O}{\|}}{C}-\overset{\overset{H}{|}}{O_+}-\overset{\overset{O}{\|}}{C}-CH_3 \qquad CH_3-\overset{\overset{O}{\|}}{C}-\overset{|}{\underset{\underset{CH_3}{\overset{|}{C=O}}}{O^+}}-\overset{\overset{O}{\|}}{C}-CH_3$$

m/z 73	m/z 103	m/z 145
9	**10**	**11**

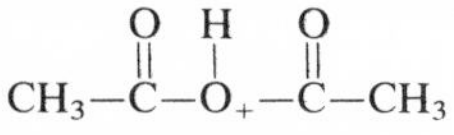

An automatic computer identification of the TMS ethers of neutral sugars was tentatively proposed. The determination is based on the relative

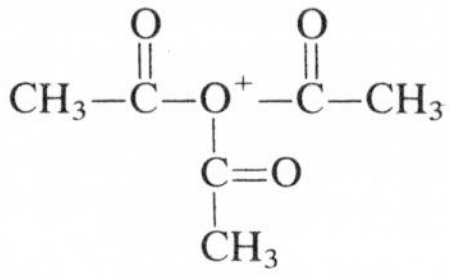

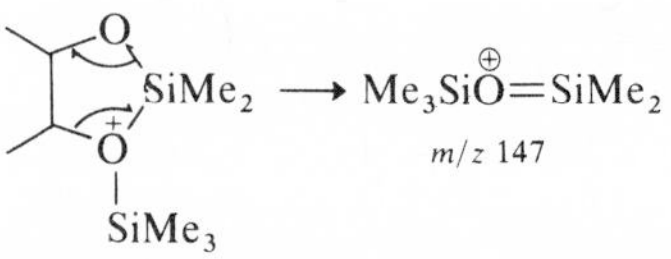

$$\longrightarrow Me_3Si\overset{\oplus}{O}=SiMe_2$$

m/z 147

FIGURE 5. Formation of m/z 147 by interaction between two vicinal TMS groups.

intensity of the five peaks at m/z 73, 147, 191 (**3**), 204 (**7**), and 217 (**6**), together with the last visible fragment at the highest mass and on the relative retention times of each derivative (Jankowski and Gaudin, 1978). However, the definition of the "last visible fragment" seems ambiguous since it is dependent on the signal/background ratio and thus on the amount of substance which is injected.

As already mentioned, the EI mass spectra of methyl ethers, acetates, trifluoroacetate, or TMS derivatives are very little dependent, if at all, on the relative stereochemistry of the hydroxyl groups. Thus the identification of sugars should be helped by the determination of precise GC parameters. The assignement of a structure by both the mass spectra and the retention time may avoid some confusion. For example, in the analysis of sugars in bacterial endotoxins, Davis *et al.* (1969) found GC peaks having the same retention time as KDO. However, none of these peaks have a mass spectrum compatible with the structure of KDO.

The general rules for the decomposition of sugar derivatives under electron impact have been used tentatively for the identification of partially methylated sugars resulting from the hydrolysis of fully methylated polysaccharides. Mass spectrometry was, in this case, a good helper to determine structural characteristics of these products, provided that a good GC separation could be achieved. In some cases, mass spectrometry could also give information about components present in an unresolved GC peak.

The derivatization of the free remaining hydroxyl groups in partially methylated sugars highly improves their separation by GC and the mass spectra allow the location of the methyl groups more certainly. In the mannose series, for example, the TMS derivatives of the different tri-O-methyl mannoses have very different retention times on polar stationary phases (Carbowax 6000: Sikl *et al.*, 1970; carbowax 20 M: Stewart *et al.*, 1968). This result was used for structural studies of some yeast mannans by the identification of sugars arising the hydrolysis of the permethylated samples. Kochetkov and Chizhov (1964) effected a deuteromethylation of the free hydroxyl groups: the assignment of the CD_3 groups by mass spectrometry leads to the determination of the structure of partially methylated saccharides. In a similar way, trimethylsilylation (Matsubara and Hayashi, 1974) or acetylation (De Jongh and Biemann, 1963) gives derivatives having characteristic mass spectra, from which it is possible to determine the number and the position of the methoxyl groups.

It would seem that the possibilities of using GC/MS for the qualitative and quantitative analysis of such mixtures has not been fully exploited. The reasons would be (a) the number of GC peaks in an already complex mixture is increased; (b) the identification of the position of the O-methyl

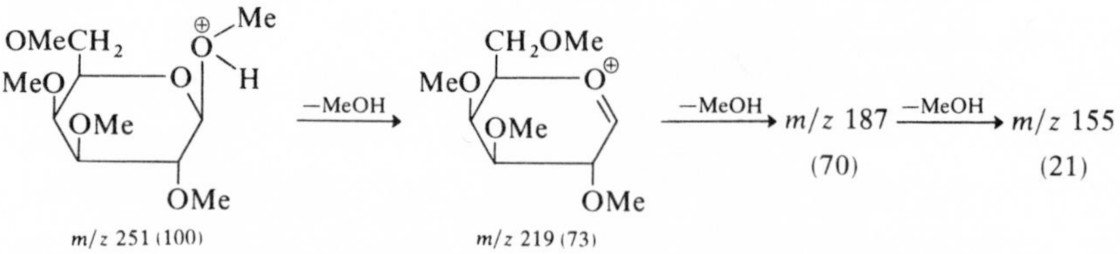

FIGURE 6. Chemical ionization decomposition pathway of methyl 2,3,4,6-tetra-O-methyl-β-D-galactopyranoside. Reactant gas: methane. Relative intensities are indicated in parentheses. (From Kadentsev *et al.*, 1968.)

group may be in some cases relatively difficult; (c) the spectra of acyclic derivatives of sugars can be interpreted more easily and speedily.

2.3. Chemical Ionization Mass Spectra of Trimethylsilyl Ethers, Methyl Ethers, and Acetate Derivatives of Monosaccharides

Under chemical ionization conditions using a protonating reactant gas such as methane or isobutane, the $(M + H)^+$ ion is generally observed with high intensity for the permethyl ether derivatives of monosaccharides. In contrast, this ion possesses a weak intensity, or is absent, for TMS derivatives and acetates. For these derivatives, the molecular weight may be deduced from the ammonia CI spectra which exhibit a $(M + NH_4)^+$ ion.

The $(M + H)^+$ ions of permethyl or peracetyl derivatives decompose by the loss of the C1 substituent, giving an oxonium ion by elimination of

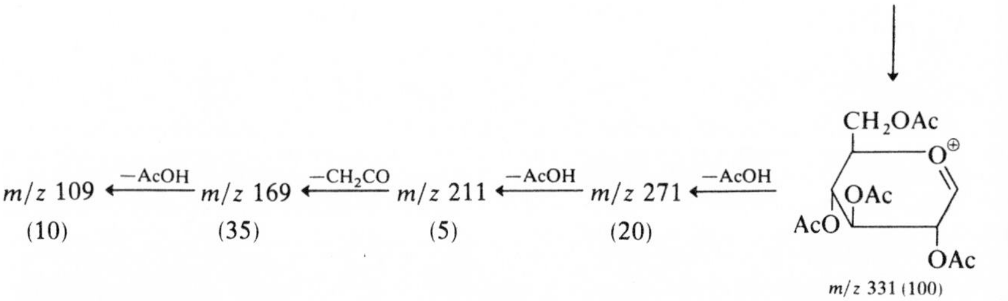

FIGURE 7. Chemical ionization spectra of methyl 2,3,4,6-tetra-O-acetyl-β-D-glucopyranoside. Reactant gas: isobutane. Relative intensities are indicated in parentheses. (From Horton *et al.*, 1974.)

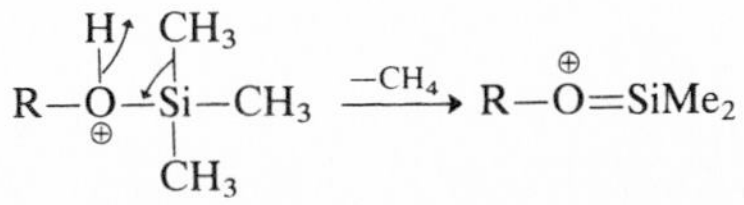

FIGURE 8. Elimination of a methane molecule from a protonated trimethylsilyloxy group, according to ICR measurements (Blair and Bowie, 1979).

$$R-\overset{\oplus}{O}-\overset{\overset{\displaystyle H}{\underset{\displaystyle CH_3}{|}}}{\underset{\displaystyle CH_3}{Si}}-CH_3 \xrightarrow{-CH_4} R-\overset{\oplus}{O}=SiMe_2$$

methanol or acetic acid, respectively, which may be decomposed further by elimination of other substituents (Kadentsev *et al.*, 1978; Horton *et al.*, 1974) (Figures 6 and 7).

The CI isobutane spectrum of TMS-derivatives is more complex. Besides the expected successive eliminations of trimethylsilanol molecules from the MH^+ ion (generally not observed in the spectrum), elimination of a methane molecule produces the formation of the $(MH - 16)^+$ ion (Schoots and Leclercq, 1979), which further decomposes by successive losses of trimethylsilanol. Moreover, some abundant ions at m/z 73, 147, 191, 204, and 217 appear, as in the EI spectra. The elimination of a methane molecule from a protonated trimethylsilyl group is a characteristic behavior of such derivatives (Blair and Bowie, 1979) (Figure 8).

Less abundant fragmentation occurs under CI ammonia conditions. In the peracetylated glycosides, the base peak is $M \cdot NH_4^+$ and only a weak fragment is formed by the loss of the substituent on C1. The spectra of the TMS-ethers remain complex (Murata and Takahashi, 1978). The MH^+ ion (which is not seen) decomposes by successive losses of trimethylsilanol molecules giving an odd-mass-number ion series (for example m/z 451, 361, 271 for the TMS-derivatives of hexoses). However, intense ions produce even-mass-number series (m/z 468, 378, 288, 198). The first member of the latter series was interpreted as the result of an elimination of a trimethylsilyl radical from the MH^+ ion at m/z 541 (Murata and Takahashi, 1978). This interpretation seems improbable, since the "even electron rule" must generally be respected under CI conditions. A most likely assumption is the intervention of an ammonolysis of a trimethylsilyl ether group, giving m/z 468, followed by successive elimination of trimethylsilanol groups (Figure 9a). The ammonolysis of functional groups was presented as a general feature by Haegele and Desiderio (1974).

Another series (m/z 396, 306, 216) was interpreted as successive losses of Me_3Si and CH_2SiMe_2 from the MH^+ ion. A more likely interpretation seems to be the elimination of (bis)trimethylsilyl ether from $M \cdot NH_4^+$ (Figure 9b).

The relative intensity of some ions in the CI spectra allows a differentiation of sugar derivatives according to their stereochemistry. For example, it was found that the relative intensity of the $M \cdot NH_4^+$ ion is much higher for the β-anomer than for the α-anomer in the CI ammonia spectra of pertrimethylsilyl derivatives of hexoses (Murata and Takahashi, 1978). In

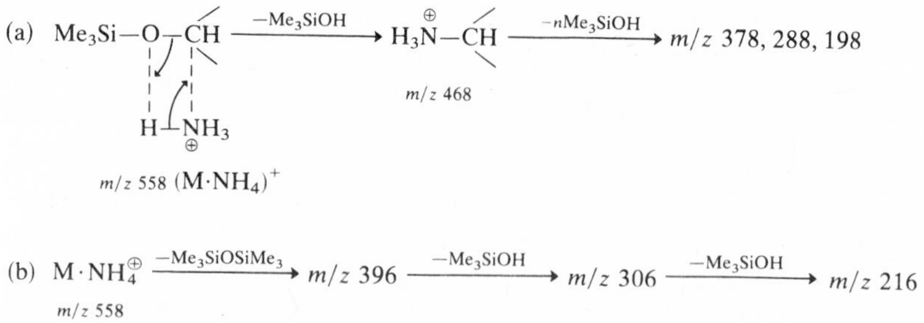

(a)
$$\text{Me}_3\text{Si}-\text{O}-\overset{\diagup}{\underset{\diagdown}{\text{CH}}} \xrightarrow{-\text{Me}_3\text{SiOH}} \overset{\oplus}{\text{H}_3\text{N}}-\overset{\diagup}{\underset{\diagdown}{\text{CH}}} \xrightarrow{-n\text{Me}_3\text{SiOH}} m/z\ 378, 288, 198$$

$$\overset{\diagup}{\underset{\oplus}{\text{H}}}-\overset{|}{\text{NH}_3}$$

m/z 468

m/z 558 $(\text{M}\cdot\text{NH}_4)^+$

(b) $\text{M}\cdot\text{NH}_4^{\oplus} \xrightarrow{-\text{Me}_3\text{SiOSiMe}_3} m/z\ 396 \xrightarrow{-\text{Me}_3\text{SiOH}} m/z\ 306 \xrightarrow{-\text{Me}_3\text{SiOH}} m/z\ 216$

m/z 558

FIGURE 9. Reactions induced by ammonia chemical ionization of TMS derivatives. (a) Ammonolysis in the gas phase of a trimethylsilyloxy group. (b) Explanation of the formation of even-mass number ions.

contrast, the ratio $(\text{MH} - \text{CH}_3\text{OH})^+/(\text{MH})^+$ is always higher for the β-anomer of permethyl hexopyranosides under CI isobutane or methane conditions. Stereoisomers of the TMS derivatives of hexoses also present reliable differences under CI isobutane conditions (Schoots and Leclercq, 1979).

The CI spectra present few fragmentations, if any, of the sugar backbone. A combination of EI and CI spectra is generally needed to obtain a reliable identification of the sugar structure. An example is presented in the analysis of aminosugars from bacterial lipopolysaccharides (Bowser *et al.*, 1978). The lipopolysaccharide from *Brucella* and *Neisseria* strains were hydrolyzed. The aminosugars were isolated by elution from a Dowex 50 H$^+$ resin, *N*-acetylated and then trimethylsilylated. The mixture was analyzed by GC/MS. Two different GC peaks were obtained for each individual aminosugar, for which similar EI or CI (methane) spectra were observed. Under EI conditions, the base peak for all amino-sugars was found to be m/z 173, indicating 2-deoxy-2-amino sugars $(\text{TMSi}-O-\text{CH}=\text{CH}-\text{NH}-\text{COCH}_3)^+$. The *N*-acetylquinovosamine, a 2-acetamido-2,6-dideoxy sugar, was readily identified by its CI spectrum (determination of its molecular weight) and EI spectrum which was very different from other 2-acetamido-2-deoxy hexose derivatives. The differentiation between *N*-acetyl glucosamine, galactosamine, and mannosamine was realized both by their relative GC retention times and their different CI spectra. Thus, the lipopolysaccharides of *Neisseria* strains were found to contain glucosamine. Some strains also contained galactosamine in addition to glucosamine. The *Brucella suis* strains contained glucosamine and quinovosamine and a *Brucella abortus* strain only contained glucosamine.

The peracetates of pentoses and hexoses were also studied by negative ion mass spectrometry (Khvostenko *et al.*, 1976; Tolstikov *et al.*, 1977) in conditions where the molecular anion dissociates through dissociative electron capture. The dissociation occurs by the loss of 43 mass units (CH_3CO) from $M^{-\cdot}$. This even electron fragment anion further decomposes by successive losses of ketene, acetic acid, or acetic anhydride. The base peak is the cluster anion m/z 119 ($CH_3COO--H^--OCOCH_3$). The intensities of the fragments are different according to the structure and the stereochemistry of the molecules.

2.4. Trimethylsilyl Derivatives of Partially Acylated Sugars

The formation of the trimethylsilyl ethers of acylated sugars can be done without isomerization of the acyl groups, provided that some precautions are taken (Promé *et al.*, 1976). The mass spectra of the TMS derivatives of mono- and diacetylated methyl hexopyranosides present characteristic differences allowing the location of the acetyl groups (Boren *et al.*, 1973). The characteristic peaks are mainly due to ions which did not contain the acetyl group. Some similar results are obtained with long-chain acylated glycopyranosides (Puzo and Promé, 1978). In the latter study, the presence of fragments containing the acyl chain was recognized early in the spectra owing to the important mass shift induced by the presence of the acyl chain. These ions appear in a mass range for which the probability of interfering ions is low, and give complementary information on both the nature of the acyl chain and its location on the sugar ring.

An interaction between the chain and the sugar moiety induces the formation of an intromolecular H-transfer which promotes the elimination of the glycosidically linked residue (Puzo *et al.*, 1978a). The fragmentation of the sugar ring produces two types of ions: those containing the acyl chain and those devoid of aliphatic residues. The ions at m/z 204, 217, and 218 allow a preliminary differentiation between the position of the acyl group: m/z 204 has a particularly low intensity in the 3-acylated isomers, m/z 217 has a low intensity in the spectrum of the 4-acylated sugar, and the ratio m/z 218/217 is higher for the 2-acylated isomer. In the 2- and 3- long-chain acylated glycosides, an intense ion *12* is present (m/z 370 for a palmitoyl ester). This ion is absent in the spectra of the 4- and 6-acylated sugars. The 4-acylated isomer forms an ion 13, which is also present in the 2-acylated sugar (m/z 456 for a palmitoyl ester). Finally, the 6-acylated isomer is characterized by an intense ion 14 (m/z 371 for a palmitoyl ester) and the absence of m/z 132. Ions 15 and 16 only contain the acyl group (Figure 10).

These characteristics were used for structural determination of acylated trehaloses from *Corynebacterium diphtheria*: 6-corynomycoloyl-α-trehalose (Puzo and Promé, 1978), 6-β-keto acyltrehaloses (Ahibo-Coffy

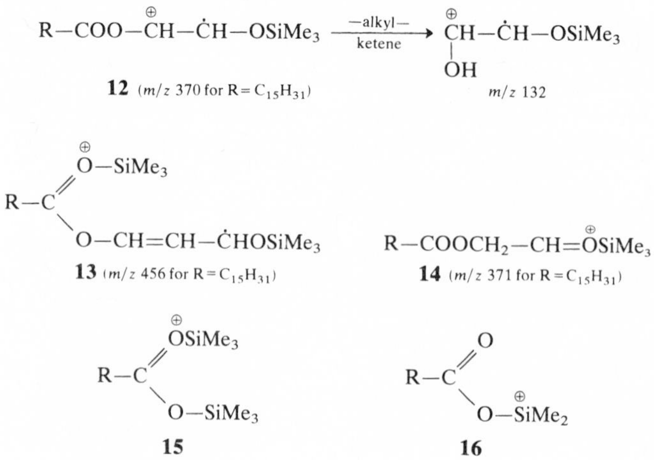

$$R-COO-\overset{\oplus}{C}H-\overset{\cdot}{C}H-OSiMe_3 \xrightarrow[\text{ketene}]{-\text{alkyl}-} \overset{\oplus}{C}H-\overset{\cdot}{C}H-OSiMe_3$$
$$\qquad\qquad\qquad\qquad\qquad\qquad\qquad\qquad\qquad\qquad\qquad\qquad\qquad\qquad | $$
$$\qquad\qquad\qquad\qquad\qquad\qquad\qquad\qquad\qquad\qquad\qquad\qquad\qquad\qquad OH$$

12 (m/z 370 for R = $C_{15}H_{31}$) $\qquad\qquad$ m/z 132

$$R-C \begin{matrix} \overset{\oplus}{O}-SiMe_3 \\ \diagup\diagup \\ \diagdown \\ O-CH=CH-\overset{\cdot}{C}HOSiMe_3 \end{matrix}$$

13 (m/z 456 for R = $C_{15}H_{31}$)

$$R-COOCH_2-CH=\overset{\oplus}{O}SiMe_3$$

14 (m/z 371 for R = $C_{15}H_{31}$)

$$R-C \begin{matrix} \overset{\oplus}{O}SiMe_3 \\ \diagup\diagup \\ \diagdown \\ O-SiMe_3 \end{matrix} \qquad\qquad R-C \begin{matrix} O \\ \diagup\diagup \\ \diagdown \\ O-\overset{\oplus}{S}iMe_2 \end{matrix}$$

15 $\qquad\qquad\qquad\qquad\qquad$ **16**

FIGURE 10. Structures of the main characteristic fragment ions in the spectra of per TMS ethers of long-chain acylated methyl glucopyranosides.

et al., 1978), bis-6,6'-hydroxy and keto esters of trehalose (Puzo et al., 1979; Puzo et al., 1978b). In such studies, the molecular weight of the glycolipid was determined by field desorption, using cationization by cesium salts (Promé and Puzo, 1978).

3. DERIVATIVES OF ACYCLIC FORMS OF CARBOHYDRATES

The presence of a free hemiacetal group in sugars led to multiple structures for the already observed derivatives, thus complicating the GC analysis. Many attempts were made to obtain a single volatile component by chemical transformation of sugars.

Reduction of the carbonyl group of aldoses gives alditol compounds, which can be derivatized further. This procedure, however, produces the loss of what was the beginning and the end of the sugar backbone, and the use of deuterated reducing agents is necessary to avoid the loss of this important information. Since the reduction reaction is not stereospecific, ketoses give two products.

The transformation of the carbonyl group into a nitrile group was another useful way to solve the problem. However, ketoses cannot give such derivatives. According to this procedure, the carbonyl group is first derivatized into an oxime function which is further transformed into a nitrile by loss of water. The oxime function was also used to avoid cyclic forms from sugars, and this derivatization can be applied both to aldoses and ketoses; however, the syn-anti isomerism of the oxime function generally led to two different GC peaks for each sugar. The explanation of how mass spectrometry is used in the identification of such derivatives is the aim of this section.

3.1. Alditol Acetates

A great simplification in the GC pattern of a mixture of sugars was obtained by their transformation into alditol acetates. These derivatives were easily synthesized by reduction of the carbonyl function by NaBH₄ or NaBD₄ followed by per acetylation of the polyol (Björndal *et al.*, 1967a). A typical procedure is the following: 1 mg of sugars were reduced in water (5 ml) with NaBH₄ (or NABD₄) for 2 h. After treatment with Dowex (H⁺) and concentration, boric acid was removed by codistillation with methanol and the product was treated with acetic anhydride pyridine 1:1 (2 ml) at 100°C for 10 min. The reactants were evaporated and the residue was analyzed by GC/MS on different types of columns. ECNSS-M is the most generally useful column-packing for alditol acetates, but other polar stationary phases give good results (for example XE-60, OV-225).

The fragmentation pathways of these derivatives under electron impact are very simple (Golovkina *et al.*, 1966). Generally, no molecular ion is present in the spectrum but $(M-CH_3COO)^+$ is found in low abundance. The mass spectra of stereoisomers are nearly identical. The base peak is the acetylium ion m/z 43 $(CH_3CO)^+$. Primary fragments are formed by cleavage of the carbon chain.

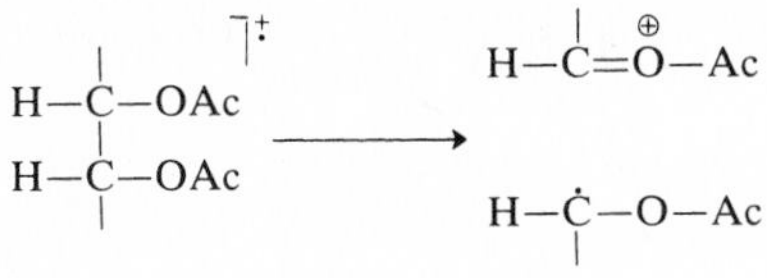

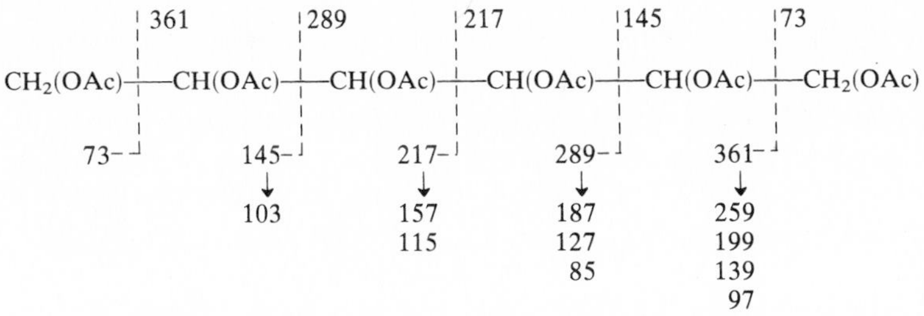

FIGURE 11. Fragmentation pattern of an hexitol peracetate.

These fragments are further degraded, mainly by elimination of acetic acid (60 mass units), ketene (42 mass units) or acetic anhydride (102 mass units) to give fragments having odd mass numbers (Figure 11).

Cleavage at a deoxy group is insignificant as exemplified by the spectrum of alditolacetate from abequose **17**. Introduction of an acetamido group enhances the relative abundance of fragments formed by cleavage at this group (Björndal *et al.*, 1970). In the nonmethylated compounds, fragmentation is governed by the acetamido group (derivative **18** from glucosamine).

```
      CH₂OH                              CH₂OAc
       |         73                        |
 - - - | - - - - - - - -            - - - - | - - - - - - -
       |       245                          |         360
      CHOAc                             HCNHAc
       |                        144 _ _ | _ _ _ _ _ _ _ _ _
      CH₂                               |
       |                              AcOCH
 AcO—CH                                  |
 - - - | - - - -                       CHOAc
       |       231                        |
       |        87                      CHOAc
      CHOAc                               |
       |                              CH₂OAc
      CH₃

        17                               18
```

However odd-electron ions are observed in the spectra of peracetylated polyols where these peaks have even mass numbers. Those at m/z 170 and 128 are the most prominent fragments of even mass for hexitol acetates (4% and 6%) and are stronger in the mass spectra of 1-deoxy hexitol acetates (8% and 11% respectively).

By the use of labeled compounds and of metastable transitions, it was found that the first step was the elimination of acetaldehyde from the molecular ion, followed by successive expulsions of acetic anhydride, acetic acid, and ketene (Jansson and Lindberg, 1980) (Figure 12).

These even-mass fragments are insignificant in the mass spectra of partially methylated alditol acetates, in which the fission of the alditol chain at a methoxylated carbon atom is the most important reaction.

The GC/MS of alditol acetates is very widely used for the determination of the structure of partially methylated sugars arising from the hydrolysis of permethylated oligo- and polysaccharides.

For such derivatives, simple fragmentation rules have been presented (Björndal *et al.*, 1976b). In the primary fragmentation, fission occurs between the two carbon atoms in the partial structures **20** and **21** in preference to structure **19**. The most favorable cleavage is observed between the carbons containing two vicinal methoxyl groups as in **21**. By cleavage of structure **20**, the methoxylated carbon atom mainly carries the

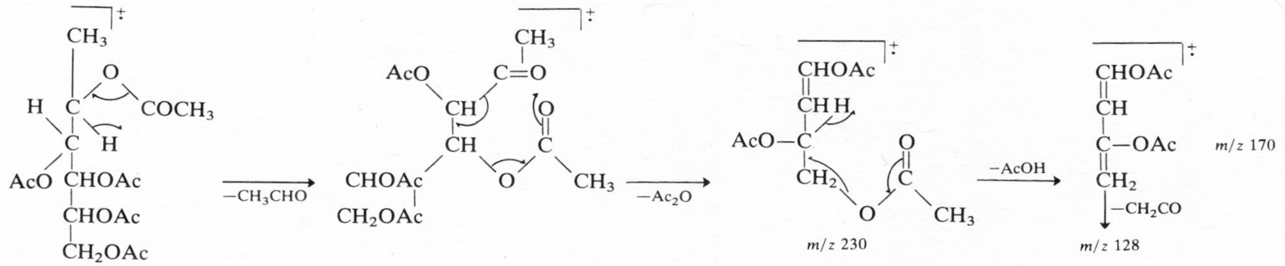

FIGURE 12. The formation of even mass number ions in the EI spectra of peracetylated alditols (from Hansson and Lindberg, 1980).

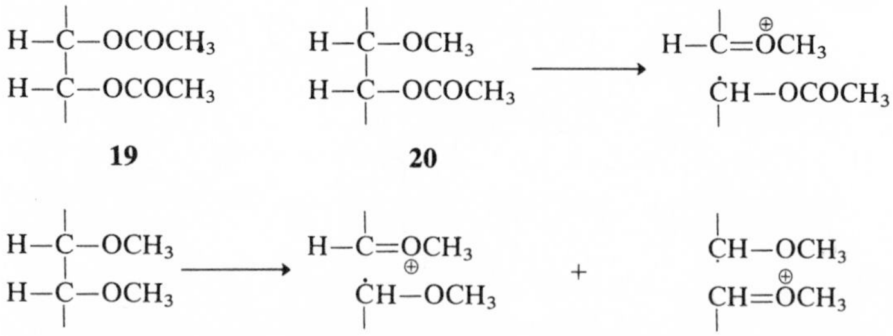

$$H-\overset{|}{C}-OCOCH_3 \qquad H-\overset{|}{C}-OCH_3 \qquad \longrightarrow \qquad H-\overset{|}{C}=\overset{\oplus}{O}CH_3$$
$$H-\overset{|}{C}-OCOCH_3 \qquad H-\overset{|}{C}-OCOCH_3 \qquad \qquad \overset{\cdot}{C}H-OCOCH_3$$

$$\mathbf{19} \qquad\qquad\qquad \mathbf{20}$$

$$H-\overset{|}{C}-OCH_3 \qquad H-\overset{|}{C}=\overset{\oplus}{O}CH_3 \qquad + \qquad \overset{\cdot}{C}H-OCH_3$$
$$H-\overset{|}{C}-OCH_3 \quad\longrightarrow\quad \overset{\cdot}{C}H-OCH_3 \qquad\qquad CH=\overset{\oplus}{O}CH_3$$

$$\mathbf{21}$$

FIGURE 13. Main fragmentations of partially methylated peracetylated alditols.

position charge, whereas by fission of structure **21**, the two fragments are observed (Figure 13).

Applications of these fragmentation rules to partially methylated hexitol acetates give the following results:

An intense ion **22** at m/z 45 is observed when C1 is methoxylated and C2 acetoxylated. The presence of the ion **23** at m/z 89 for 1,2-*O*-methyl hexitol is an exception to the preceding rules; this fragment is more stabilized than the expected ion **22**. Ion **24** is observed when C1 is acetoxylated and C-2 methoxylated.

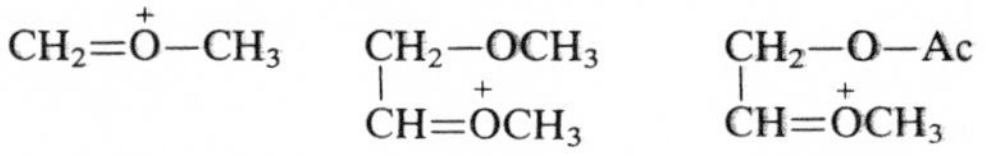

$$CH_2=\overset{+}{O}-CH_3 \qquad \begin{matrix}CH_2-OCH_3\\ |\\ \overset{+}{CH}=OCH_3\end{matrix} \qquad \begin{matrix}CH_2-O-Ac\\ |\\ \overset{+}{CH}=OCH_3\end{matrix}$$

$$\underset{\mathbf{22}}{m/z\ 45} \qquad \underset{\mathbf{23}}{m/z\ 89} \qquad \underset{\mathbf{24}}{m/z\ 117}$$

The cleavage of the C3–C4 bond only gives an intense ion when C3 is methoxylated. However, when C2 is methoxylated, the corresponding ion possesses a low intensity (i.e., the C2–C3 cleavage is preferred), unless C4 is also methoxylated.

Similar features occur for the formation of fragments containing four carbon atoms of the alditol **25**.

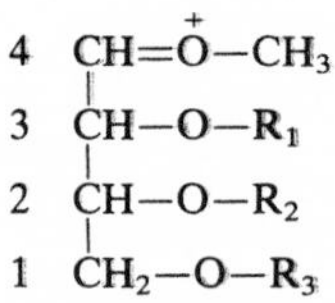

$$\begin{matrix}4 & CH=\overset{+}{O}-CH_3\\ 3 & CH-O-R_1\\ 2 & CH-O-R_2\\ 1 & CH_2-O-R_3\end{matrix}$$

$$\mathbf{25}$$

The ion **25** possesses a low intensity when $R_1=R_2=R_3=CH_3$, except if C5 is methoxylated. The ions corresponding to $R_1=COCH_3$, and

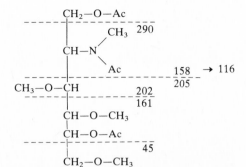

FIGURE 14. Main fragments observed in the spectra of the peracetyl ester of a reduced and partially methylated N-acetyl glucosamine issued from Hakomori methylation reaction of a N-acetyl glucosaminyl oligosaccharide.

$R_2=R_3=CH_3$ or to $R_1=R_3=CH_3$ and $R_2=COCH_3$ only possess a high intensity if C5 is methoxylated. The acetyl moiety is further expelled from the ion as an acetic acid molecule. The ions corresponding to $R_1=R_2=COCH_3$, $R_3=CH_3$, and $R_1=R_2=R_3=COCH_3$ are relatively intense and can lose two molecules of acetic acid.

For acetamido sugars, the Hakomori methylation procedure methylates both hydroxyl and acetamido groups. The resulting sugars possess a N-methyl N-acetyl group. The main fragments of the alditol acetate resulting from the presence of a N-acetyl hexosamine in a polysaccharide are shown in Figure 14. Preferential cleavages occur between methoxy and N-methyl acetamido groups, and also between two adjacent methoxyl groups. The nitrogen-containing fragments are easily identified by the even mass numbers of these intense ions (Figure 14).

As pointed out before in the analysis of alditol acetates, it is essential to differentiate the hydroxyl group which is produced by reduction of the carbonyl group. By using sodium borodeuteride instead of sodium borohydride for the reduction step, a deuterium atom is introduced into the resulting alditol and the localization of such a label leads to the determination of the carbon originating from the carbonyl function. From aldoses, 1-d^1-alditols are thus synthesized and the fragments containing the deuterium are shifted by one mass unit.

For example, the 2,3 and 3,4-di-O-methyl-D-xyloses give, by $NaBH_4$ reduction, identical hexitols. By the use of $NABD_4$, two different hexitols, **26** and **27**, are obtained. Although their acetate derivatives remain inseparable by GC analysis, different mass spectra are obtained.

Similar results are obtained in the reduction of ketoses. However, the problem is complicated by the introduction of a supplementary asymmetric carbon atom by the reduction of the ketonic function. It was pointed out (Lindberg *et al.*, 1973) that the reduction of a mixture of partially methylated fructoses (resulting from the hydrolysis of a permethylated levan)

```
            CHD—O—Ac                    CHD—O—Ac
            H—C—OCH₃                    H—C—OAc
   118      ----|-----------          CH₃O—C—H
  ------          |                     ----|-----------  190
   189      CH₃OC—H                           HC—OCH₃      117
            HC—OAc                            CH₂OAc
            CH₂OAc
              26                              27
```

gives pairs of partially methylated D-glucitol and D-mannitol, but their retention times are so similar that they are not separated by GC. The chromatogram remains simple and the distinction between the 1,3,4- and 3,4,6-tri-O-methyl isomers of D-fructose could be made after reduction by a deuterated reducing agent (**28, 29**).

```
      CH₂—OCH₃    45          CH₂—OAc
   ---|-----------           CD—OAc
      CD—OAc               CH₃O—CH
   CH₃O—CH                  ----|-----------  190
   ---|-----------  162           |            161
       |            189      HC—OCH₃
      HC—OCH₃               CH—OAc
      CH—OAc            ----|-----------
      CH₂OAc                CH₂OAc        45

   1,3,4 isomers              3,4,6 isomers
       28                         29
```

Aldonic and uronic acids could be easily transformed into alditol acetates using lithium aluminum hydride reagent. The use of AlLiD$_4$ permits their distinction among the reduction products, since two deuterium atoms are incorporated into the polyol by reduction of aldonic acids and three deuterium atoms by reduction of uronic acids. This labeling technique was used for the methylation analysis of a fungal glucuronoglucan (Björndal and Lindberg, 1970) and also for the distinction between aldonic and uronic acids, allowing their quantitative determination by GC/MS (Sjöström et al., 1974).

In some cases the chromatograms of partially methylated alditol acetates have been difficult to interpret because of overlapping peaks. A new derivative, the partially ethylated alditol acetate, was synthesized by replacing the methyl iodide used in the Hakomori methylation procedure of polysaccharides, by ethyl iodide. These derivatives give sharp GC peaks on different types of columns allowing more extensive separation of components than the methylated analogs do (Sweet et al., 1975). The fragmentation under electron impact of the ethylated alditol acetates is analogous to those of the partially methylated alditol acetates and permits an easy location of the ethoxyl groups (Sweet *et al.*, 1974).

Since the stereochemistry has no influence on the decomposition of alditol acetates upon electron impact, the assignment of structure to *O*-acetyl *O*-methyl alditols is generally based both on the pattern of the EI mass spectra and on the GC retention times. The latter parameter permits not only the identification of stereoisomers but often aids in determining the number of *O* methyl groups. This is useful because molecular ions produced by EI ionization are rarely stable enough to be observed. The ambiguities inherent in this method could be ruled out by the use of chemical ionization mass spectrometry. For example, the use of isobutane as a reagent gas allowed an easy observation of the (M + H) peaks of the different methylated hexitol-2-*d*-acetates obtained by the methylation–hydrolysis–reduction–acetylation degradation procedure of a levan which is produced by a *Streptococcus salivarius* strain (Hancock *et al.*, 1976).

The fragmentation pathway under CI isobutane conditions strongly differs from the EI conditions. The main fragment ions are formed either by the loss of methanol (32 mass units) or acetic acid (60 mass units) from the $(M + H)^+$ ion. When selected and reproducible conditions are used, it has been observed that methylated hexitol acetates steroisomers having the same arrangement of *O*-methyl and *O*-acetyl groups show markedly different CI spectra (McNeil and Albersheim, 1977). Although structural determinations of partially methylated hexitol acetates cannot be obtained by the use of CI spectra alone (since there is an absence of cleavages of carbon–carbon bonds), such spectra give additional information about the relative stereochemistry of different groups. This mode of analysis was used to identify the methylated hexitol acetates derived from terminal hexoses. The derivatives originating from terminal glucose and mannose possess similar retention times on many chromatographic columns. They can be slightly separated on some peculiar stationary phases, but the separation is not suficient to afford two peaks. Hence, it is difficult to determine whether a sample contains one or both of the terminal aldohexoses. Using isobutane CI mass spectrometry and monitoring the (M + 1), [(M + 1) − 32] and [(M + 1) − 60] ions of the methylated alditol acetates, it can be seen that the relative three-ion intensities vary significantly during the elution of the chromatographic peak which is made up of a mixture of glucose and mannose derivatives. The leading edge of the peak is enriched in the glucose derivative [higher relative intensity of the (MH) and (MH-32) ions] whereas the trailing edge is enriched in the mannose derivative. To make such experiments, it is necessary to use carefully controlled experimental conditions because the relative decomposition rate constants of the ions are very sensitive to slight changes in the ion source conditions.

In recent years, a large number of structures of oligosaccharides have been determined by using GC/MS analysis of partially methylated alditol acetates. Some characteristic examples are the following.

The core of the lipopolysaccharide of *Proteus mirabilis* was found to contain D-glucosamine, D-glucose, D-glycero-D-mannoheptose, and L-glycero-D-mannoheptose in a ratio of about $1:1:1:1:2$, and additional 3-deoxy-D-mannooctulosonic acid (KDO). In the R4 mutant of a strain of *P. mirabilis*, the lipopolysaccharide only contains D-glucose, L-glycero-D-mannoheptose, and KDO, in addition to lipid A. The lipopolysaccharide of this mutant was investigated in order to elucidate the structural details of the *P. mirabilis* deeper R-core region (Radziejewska-Lebrecht *et al.*, 1980). The core oligosaccharide was obtained by acetic acid hydrolysis of the whole lipopolysaccharide. A part of this sample was dephosphorylated by treatment with aqueous hydrofluoric acid. The aliquot that was treated with H_2F_2 as well as the untreated aliquot were methylated by the Hakamori procedure (using CH_3I). The products were hydrolyzed, reduced by $NaBD_4$, and acetylated. The partially methylated alditol acetates were separated on a capillary column CP-Si15, and their structures were identified by mass spectrometric analysis (Radziejewska-Lebrecht *et al.*, 1979). The untreated oligosaccharides produced the alditol acetates **30, 31, 32** in the ratio

1,5-di-O-acetyl-2,3,4,6-tetra-O-methyl-d^1-glucitol	**30**
1,5-di-O-acetyl, 2,3,4,6,7-penta-O-methyl-d^1-heptitol	**31**
1,3,4,5-tetra-O-acetyl-2,6,7-tri-O-methyl-d^1-heptitol	**32**

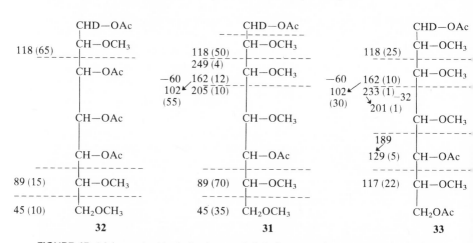

FIGURE 15. Main saccharide derivatives, and their fragmentation patterns, obtained after different chemical treatments of the Core lipopolysaccharide of the P-mirabilis R4 mutant (see text). The relative intensities of the different fragment ions are indicated in parentheses.

1 : 0.2 : 1 (the relative intensities of the characteristic peaks were indicated in parentheses) (Figure 15).

When the methylation was conducted on the dephosphorylated oligosaccharide, the same alditol acetates **30, 31, 32** were identified, but with a different relative molar ratio (1.4 : 1.2 : 1).

When the untreated permethylated oligosaccharide was further dephosphorylated and hydrolyzed, a new alditol acetate was obtained, namely, the 1,5,6-tri-O-acetyl-2,3,4,5-tetra-O-methyl-d^1-heptitol **33** [relative molar ratio for compounds **30–31–32–33** (0.9 : 0.3 : 1 : 1)].

These experiments (which were confirmed by subsequent ones) are consistent with the presence of terminal glucose, a branched heptose, and a terminal heptose which partially possesses a phosphate group on the C7 position. Some further experiments (^{1}H-NMR; no coprecipitation with Concana valin A) indicated the anomeric configurations of this oligosaccharide and the partially resolved structure shown in Figure 16 was proposed.

A similar study using successive labeling was made on the heptose region of the lipopolysaccharide of *Escherichia coli* K12 CR34 (Blache *et al.*, 1981).

These procedures are widely used and may differ slightly from one to another. A generalization of the methods for structural studies of polysaccharides was proposed (Valent *et al.*, 1980).

The strategy can be summarized as follows: the polysaccharide is first fully methylated. A partial hydrolysis is then done giving rise to partially methylated oligosaccharides. This mixture is reduced with sodium borodeuteride and ethylated. The resulting peraklylated oligosaccharides are separated by liquid chromatography. Each individual oligosaccharide is then fully hydrolyzed, reduced, and acetylated. The resulting partially methylated and ethylated alditol acetates are identified by GC/MS.

In this way, methoxyl groups are introduced in place of the free hydroxyl groups in the polysaccharide and ethoxyl groups are present both at (a) the position where other glycosyl residues were originally attached to that particular oligosaccharide, when the oligosaccharide was pert of the unfragmented polysaccharide and (b) the C1 and C5 (or C4 for furanosyl residues) of each alditol present at what was the reducing end of each internal oligosaccharide. Finally, the acetoxy groups replaced the free

D-glucose (β1 → 3 or 4)L-α-D-heptose (1 → ?) KDO
$\uparrow$(1 → 4 or 3)
L-α-D-heptose-7-P

FIGURE 16. Partial structure of the deeper R-core region of a lipopolysaccharide obtained after methylation analysis and GC/MS identification of partially methylated sugars.

hydroxyl groups which were liberated by the hydrolysis of the fully alkylated oligosaccharide.

Thus, the structure of oligosaccharides (mainly disaccharides and trisaccharides) may be determined unambiguously by GC/MS analysis of alditol acetates, Since these oligosaccharides overlap within the complex carbohydrate, and in conjunction with the knowledge of the glycosyl linkage of the complex carbohydrate, the structure of the latter can be determined.

An example of the utilization of this method is presented in the structural analysis of xanthan (Valent *et al.*, 1980). This polysaccharide is secreted by *Xanthomonas campestris*. The glycosyl composition shows that the polysaccharide is composed of glucosyl, mannosyl, and glycosyl uronic acid residues in a ration of 2:2:1. Moreover, it has been shown to contain acetal-linked pyruvic acid and O-acetyl groups.

Step 1. The polysaccharide is methylated by a modification of the Hakomori method, using potassium dimethylsulfinyl anion (instead of sodium dimethyl sulfinyl anion) and methyl iodide.

Step 2. The uronic acid carboxyl groups are reduced by sodium borodeuteride in ethanol-tetrahydrofuran (this reduction procedure was found to be more effective than the one employing lithium aluminum deuteride).

Step 3. A small portion of this methylated polysaccharide is fully hydrolyzed and the resulting partially methylated carbohydrates are analyzed as alditol acetates by GC/MS. This step allows the determination of the nature of the glycosyl linkages.

Step 4. An incomplete formic acid hydrolysis of the methylated, carboxyl-reduced xanthan affords good yields of di-, tri-, and tetrasaccharides. It was found that the hydrolysis cleaved preferentially all terminal mannose residues that were not substituted with pyruvic group. Thus, in the mixture of oligosaccharides obtained by partial formolysis of xanthan, oligosaccharides which contained a pyruvic-acid-substituted terminal mannose residue could be identified.

Step 5. Location of the terminal mannosyl residues. A very mild formolysis is performed which cleaves mainly the acid-labile terminal mannosyl residues and not the other glycosyl residues. The mixture is ethylated, then completely hydrolyzed and converted to alditol acetates which are then analyzed by GC/MS. Two major products containing the ethoxyl groups are obtained; the first one derives from a 6,6 dideuterio, terminal glucosyl residue having ethoxyl groups linked to both C4 and C6; the second one comes from a 6,6-dideuterio-4-linked glucosyl residue having a single ethoxyl group attached to C6. Since two deuterium atoms were incorporated, these derivatives are formed from glucuronic residues. The ethoxyl group on the C4 position indicated the position of linkage of the terminal mannose.

Step 6. The mixture of di-, tri-, and tetraoligosaccharides arising from the formolysis of step 4 is reduced (NaBD$_4$) and ethylated. The mixture is separated by liquid chromatography (reverse phase).

Each individual fraction is fully hydrolyzed, reduced (NaBD$_4$), and acetylated. The analysis is made by GC/MS for the determination of the position of methoxy, ethoxy, and acetoxy groups.

Step 7. Interpretation of the results. The determination of the structure of trisaccharide which produces the alditol acetates **34**, **35**, and **36**, is presented as an illustration.

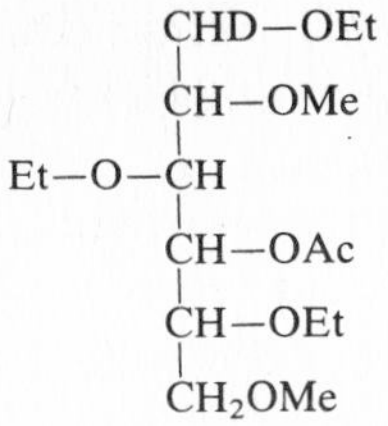

CHD—OEt	CHD—OAc	CHD—OAc
CH—OMe	CH—OMe	CH—OMe
Et—O—CH	Et—O—CH	MeO—CH
CH—OAc	CH—OEt	CH—OAc
CH—OEt	CH—OAc	CH—OAc
CH$_2$OMe	CH$_2$OMe	CH$_2$OMe
34	**35**	**36**

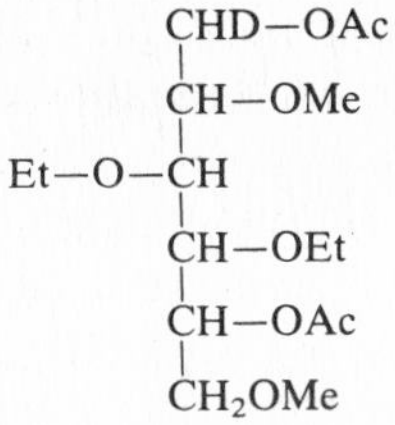

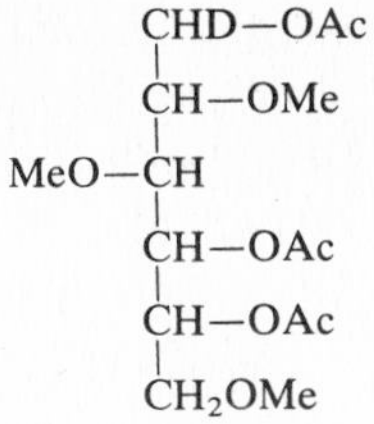

An ethoxyl group on C1 (such as in **34**) indicates the reducing part of the oligosaccharide. In this alditol acetate, a second ethyl group corresponds to the hydroxyl group linked through the semiacetalic function (C5). Thus the third ethoxyl group on C3 corresponds to a branching point in the polysaccharide. The acetyl group on C4 indicates the location of the glycosyl linkage in the oligosaccharide.

The absence of ethoxy groups in **36** is an indication that the corresponding sugar is an internal residue in the oligosaccharide. The *O*-acetyl groups on C1, C4, and C5 correspond to the linkages indicated in Figure 17. Lastly, the remaining alditol would correspond to the nonreducing end of the oligosaccharide. The ethoxyl groups correspond to the branching points in the polysaccharide. Thus, this oligosaccharide possesses the structure presented in Figure 17 in which arrows indicate the linkage points in the intact polysaccharide.

The pyruvic acid-linked mannose residues are found in some oligosaccharides. The analysis showed that pyruvic acids are linked ketosidically to the oxygen atoms on C4 and C6 of terminal mannose. Moreover, such

$$\downarrow \qquad\qquad\qquad\qquad \downarrow$$
$$(3) \qquad\qquad\qquad\qquad (3)$$

$(x \rightarrow 4)$ Glucose $(1 \rightarrow 4) -$ Glucose $(1 \rightarrow 4) -$ Glucose $(1 \rightarrow y)$

$(\rightarrow \textbf{35}) \qquad\qquad (\rightarrow \textbf{36}) \qquad\qquad (\rightarrow \textbf{34})$

FIGURE 17. Structure of an internal oligosaccharide obtained from a Xanthan and linkage points to other saccharides in the intact polysaccharide, as determined by the Valent's procedure and GC/MS analysis.

analysis also indicated that mannosyl, glucosyl, and glucosyl uronic residues are in the pyranoid ring form.

This whole analysis was made on about 20 mg of purified xanthan, which demonstrates the sensitivity of the method, to which the performance of the GC/MS analysis is the major contribution.

3.2. Other Derivatives of Polyols

The mass spectra of TMS-derivatives of deuterated polyols, obtained by reduction of sugars by means of a deuterated reagent and further trimethylsilylated, allow firm assignment of the number and position of the deuterium atoms, and hence of the nature of the starting sugar (aldose, ketose, lactones, etc.) (Dizdaroglu et al., 1974).

The trifluoroacetate derivatives of polyols were used for the determination of the components of oligosaccharides (Kamei et al., 1976). The deuterium-labeled hexitols were analyzed by mass fragmentography, monitoring the ion $(M - CF_3COO)^+$ for each saccharide. These fragment ions are highly intense and are not affected by other peaks because of the absence of fragment ions in the adjacent area. The fragmentation ways are similar to those of the alditol acetates except that the secondary fragments are only formed by elimination of trifluoroacetic acid or trifluoroacetic anhydride (Chizhow et al., 1969).

3.3. Aldononitrile Peracetates

Aldoses were rapidly and easily converted into peracetylated aldononitriles (PAAN) by successive reactions with hydroxylamine and acetic anhydride. These derivatives allowed excellent separations of monosaccharides on polyester stationary phases. (Lance and Jones, 1967). A typical derivatization procedure is the following: 12 mg of an aldose mixture are mixed with 12 mg hydroxylanine HCl in 0.6 ml pyridine. After heating in a sealed tube (15 min, 90°C) 1.8 ml of acetic anhydride is added and the mixture is heated for an additional 30 min. After evaporation of the solvent, the residue is dissolved in 0.1 ml dry chloroform and analyzed by GC (Dimitriev et al., 1971).

The aldonitriles retain terminal dissymmetry of the starting sugar. The mass spectra of the PAAN derivatives are easy to interpret. Although the molecular ion is not observed under EI conditions, the ions produced by the loss of CH_2OAc $(M - 73)$ and $AcO-CHCN$ $(M - 98)$ allow the determination of the molecular weight (Dmitriev et al., 1971). The cyano group protects the derivative from undergoing C1, C2 cleavage, but all the fragmentation pathways are similar to those of the peracetylated polyols.

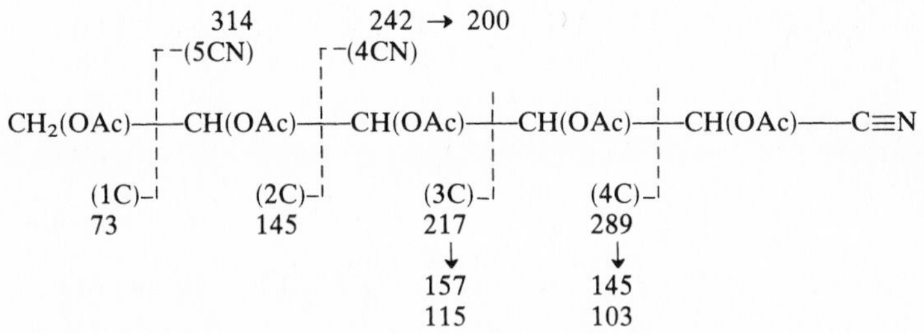

FIGURE 18. Fragmentations of a PAAN derivative of hexose.

Direct cleavages of the C–C bonds, followed by the loss of ketene, acetic acid, or acetic anhydride give rise to two series of ions. The "C" series, which gives the most prominent peaks, do not contain the cyano group. The "CN" series contains this group and produces even-mass-number ions (Figure 18). Detailed fragmentation studies were presented by several authors (Seymour et al., 1975, Dimitriev et al., 1971, Szafranek et al., 1974, Seymour et al., 1979). For unmethylated sugars, the (3CN) and (2CN) series are absent.

In the partially methylated PAAN, the ions of the CN series are not observed when both C2 and C3 bear methoxyl groups. In these cases, fragmentation is simpler, as only one end of the molecule contributes to intense, primary fragments (Seymour et al., 1975). However, an acetyl group on C2 or C3 stabilizes the ions arising from the nitrile end. For example, in the 2,5-di-O-acetyl-3,4,6-tri-O-methyl mannonitrile, ions of the 3CN and 4CN series are observed (Figure 19).

The GC/MS of PAAN derivatives was used to determine the structure of bacterial dextrans by methylation analysis (Seymour et al., 1977; Morrison, 1975). It was noticed (Szafranek et al., 1973; Pfaffenberger et al., 1975) that the GC properties of alditol acetates and PAAN derivatives are similar and that both derivatives of the same saccharide could be cochromatographied. Hence, when a reducing oligosaccharide was subjected

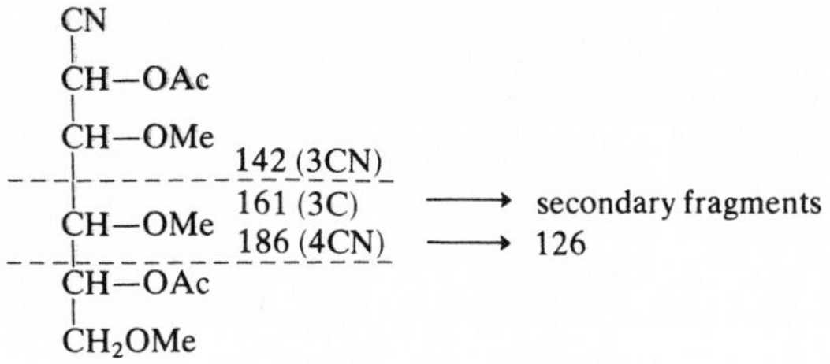

FIGURE 19. Fragmentation of the PAAN derivative obtained from 3,4,6-tri-O-methyl mannose.

to the sequence of (a) reduction, (b) hydrolysis, and (c) formation of PAAN derivatives of the saccharides in the hydrolyzate, the chromatogram contained one peak for the acetylated alditol and other peaks for the PAAN derivatives. Thus the degree of polymerization of the oligosaccharides would be represented by [(1% of PAAN/% alditol acetate) + 1]. Varna and Varna (1977) empolyed this combined alditol–aldononitrile method to evaluate the chain lengths of glycosaminoglycans.

However, like alditol acetates the aldononitrile peracetates seldom yield the molecular ion upon electron impact. In contrast, under CI (isobutane) conditions, the MH^+ ion is observed in each instance (Li *et al.*, 1977) and the base peak is $(MH - 60)^+$ resulting from the loss of acetic acid. The very low number of fragment ions allows the easy distinction of different components which are contained in unresolved GC peaks such as for fucose and ribose derivatives. With ammonia as reactant gas (Seymour *et al.*, 1979, the $(M + NH_4)^+$ ion is obtained, together with MH^+ and $(MH - 60)^+$. These very simple spectra allow an easy determination of the molecular weight of the derivatives. Using methane CI conditions, the MH^+ ion is not seen, but a prominent peak, corresponding to $[(M + H) - 60 - 42]^+$, is found together with other peaks due to the successive stripping of acetoxyl groups either as acetic acid or ketene molecules.

3.4. Oxime Derivatives

The peracetylated oxime derivatives of ketoses are produced by the same procedure which converts aldoses to aldononitrile acetates, because the ketoximes are not transformed into dehydration products. Such PAKO derivatives (percetylated ketone oximes) are thus obtained together with

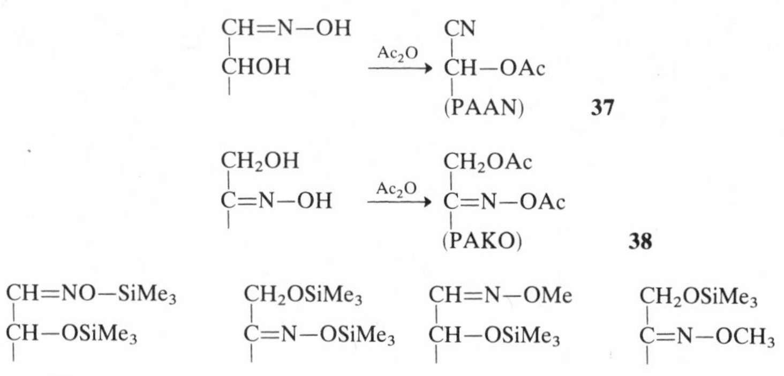

FIGURE 20. Different partial structures of PAAN, PAKO, and MO–TMS derivatives from aldoses and ketoses.

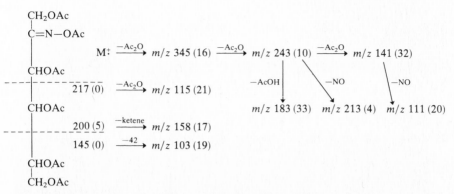

FIGURE 21. Mass spectral decomposition of the PAKO derivative of fructose. Intensities are indicated in parentheses

the PAAN derivatives for aldoses when a mixture of these sugars is treated by hydroxylamine and then acetylated (**37, 38**).

Alternatively, other oxime derivatives can be synthesized, mainly per-Me$_3$Si oxime **39, 40** and per-Me$_3$SiO-methyl oxime derivatives **41, 42** (MO-TMS) (Figure 20).

The EI and CI mass spectra of the PAKO derivatives were studied by Seymour *et al.*, (1980). For most of them, the backbone cleavage is not the prominent decomposition process under EI conditions. Only weak fragments of the IC, 2C, ..., series are observed. Some (CN) fragments are observed but they arise from cleavages far from the oxime function. The most prominent peaks contain nitrogen, and come from successive splittings of the acetoxy groups from the molecular ion. The EI spectrum of the PAKO derivatives of a 2-hexulaose such as fructose is presented in Figure 21.

For sugars possessing larger backbones, such as 2-heptuloses, some cleavages of C—C bonds contribute to higher intensity ions of the (C) and (CN) series.

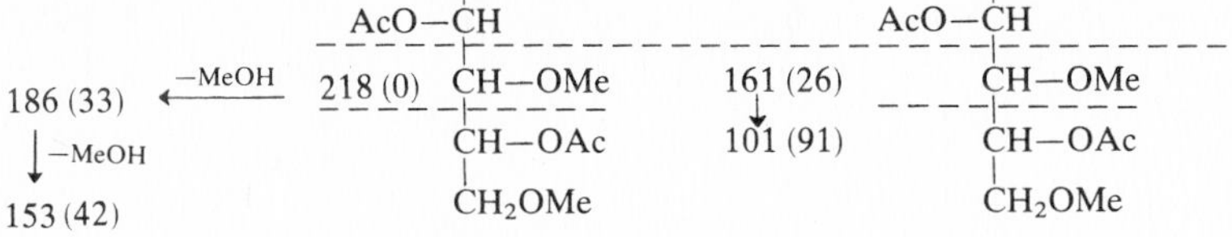

		CH$_2$OMe		CH$_2$OAc
		C=N—OAc		C=N—OAc
		AcO—CH		AcO—CH
186 (33) ←$^{-MeOH}$	218 (0)	CH—OMe	161 (26)	CH—OMe
↓$^{-MeOH}$		CH—OAc	101 (91)	CH—OAc
153 (42)		CH$_2$OMe		CH$_2$OMe
		43		**44**

FIGURE 22. Mass spectral decompositions of the PAKO derivatives obtained from a 1,4,6-tri-*O*-methyl-2-hexulose and a 4,6-di-*O*-methyl-2-hexulose. Intensities are in parentheses.

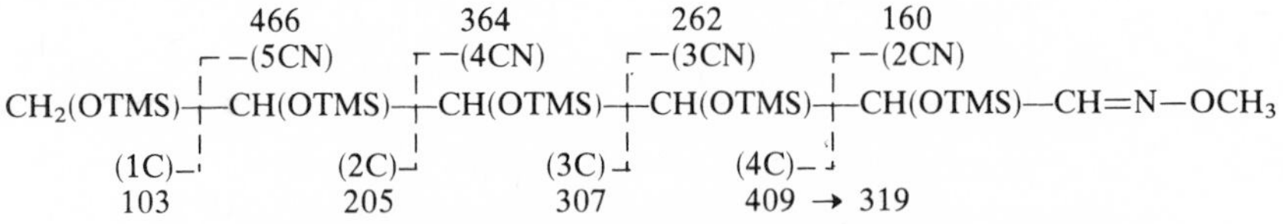

```
          466           364           262           160
        r -(5CN)      r -(4CN)      r -(3CN)      r -(2CN)
CH₂(OTMS)-+-CH(OTMS)-+-CH(OTMS)-+-CH(OTMS)-+-CH(OTMS)—CH=N—OCH₃
        |           |           |           |
      (1C)-¦      (2C)-⌋      (3C)-⌋      (4C)-⌋
       103         205         307         409 → 319
```

FIGURE 23. Fragmentations of MO–TMS derivatives of aldohexoses.

In contrast, the EI mass spectra of highly methylated PAKO derivatives contain high intensity ions resulting from the backbone cleavage. Little or no cleavage adjacent to the oxime group is observed, and many O-methylated PAKO derivatives only give (C) series (Figure 22). However, in some instances, ions of the (CN) series are present, which can suppress the ambiguities in structure assignments (compare **43, 44**).

Under CI conditions (Ammonia), relatively weak MH^+ and $M \cdot NH_4^+$ ions are observed. A set of peaks $(MH - n \times 60 \text{ mass units})^+$ is formed $(n = 1, 2, 3 \ldots)$ together with $[(M - 57) - n \times 60 \text{ mass units}]^+$ $(n = 1, 2, \ldots)$. No explanation was presented for the formation of the last series which exhibits very high intensities.

The other type of oxime derivative which is frequently synthesized is the methyl oxime, O-TMS derivative (MO-TMS). They are easily prepared by the method of Laine and Sweeley (1973): 1 mg of methoxylamine hydrochloride, 1 mg of sugar, and 50 μl of dry pyridine are allowed to react 2 hours at 80°C. Then 50 μl of bis(trimethylsilyl) trifluoroacetamide (BSTFA) are added and the mixture is allowed to react 15 min at 80°C. Such derivatives have good chromatographic properties, and can be separated on nonpolar stationary phases (Laine and Sweeley, 1971). The methoxime-TMS derivatization can be applied to ketoses and aldoses. However, each aldose or ketose derivative often gives rise to two peaks on capillary columns corresponding to the syn–anti isomerism of the oxime

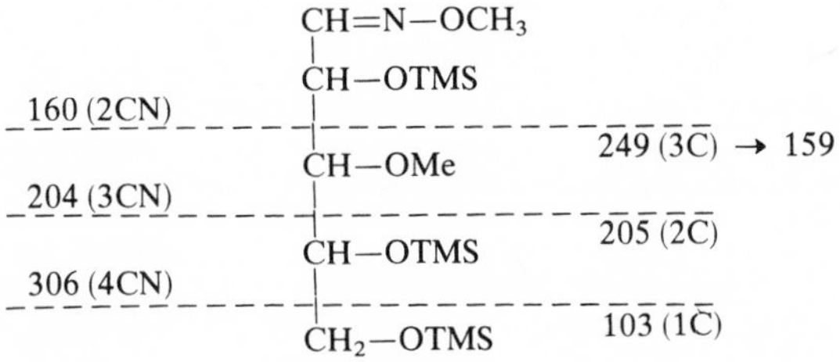

```
                        CH=N—OCH₃
                        |
                        CH—OTMS
        160 (2CN)       |
        _____|_____ 249 (3C) → 159
                        CH—OMe
        204 (3CN)       |
        _____|_____ 205 (2C)
                        CH—OTMS
        306 (4CN)       |
        _____|_____ 103 (1C)
                        CH₂—OTMS
```

FIGURE 24. Ions observed in the spectra of a MO–TMS derivative of a 3-O-methyl aldohexose.

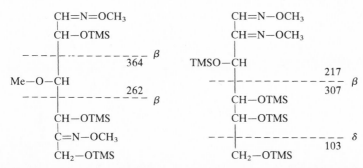

FIGURE 25. Ions observed in the spectra of MO–TMS derivatives of some aldosuloses.

group (Størset *et al.*, 1978). The fragmentation under EI is very simple (Laine and Sweeley, 1973). For aldose-MO-TMS, fission did not occur between C1 and C2 but all other C—C bonds are cleaved. The (CN) series gives even-mass-number ions since they contain a nitrogen atom. The (C) series gives odd-mass-number ions, which can lose a trimethylsilanol molecule (90 u) (Figure 23). In the high mass range, a weak (M − 15) peak is observed.

With desoxy compounds, a vicinal cleavage to the methylene group is not likely, because of the lack of alkoxyl function. Such simple fragmentation rules are also observed for the derivatives of partially methylated sugars allowing an easy determination of the location of the methoxyl groups (Figure 24).

In the derivatives of ketoses, the lack of fragmentation adjacent to the oxime carbon permits their differentiation (absence of the ions 2CN and 3C in the spectra of the MO-TMS derivative of 2-keto hexoses). The mass spectra of the MO-TMS of 2-acetamido 2-deoxy aldoses were reported and their behavior is similar (Orme *et al.*, 1974).

In the MO-TMS derivatives of aldosuloses, dialdoses and diuloses, preferential β and δ cleavages to the oxime groups allow the location of the oxime groups (Dizdaroglu *et al.*, 1977) (Figure 25).

3.5. Derivatives of Thioacetals and Thioketals

The mass spectra of peracetylated thioacetals and thioketals of aldoses and ketoses exhibit weak molecular ions. The base peak for the peracetylated diethylthioacetal derivatives of aldoses results from a C1, C2 cleavage (De Jongh, 1964a) (Figure 26).

This cleavage is not observed for 2-deoxy sugar derivatives in which the radical which would be formed by this fission is not stabilized by a

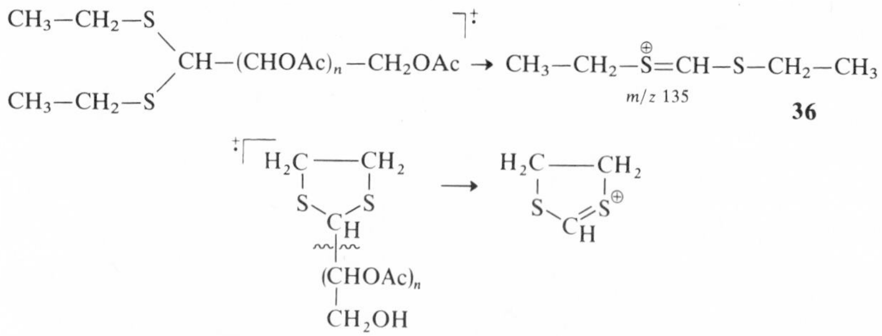

FIGURE 26. C_1–C_2 cleavage of peracetylated thioacetals.

neighboring acetoxy group. In this case, a strong peak resulting from successive losses of thioethyl and acetoxyl radical, acetic acid, and ketene from the molecular ion is formed (m/z 155 for 2-deoxy hexose derivatives). The mass spectra of the same derivatives of ketoses are strongly different as only weak peaks are observed above m/z 100. These ions are mainly formed by multiple losses of acetic acid and ketene residues from $(M-C_2H_5S)^+$ ion.

Ethylene thioacetals behave similarly to diethyldithioacetal derivatives (De Jongh, 1964b); the base peak is due to the cleavage of the C1, C2 bond with charge retention on the sulfur-containing fragment (Figure 26).

Under CI conditions, the diethyldithioacetals easily lose one or two molecules of ethane thiol from either the protonated molecular ion (isobutane CI) or the ammonium complex (ammonia CI). It is proposed that these decompositions are produced by intramolecular cyclization-elimination reactions, thus giving an explanation for the intervention of stereochemical dependence which was observed (Blanc-Muesser *et al.*, 1980).

4. CYCLIC DERIVATIVES OF MONOSACCHARIDES AND POLYOLS

The use of difunctional reagents might be expected to be helpful for the differentiation of sugars, as steric requirements for the derivatization reactions could lead to the preponderant or exclusive formation of one derivative from an anomeric mixture of sugars. Since the formation of constitutional isomers from diastereoisomers could occur, the unequivocal identification by mass spectrometry might be possible. Two main series of cyclic derivatives were used in GC/MS: the boronic esters and the cyclic acetals or ketals.

4.1. Boronic Esters

The synthesis of these derivatives is simply obtained by the reaction of a suitable boronic acid with polyhydroxy compounds (Wood *et al.*, 1975).

Some pentoses (fucose, arabinose, xylose) give a single derivative. Derivatives which possess free hydroxyl groups could be derivatized further to obtain best GC properties. For this reason, the formation of boronic esters is often followed by trimethylsilylation or acetylation.

In a typical procedure, for the synthesis of TMS derivatives of alkyl-boronates, 200 μg of sugar sample are mixed with ca. 20 mole equivalent of alkylboronic acid in pyridine and the mixture is heated 30 min at 110°C. Then 50 μl BSTFA/TMCS (1 : 1, v/v) are added and the mixture is heated at 110°C for an additional 10 min.

A review of the formation and properties of carbohydrate boronates was published recently (Ferrier, 1978). The mass spectrometric analysis of benzene butane-, and methane-boronate derivatives show the main similar features (Reinhold *et al.*, 1974). In the MS of nonsilylated derivatives the cyclic boronate group generates characteristic ions from which the ring size and the relationship (2,3 or 4,6 cyclic boronate) of the hydroxyl groups involved can be deduced. For example, methyl-α-D-mannopyranoside possesses two pairs of hydroxyl groups suitably spaced to form two boronate cyclic esters—a property shared by no other naturally occurring hexose glycoside. The molecular ion was easily discernible and the derivative showed the absence of TMS groups after silylation (lack of ions m/z 73 and 75). Two ions of high intensity were found (**45, 46**) corresponding to cleavages at both sides of the six- and five-membered ring-cyclic esters (m/z 98 and 84 for methane boronates).

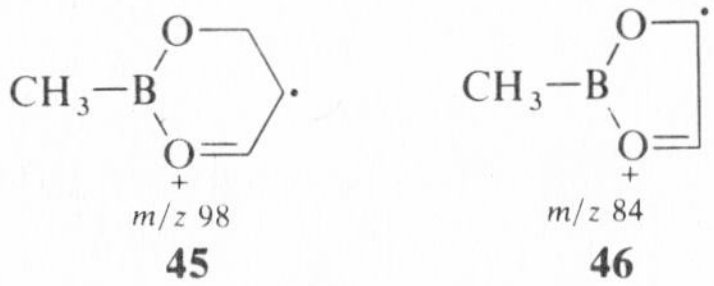

m/z 98 m/z 84

45   **46**

However, the presence of trimethylsilyl groups completely obscures the appearance of fragments containing boron. For example, the trimethyl-silyl methane boronate derivative of methyl-α-D-glucopyranoside gave a

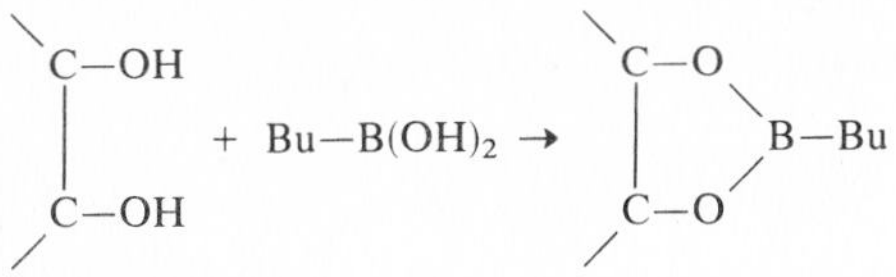

FIGURE 27. Formation of alkyl boronic esters.

single GC peak. The molecular ion is very weak, and a $(M - 15)$ ion of low intensity appears at m/z 347, indicating the presence of a single cyclic boronate with two TMS groups. The base peak is m/z 204 (indication of two adjacent TMS groups—see Section 2.1) and the corresponding structure of the boronate is **47**.

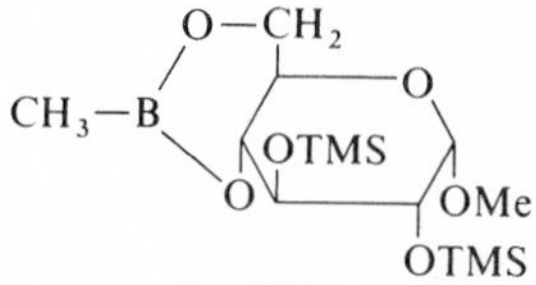

47

The methyl-α-D-galactopyranoside gives two distinct TMS-boronate derivatives. The MS of the 4,6 cyclic boronate is very similar to the preceding one. The 3,4-cyclic boronate is characterized by modified intensities of the ions which contain trimethylsilyl groups: m/z 204 possesses a lower intensity, but m/z 191 and 218 are more intense.

The acetylation of the remaining hydroxyl groups is an alternative to the borosilylation procedure. The mass spectra of the acetate boronates present interesting features (Wiecko and Sherman, 1976). The main feature is the interaction of a suitably placed acetoxy group with boron, to induce the loss of the alkyl radical, as in the derivative of D-glucofuranose (Figure 28).

This interaction may be used as a sensitive stereochemical probe for structure. For instance, D-idose possesses the reverse configuration of D-glucose at all carbons except C5. The 1,2,3,5-bis(butane boronate)-6-acetate of D-idose shows a drastic reduction of the loss of the butyl radical; the same result is observed in the D-mannofuranose derivative. For these derivatives, steric hindrance does not permit an easy interaction of the 6-acetate with 3,5-cyclic boronate. For aldopyranoses, the 8-membered ring for this interaction is less favorable, and the corresponding ion is weak. But the fragment ions resulting from the elimination of a part of the sugar

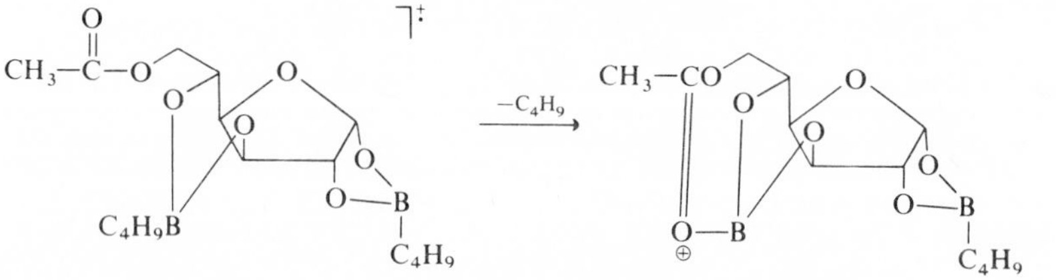

48

FIGURE 28. Elimination of a butyl group by intramolecular interaction in the 1,2-3,5-bis(butane boronate)-6-acetate of D-glucofuranose.

ring could successively eliminate the butyl radical by an acetate–boron interaction. This decomposition is also under stereochemical control. A similar interaction was demonstrated for cyclic alkyl boronates of sugar phosphates (as methyl esters) (Wiecko and Sherman, 1975).

4.2. Cyclic Acetals and Ketals

Treatment of free sugars with acetone containing an acid catalyst converts them into O-isopropylidene derivatives, which can be analyzed by GC/MS. These derivatives could be prepared as follows: 1.2 mg of sugar are mixed with 1.5 ml acetone containing 1% H_2SO_4 (v/v). After shaking 2 h, the mixture is neutralized by sodium bicarbonate and the supernatant analyzed by GC (Morgenlie, 1975). Alternatively, cupric sulfate may be used as a catalyst, but the reaction takes more time (steam bath, 18 h) (De Jongh and Biemann, 1964). Finally, the sugar acetonides may be prepared by an acid-catalyzed exchange reaction using 2,2-dimethoxy-propane (Kiso and Hasagawa, 1976). In some cases, more than one product is formed. The mass spectra interpretation was discussed by De Jongh and Biemann (1964) and Buchs *et al.*, (1975). Since the formation of such derivatives is sensitive to the configuration of the diol groups, structurally different isomers are obtained for stereochemically different isomers. Thus, major differences in the mass spectra are observed according to the stereochemistry of the starting sugar. An abundant $(M - 15)^+$ fragment is generally formed, allowing the determination of the molecular weight.

The fragmentation pathway of the benzylidene-acetals of hexopyranosides has been presented and discussed (Chizhov *et al.*, 1968; Bosso *et al.*, 1977). Under CI conditions, the isopropylidene acetals give intense MH^+ ions (isobutane reagent gas) or both MH^+ and $M \cdot NH_4^+$ ions (ammonia reagent gas). The main fragmentation pathway is the loss of acetone (Hortone *et al.*, 1974).

5. OLIGOSACCHARIDE DERIVATIVES

5.1. Permethylated Oligosaccharides

The EI mass spectra of the permethyl ethers of some oligosaccharides have been discussed by Chizhov and Kochetkov (1966). More recent studies (Moor and Waight, 1975; De Jong *et al.*, 1979a) report some complementary investigations.

The cleavage of the glycosidic bond produced glycosyl cations which can lose further one or two molecules of methanol. Monosaccharidyl cations

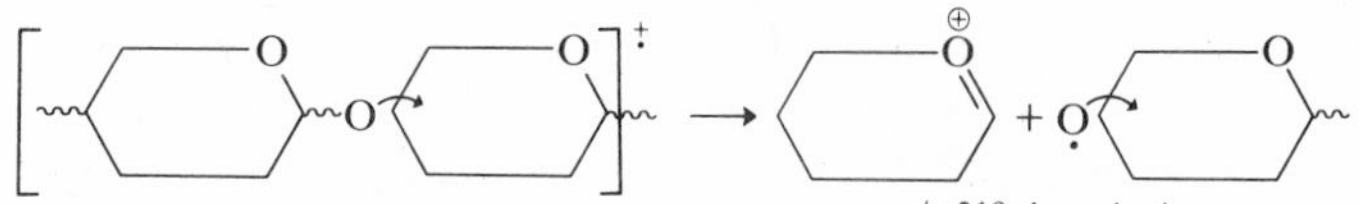

m/z 219 glycosylcation
m/z 423 diglycosylcation

FIGURE 29. The glycosidic cleavage in oligosaccharide derivatives.

are obtained from disaccharides. Mono- and disaccharidyl cations are formed from trisaccharides (Figure 29).

Characteristic differences appear mainly in the intensities of fragment ions, according to the mode of linkage. In aldohexose derivatives, the following observations have been made: (a) the $(1 \to 1)$ glycosidic linkage is characterized by an intense peak at m/z 101, together with relatively weak peaks at m/z 88, 219, 279; (b) the $(1 \to 2)$ and $(1 \to 4)$ linkages are characterized by peaks at m/z 380 and 305 (Figure 30).

A weak peak at m/z 380 is also found for the $(1 \to 6)$ linked disaccharides, but the process which gives rise to m/z 305 is due to the migration of a methoxy group to C-1 and did not occur for $(1 \to 6)$ linked sugar. The isobaric ion at m/z 380 from the $(1 \to 4)$ isomer also fragments through another way, by elimination of the nonreducing sugar moiety, giving m/z 161 (**52**) (Figure 31). This process is also observed for the $(1 \to 5)$ linked sugars. In the $(1 \to 2)$ isomer, the ion at m/z 380 loses a methyl radical to give m/z 365.

(c) A relatively abundant ion at m/z 353 is characteristic for $(1 \to 6)$ linked sugars (**53**).

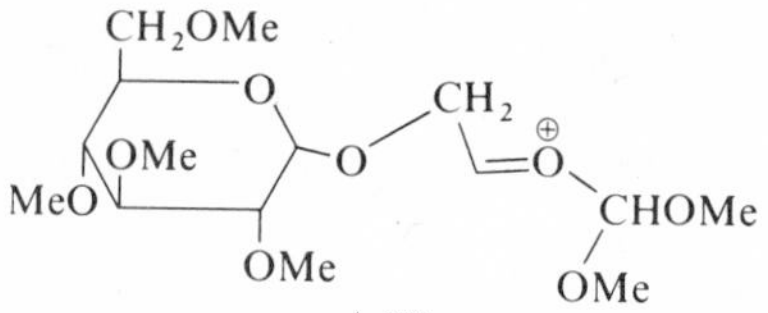

m/z 353
53

(d) The $(1 \to 3)$ linked disaccharides show relatively intense peaks at m/z 145 (**54**) and 159 (**55**).

(e) In the $(1 \to 5)$ linked disaccharides, an intense and characteristic peak is observed at m/z 325 (**56**). In addition, a relatively intense peak is found at m/z 155.

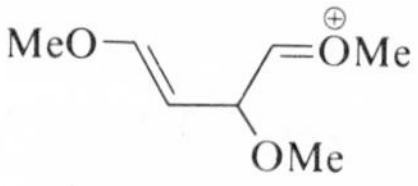

m/z 145
54

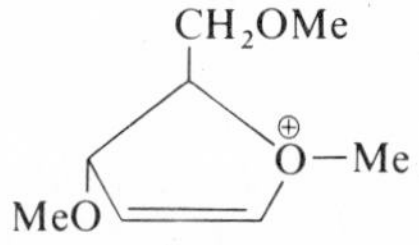

m/z 159
55

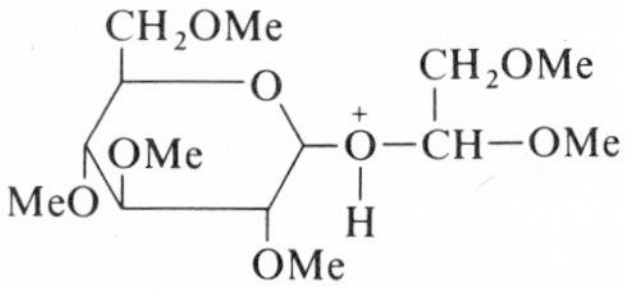

m/z 325
56

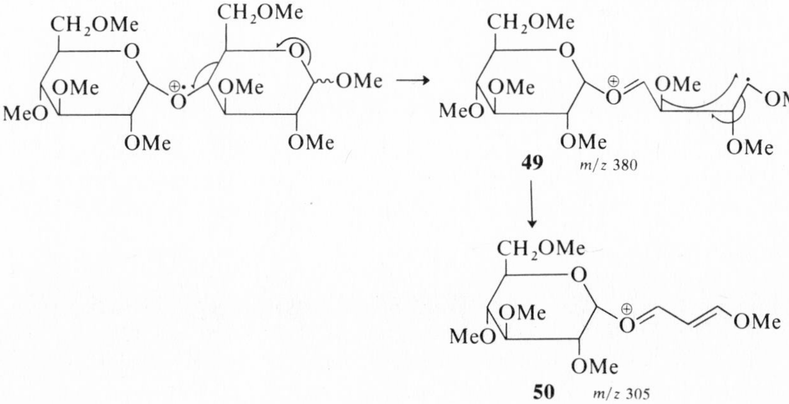

FIGURE 30. The formation of m/z 380 and m/z 305 in the permethyl ethers of methyl glucopyranos ~ yl (1 → 4) glucopyranoside.

For ketohexoses containing oligosaccharides, elimination of a methoxy methyl radical gives a stabilized oxonium ion, and this process is far more favorable than the elimination of the radical containing C6 (Figure 32).

This observation allows the distinction of ketohexoses linked through C1. The elimination of the C1 methoxy methyl radical is also influenced by the presence of an interglycosidic linkage at C2.

The cleavage of the glycosidic linkage gives very intense ions when a ketohexofuranosyl group is present, because of the higher stability of the ketohexofuranosyl cation compared with the hexopyranosyl cation derived from aldohexosyl residues.

In oligosaccharides containing a terminal fructofuranose residue linked through C1 or C6, simple cleavage between C1–C2 or C5–C6, respectively, produces an intense ion at m/z 205 (Figure 33). A similar ion at m/z 205 is also found for derivatives having a terminal function linked through C2, but it is formed through a different process involving the migration of a methyl group or a methoxyl group from a saccharide moiety to the furanose ring (Das and Thayumanavan, 1972).

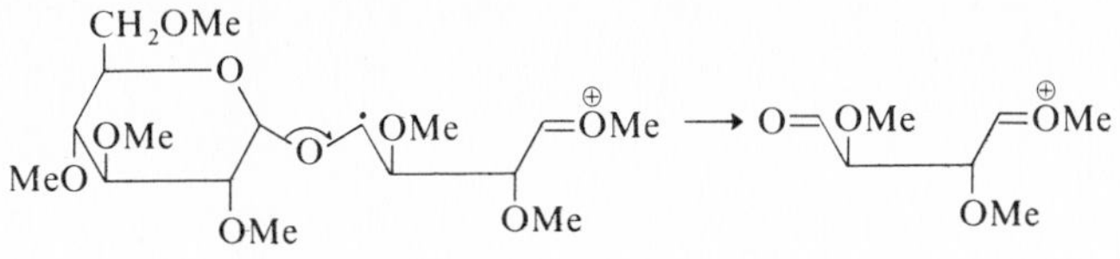

FIGURE 31. The formation of m/z 161 from m/z 380 in the permethyl ether of methyl glucopyranos ~ yl (1 → 4) glucopyranoside.

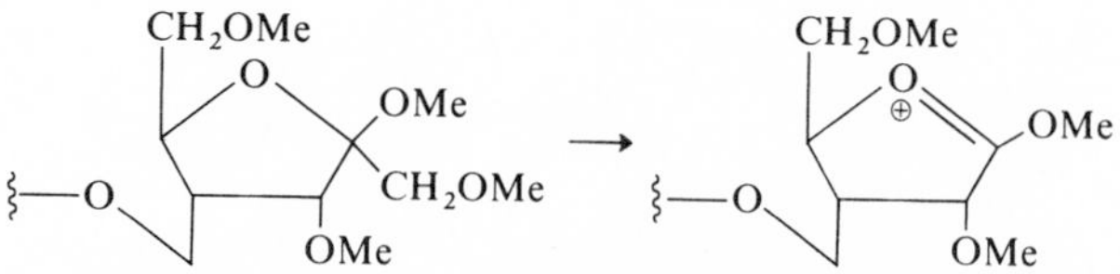

FIGURE 32. The loss of the C_1 side chain in a terminal fructo–furanosyl moiety linked through C4 (permethyl ether derivatives).

The CI spectra of permethylated aldohexopyranosyl $(1 \to 6)$ aldo-hexoses with methane and isobutane show more pronounced differences related to different glycosidic linkages than the corresponding electron impact spectra (De Jong et al., 1979b). A reliable difference between $(1 \to 2)$ and $(1 \to 4)$ linked disaccharides could be achieved. Moreover, for $(1 \to 2)$ and $(1 \to 6)$ linked saccharides, the α and β anomers could be differentiated also.

The metastable decomposition MIKE spectra and the collision activation spectra of the MH^+ ion, from the different permethylated aldohexopyranosyl aldohexoses generated under CI conditions, show no difference. However, with the use of ammonia or trimethylamine as reagent gas, differences are apparent in the collision activation spectra of the corresponding adducts $(M \cdot NH_4^+$ or $M \cdot Me_3NH)^+$ and thus enable differentiation between the anomeric disaccharides (De Jong et al., 1980a).

5.2. Pertrimethylsilylated Disaccharides

Mass spectrometric investigations have been conducted on a large number of trimethylsilylated disaccharides, representing all types of linkages between two sugar residues (see for example Kochetkov et al., 1968; Kamerling et al., 1972).

The mass spectra of the TMS of aldohexylaldohexose disaccharides may be divided into three groups, depending on the type of linkage.

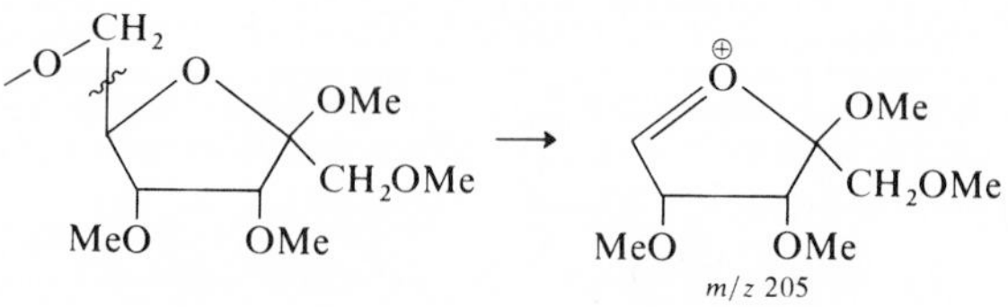

FIGURE 33. Fragmentation at the C_5–C_6 bond in a terminal 6-linked fructofuranose moiety (permethyl ether derivatives).

$$(1 \rightarrow 5)\,\text{linked} \rightarrow G1 - \overset{\oplus}{O} = \overset{5}{C}H - \overset{6}{C}H_2O - SiMe_3$$

m/z 583

$$(1 \rightarrow 6)\,\text{linked} \rightarrow G1 - O1 - \overset{6}{C}H_2 - \overset{5}{C}H = \overset{\oplus}{O}SiMe_3$$

FIGURE 34. Typical fragments arising from C_4–C_5 cleavage in $(1 \rightarrow 5)$- and $(1 \rightarrow 6)$-linked aldohexosyl aldohexoses (TMS derivatives).

The $(1 \rightarrow 1)$ linked aldohexoses give three typical peaks at m/z 565, 553, and 540. The structure **58** was proposed for the latter ion, and its formation involves an intramolecular transfer of a TMS group concomitant with the cleavage of the interglycosidic linkage.

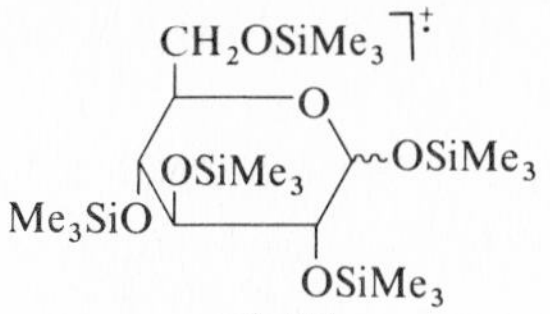

CH_2OSiMe_3

m/z 540

58

The mass spectra of the $(1 \rightarrow 2)$, $(1 \rightarrow 3)$ and $(1 \rightarrow 4)$—linked disaccharides are similar. It was suggested that the ion m/z 683 is typical for these three modes of linkage but this ion was further found in the spectra of $(1 \rightarrow 5)$ and $(1 \rightarrow 6)$ linked disaccharides.

The disaccharides $(1 \rightarrow 5)$ and $(1 \rightarrow 6)$ give almost identical spectra with a typical fragment at m/z 583, arising from the C4–C5 cleavage of the moiety at the reducing end (Figure 34).

The TMS ether of disaccharides containing a 2-acetamido-2-deoxy-hexose residue give characteristic spectra in which a 2-acetamido-2-deoxy sugar at the reducing end could be determined by the fragment at m/z 538 (**59**).

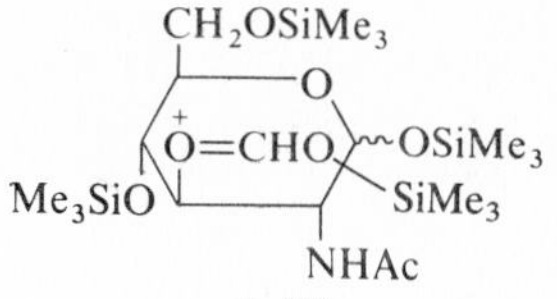

CH_2OSiMe_3

NHAc

m/z 538

59

Differentiation between $(1 \rightarrow 2)$, $(1 \rightarrow 3)$, and $(1 \rightarrow 4)$ linked 2-acetamido-2-deoxy aldosylaldoses is difficult. The $(1 \rightarrow 6)$ linked disaccharide of this series possesses an ion at m/z 552, which is the analog of

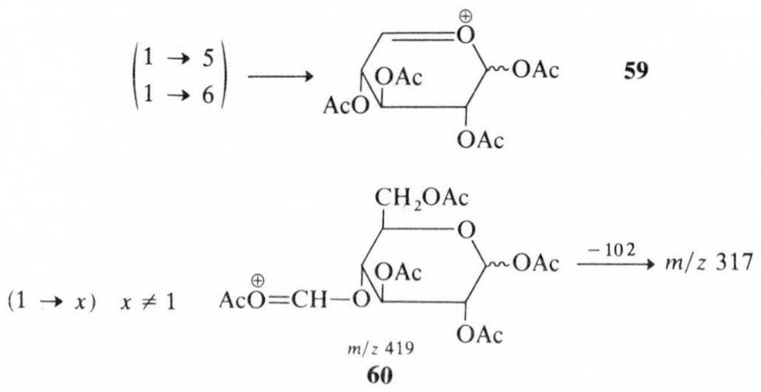

FIGURE 35. Characteristic fragmentations in peracetyl derivatives of aldohexopyranosyl-aldohexopyranoses.

m/z 583 found in the spectra of the per-TMS of $(1 \rightarrow 5)$ and $(1 \rightarrow 6)$ linked aldosyl-aldoses. This ion is absent when the amino sugar is present at the reducing end of the $(1 \rightarrow 6)$ linked disaccharide.

5.3. Peracetates of Disaccharides (Aldohexosyl-Aldohexoses)

The cleavage of the glycosidic bond gives a relatively intense oxonium ion for all isomers. It decomposes by losses of acetic acid, ketene, and acetic anhydride. As for the preceding types of derivatives, the characteristic peaks divide the disaccharides into three classes. The $(1 \rightarrow 1)$ linked disaccharides, the $(1 \rightarrow 2)$, $(1 \rightarrow 3)$, and $(1 \rightarrow 4)$ linked disaccharides, and lastly the $(1 \rightarrow 6)$ linked sugars (Das and Thayumanavan, 1975). The ion at m/z 317 is present in all isomers, except trehalose octaacetate. It was suggested that two competitive processes are operative: a C5–C6 bond cleavage of the reducing moiety (for $1 \rightarrow 5$ and $1 \rightarrow 6$ linked disaccharides) and a C1–C2 cleavage of the reducing moiety (for $1 \rightarrow 5$ and $1 \rightarrow 6$ linked disaccharides) and a C1–C2 cleavage for all isomers but the $(1 \rightarrow 1)$ (Figure 35).

By ^{13}C-labelling, this fragmentation pathway was confirmed and the structure **61** was proposed for the ion at m/z 317 in the spectra of $(1 \rightarrow 4)$ linked sugars (Bosso *et al.*, 1978).

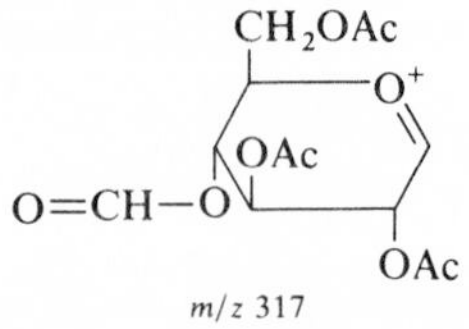

Another discrimination between the disaccharidic isomers may be obtained by the examination of the IKE spectra. In such an analysis, it was found that most of the metastable ions are due to the decomposition of the oxonium ion at m/z 331, which comes from the breakdown of the interglycosidic linkage. Since the reproducibility of the IKE spectra is almost independent of the ion source conditions, the differences appear reliable. The three groups of sugars already classified according to the EI spectra give different IKE spectra. But the $(1 \rightarrow 3)$ and $(1 \rightarrow 4)$ linked disaccharides give similar IKE spectra. However, the configuration of the glycosidic linkage does not affect the metastable decomposition pathway of m/z 331, and the slight differences which are observed are attributed to the way of formation of this ion, which could be produced in energetically different states.

5.4. Derivatives of Oligosaccharide Alditols

In contrast to the derivatives of reducing oligosaccharides, the corresponding alditol derivatives give a single peak on GC and their mass spectra are less complex. Each sugar moiety has a typical fragmentation pattern

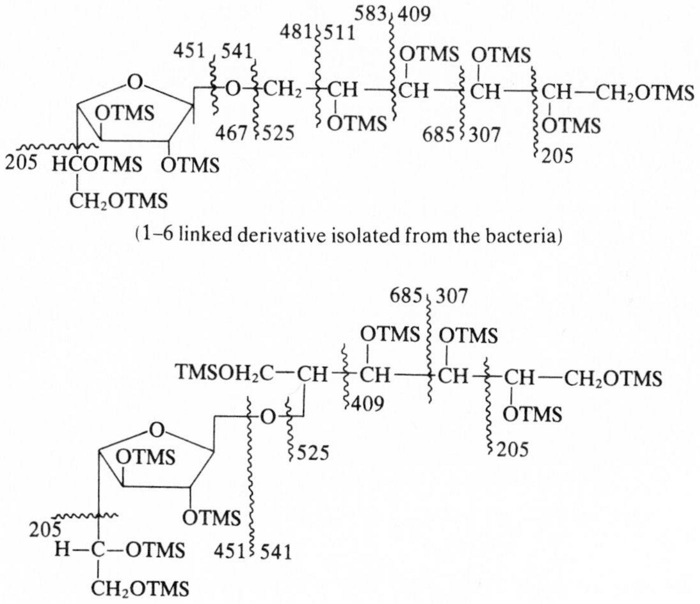

(1–6 linked derivative isolated from the bacteria)

FIGURE 36. Fragmentations in the TMS derivatives of two isomers of hexofuranosidyl alditols. Top: derivative obtained from the galactose-containing disaccharide in *M. tuberculosis* $(1 \rightarrow 6)$-linked. Bottom: expected fragmentations from a $(1 \rightarrow 5)$-linked isomer.

resembling its monosaccharide counterpart. Using permethylated derivatives, it is often difficult to determine the position of substitution on the alditol unit, because the fragments obtained by fission of the alditols are weak compared to those from the sugar part. This information is more easily obtained from TMS derivatives.

An example of such an investigation is presented in the structural determination of a galactose-containing disaccharide in fractions of *Mycobacterium tuberculosis* (Vilkas *et al.*, 1973). An intense peak at m/z 583 in the spectra of the TMS derivative of the aldosylaldose is consistent with $(1 \rightarrow 5)$ and $(1 \rightarrow 6)$ linkages. The ratio 205/204 and 217/204 in the TMS of both the disaccharide and its reduced alditol indicates an hexofuranosidyl moiety. The distinction between the $(1 \rightarrow 5)$ and $(1 \rightarrow 6)$ modes of linkage is obtained by the examination of the spectra of the TMS-glycosyl alditols (Figure 36).

In the same way, the presence and position of a 2-acetamido 2-deoxy moiety in a hexosamine-containing disaccharide could be determined by mass spectrometry (Finne *et al.*, 1977).

Using permethylated derivatives, it was possible to determine the presence of hydroxy-acyl amido groups on a glucosaminyl-glucosamine derivative which had been isolated by partial hydrolysis of a lipopolysaccharide (Hase and Rietschel, 1977) (Figure 37).

The determination of the nature of the glycosidic linkage in some glucosaminyl-glucosaminitol derivatives can also be obtained by EI–MS of the permethyl derivatives (Jensen *et al.*, 1979).

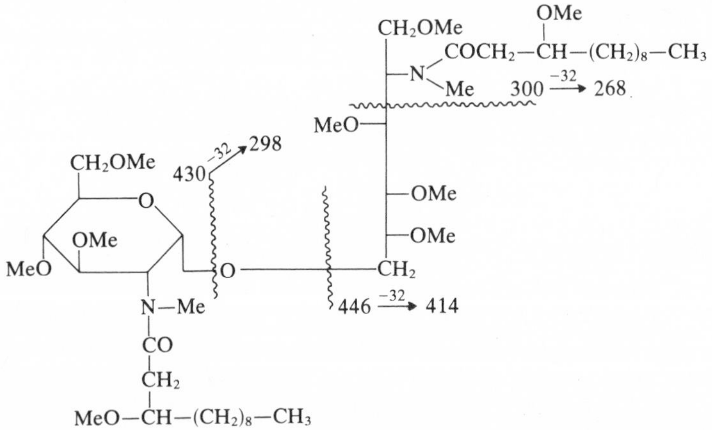

FIGURE 37. Determination of the existence of 2-hydroxy acyl amido groups in glucosaminyl-glucosamine disaccharides isolated from a lipopolysaccharide (per methyl derivatives of the glucosaminyl-glucosaminitol).

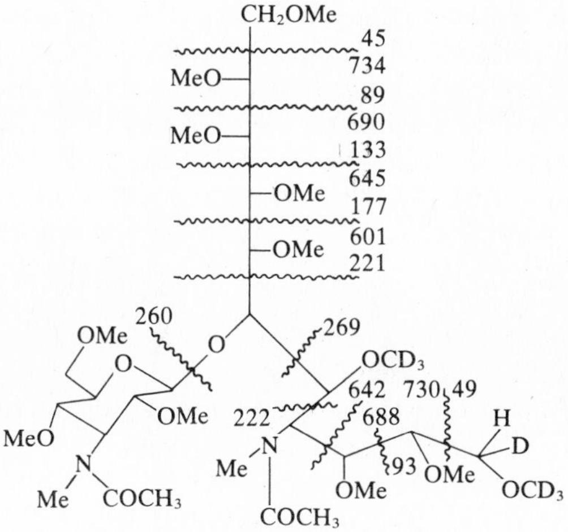

FIGURE 38. Mass spectrometric structural study of an oligosaccharide from anthelmycine. For derivatization, see text.

Similar studies were used to determine the structure of the oligosidic moiety of antibiotics. An example is the structural study of anthelmycine, an antibiotic isolated from *Streptomyces lingissimus* (Ennifar *et al.*, 1977). The reduced permethylated and partially labeled disaccharide, which contains a 3-amino-3-deoxy-β-D-glucopyranose linked to an amino-undecose (a sugar containing a C11 backbone), gives the mass spectral fragmentation shown in Figure 38.

The CD_3 groups are introduced by deuteromethylation of the free hydroxyl groups liberated by the selective hydrolysis of the aglycone part of the permethylated antibiotic followed by $NaBD_4$ reduction of the oligosaccharide. The 4-amino undecose was linked glycosidically to cytosine.

6. CONCLUSION

The interpretation of the mass spectra of carbohydrate derivatives clearly gives sufficient information to permit the determination of their structural properties (molecular weight and the nature and location of functional groups). The distinction between stereoisomers appears more difficult in spite of the numerous attempts to employ chemical ionization conditions or to synthesize structurally different derivatives from stereoisomeric sugars.

Thus the precise determination of gas chromatographic parameters is very useful to completely identify the structures of sugars. Obviously, the identification by means of chromatographic parameters requires the synthesis of reference samples with known stereochemistry and structure. For the identification of partially alkykated sugars, which is one of the most widely found problems, the reference samples can be obtained easily by an incomplete alkylation of a sugar of known configuration followed by the determination of both retention times and structure assignments of each component by GC/MS analysis.

Another problem which is easily solved by the use of GC/MS is the identification of sugars in complex mixtures containing noncarbohydrate components. The derivatization procedures often lead to the formation of volatile compounds from nonsugar components (for example, amino acids) which may be difficult to differentiate by the use of GC only. Fortunately, their mass spectra are generally so different that they can be identified with no difficulty.

Thus, the GC/MS analysis of carbohydrates is a powerful method for structural determination and identification, and its use continues to increase in microbiology, particularly in the field of complex polyasccharides.

REFERENCES

Ahibo-Coffy, A., Aurelle, H., Lacave, C., Promé, J. C., Puzo, G., and Savagnac, A., 1978, Isolation, structural studies and chemical synthesis of a palmitone lipid from *Corynebacterium diphtheriae*, *Chem. Phys. Lipids* **22**:185.

Aspinall, G. O., 1976, Polysaccharide Methodology, *Int. Rev. Sci. Org. Chem. Ser. 2* **7**:201.

Biemann, K., De Jongh, D. C., and Schnoes, H. K., 1963, Application of mass Spectrometry to structure problems. XIII Acetates of pentoses and hexoses. *J. Am. Chem. Soc.* **85**:1763.

Björndal, H., Hellerqvist, C. G., Lindberg, B., and Svensson, S., 1970, Gas–liquid chromatography and mass spectromety in methylation analysis of polysaccharides, *Angew. Chem. Int. Ed.* **9**:610.

Björndal, H., and Lindberg, B., 1970, Polysaccharides elaborated by *Polyporus fomentarius* and *Polyporus igniarius*. Part II. Water soluble, acidic polysaccharides from the fruit bodies, *Carbohydr. Res.* **12**:29.

Björndal, H., Lindberg, B., and Svensson, S., 1967a, Gas–liquid chromatography of partially methylated alditols as their acetates, *Acta Chem. Scand.* **21**:1801.

Björndal, H., Lindberg, B., and Svensson, S., 1967b, Mass spectrometry of partially methylated alditol acetates, *Carbohydr. Res.* **5**:433.

Blache, D., Bruneteau, M., and Michel, G., 1981, Structure de la région heptose du lipopolysaccharide de *Escherichia coli* K12 CR34, *Eur. J. Biochem.* **113**:563.

Blair, I. A., and Bowie, J. H., 1979, Ion Cyclotron resonance studies of alkyl-silyl ions. II. The reaction of ketones, carboxylic acids and esters with the trimethylsilyl cation, *Aust. J. Chem.* **32**:1389.

Blanc-Muesser, M., Defaye, J., Folz, R. L., and Horton, D., 1980, Chemical ionization mass spectrometry of dimethylacetals and diethylthioacetals of aldopentoses and aldohexoses and the influence of stereochemical configuration, *Org. Mass Spectrom.* **15**:317.

Boren, H. B., Garegg, P. J., Kenne, L., Pilotti, A., Svensson, S., and Swahn, C. G., 1973, Analysis of acylated methyl glycosides as their trimethylsilyl derivatives by gas–liquid chromatography and mass spectrometry, *Acta Chem. Scand.* **27**:3557.

Bosso, C., Defaye, J., Gardelle, A., and Ulrich, J., 1977, Spectrométrie de masse de benzyl-idène-acétals d'hexopyranosides, *Org. Mass Spectrom.* **12**:493.

Bosso, C., Taravel, F., Ulrich, J., and Vignon, M., 1978, Utilisation du ^{13}C en spectrométrie de masse: Étude de la fragmentation de disaccharides, *Org. Mass Spectrom.* **13**:477.

Bowser, D. V., Teece, R. G., and Somani, S. M., 1978, Identification of amino-sugars from bacterial lipopolysaccharides by gas chromatography—electron impact and chemical ionization mass spectrometry, *Biomed. Mass Spectrom.* **5**:627.

Buchs, A., Glangetas, A., and Tronchet, J. M. J., 1975, Etude en spectrométrie de masse de la fragmentation de di-*O*-isopropylidène-1,2,5,6-α-D-hexofurannoses, *Org. Mass Spectrom.* **10**:970.

Chizhov, O. S., Dmitriev, B. A., Zolotarev, B. M., Chernyak, A. Y., and Kochetkov, N. K., 1969, Mass spectra of alditol trifluoroacetates, *Org. Mass Spectrom.* **2**:947.

Chizhov, O. S., Golovkina, L. S., and Wulfson, N. S., 1968, Mass Spectrometric study of carbohydrates. A new fragmentation reaction induced by electron impact: "h rupture," *Carbohydr. Res.* **6**:138.

Chizhov, O. S., and Kochetkov, N. K., 1966, Mass spectrometry of carbohydrate derivatives, *Adv. Carbohydr. Chem. Biochem.* **21**:29.

Coduti, P. L., and Bush, C. A., 1977, Structural determination of *N*-acetyl aminosugar derivatives and disaccharides by gas chromatography–mass spectrometry, *Anal. Biochem.* **78**:21.

Das, K. G., and Thayumanavan, B., 1972, Mass spectral studies of some fructose containing oligosaccharide permethyl ethers, *Org. Mass Spectrom.* **6**:1063.

Das, K. G., and Thayumanavan, B., 1975, E. I. and I. K. E. spectra of some aldosyl disaccharide peracetates, *Org. Mass Spectrom.* **10**:455.

Davis, C. E., Freedman, S. D., Douglas, H., and Brandz, A. I., 1969, Analysis of sugars in bacterial endotoxins by gas-liquid chromatography, *Anal. Biochem.* **28**:243.

De Jong, E. G., Heerma, W., Haverkamp, J., and Kamerling, J. P., 1979a, Electron impact mass spectrometry of permethylated disaccharides, *Biomed. Mass Spectrom.* **6**:72.

De Jong, E. G., Heerma, W., and Sicherer, C. A. X. G. F., 1979b, Discrimination of the permethylated dissacharides by their chemical ionization spectra, *Biomed. Mass Spectrom.* **6**:243.

De Jong, E. G., Heerma, W., and Dijkstra, G., 1980a, The use of collisional activation spectra in the discrimination of stereoisomeric permethylated disaccharides, *Biomed. Mass Spectrom.* **7**:127.

De Jong, E. C., Heerma, W., Sicherer, C. A. X. G. F., and Dijkstra, G., 1980b, The structure of the *m/z* 101 in the spectra of permethylated saccharides, *Biomed. Mass Spectrom.* **7**:122.

De Jongh, D. C., 1964a, Mass spectrometry in carbohydrate chemistry: di-ethyl dithioacetal and dithioketal peracetates, *J. Am. Chem. Soc.* **86**:3149.

De Jongh, D. C., 1964b, Mass spectrometry in carbohydrate chemistry: Ethylene dithioacetal peracetates, *J. Am. Chem. Soc.* **86**:4027.

De Jongh, D. C., and Biemann, K., 1963, Application of mass spectrometry to structure problems: XIV. Acetates of partially methylated pentoses and hexoses, *J. Am. Chem. Soc.* **85**:2289.

De Jongh, D. C., and Biemann, K., 1964, Mass spectra of O-isopropylidene derivatives of pentoses and hexoses, *J. Am. Chem. Soc.* **86**:67.

De Jongh, D. C., Radford, T., Hribar, J. D., Hanessian, S., Bieber, M., Dawson, G., and Sweeley, C. C., 1969, Analysis of trimethylsilyl derivatives of carbohydrates by gas chromatography and mass spectrometry, *J. Am. Chem. Soc.* **91**:1728.

Diekman, J., Thomson, J. B., and Djerassi, C., 1968, Mass spectrometry in structural and stereochemical problems. CLV. Electron impact induced fragmentations and rearrangements of some trimethylsilyl ethers of aliphatic glycols and related compounds, *J. Org. Chem.* **33**:2271.

Dizdaroglu, M., Henneberg, D., and Von Sonntag, C., 1974, The mass spectra of TMS-ethers of deuterated polyalcohols. A contribution to the structural investigation of sugars, *Org. Mass Spectrom.* **8**:335.

Dizdaroglu, M., Henneberg, D., Von Sonntag, C., and Schuchman, M. N., 1977, Mass spectra of trimethylsilyl-di-O-methyloximes of aldosuloses and dialdoses, *Org. Mass Spectrom.* **12**:772.

Dmitriev, B. A., Backinovsky, L. V., Chizhov, O. S., Zolotarev, B. M., and Kochetkov, N. K., 1971, Gas liquid chromatography and mass spectrometry of aldononitrile acetates and partially methylated aldononitrile acetates, *Carbohydr. Res.* **19**:432.

Dougherty, R. C., Horton, D., Philips, K. D., and Wander, J. D., 1973, The high resolution mass spectrum of 2-acetamido-1,3,4,6-tetra-O-acetyl-2-deoxy-α-D-glucopyranose, *Org. Mass Spectrom.* **7**:805.

Dutton, G. G. S., 1973, Application of gas–liquid chromatography to carbohydrates, part I, *Adv. Carbohydr. Chem. Biochem.* **28**:11.

Dutton, G. G. S., 1974, Application of gas–liquid chromatography to carbohydrates, part II, *Adv. Carbohydr. Chem. Biochem.* **30**:9.

El-Dash, A. A., and Hodge, J. E., 1971, Determination of 3-deoxy-D-erythrohexosulose (3-deoxy-D-glucosone) and other degradation products of sugars by borohydride reduction and gas–liquid chromatography, *Carbohydr. Res.* **18**:259.

Ennifar, S., Das, B. C., Nash, S. M., and Nagarajan, R., 1977, Structural identity of anthelmycin with hikizimycin, *J. Chem. Soc., Chem. Commun.*, p. 41.

Ferrier, R. J., 1978, Carbohydrate boronates, *Adv. Carbohydr. Chem. Biochem.* **35**:31.

Finne, J., Mononen, I., and Kärkkäinen, J., 1977, Analysis of hexosaminitol-containing disaccharide alditols from rat brain glycoproteins and gangliosides as O-trimethylsilyl derivatives by gas chromatography–mass spectrometry. *Biomed. Mass Spectrom.* **4**:281.

Golovkina, L. S., Chizhov, O. S., and Vul'fson, N. F., 1966, Mass spectrometer investigation of carbohydrates. 9. Acetates of polyols, *Bull. Acad. Sci. USSR*, 1853.

Gray, G. R., and Ballou, C. E., 1971, Isolation and characterization of a polysaccharide containing 3-O-methyl-D-mannose from *Mycobacterium Phlei*. *J. Biol. Chem.* **246**:6835.

Haegele, K. D., and Desiderio, D. M., 1974, Occurrence of gas phase ammonolysis during chemical ionization mass spectrometry, *J. Org. Chem.* **39**:1078.

Hancock, R. A., Marshall, K., and Weigel, H., 1976, Structure of the levan elaborated by *Streptococcus salivarius* strain 51; An application of chemical ionization mass spectrometry, *Carbohydr. Res.* **49**:351.

Hase, S., and Rietschel, E., 1977, The chemical structure of the lipid A component of lipopolysaccharide from *Chromobacterium violaceum* NCTC 9694 *Eur. J. Biochem.* **75**:23.

Horton, D., Wander, J. D., and Foltz, R. L., 1974, Analysis of sugar derivatives by chemical-ionization mass spectrometry, *Carbohydr. Res.* **36**:75.

Hoshida, K., Honda, N., Iino, N., and Kato, K., 1969, Gas–liquid chromatography of hexopyranosides and hexofuranosides, *Carbohydr. Res.* **10**:333.

Hurst, R. E., 1973, The trimethylsilylation reactions of hexosamines, and gas-chromatographic separation of the derivatives, *Carbohydr. Res.* **30**:143.

Jankowski, K., and Gaudin, D., 1978, Critical examination of the use of gas-chromatography–mass spectrometry in the identification of persilylated saccharides, *Biomed. Mass Spectrom.* **5**:371.

Jansson, P. E., and Lindberg, B., 1980, Mass spectrometry of alditol acetates: Origin of the fragment having even mass numbers, *Carbohydr. Res.* **86**:287.

Jensen, M., Borowiak, D., Paulsen, H., and Rietschel, E. T., 1979, Analysis of permethylated glucosaminyl-glucosaminitol disaccharides by combined gas–liquid chromatography–mass spectrometry, *Biomed. Mass Spectrom.* **6**:559.

Kadentsev, V. I., Solov'yov, A. A., and Chizhov, O. S., 1978, Chemical ionization mass spectra of permethylated methyl hexopyranosides, *Adv. Mass Spectrom.* **7B**:1465.

Kamei, A., Yoshizumi, H., Akashi, S., and Kagabe, K., 1976, A new method for the simultaneous determination of a reducing end group and component hexoses in oligosaccharides by gas chromatography–mass spectrometry, *Chem. Pharm. Bull. (Jpn)* **24**:1108.

Kamerling, J. P., Vliegenthart, J. F. G., Vink, J., and De Ridder, J. J., 1972, Mass spectrometry of pertrimethylsilyl oligosaccharides contining fructose units, *Tetrahedron* **28**:4375.

Kärkkäinen, J., and Vihko, R., 1969, Characterization of 2-amino-2-deoxy-D-glucose, 2-amino-2-deoxy-D-galactose and related compounds, as their trimethyl silyl derivatives by gas liquid chromatography–mass spectrometry, *Carbohydr. Res.* **10**:113.

Kiso, M., Hasegawa, A., 1976, Acetonation of D-glucose with 2,2-dimethoxypropane, *N,N*-dimethylformamide, *p*-toluene-sulfonic acid, *Carbohydr. Res.* **52**:87.

Khvostenko, V. I., Baltima, L. A., Fal'ko, V. S., and Tolstikov, G. A., 1976, Stereospecificity of mass spectra of negative ions of hexopyranose peracetates, *Bull. Acad. Sci. USSR, Chem. Sci.* 1587.

Kochetkov, N. K., and Chizhov, O. S., 1964, Application of mass spectrometry to methylated monosaccharides identification. *Biochim. Biophys. Acta* **83**:134.

Kochetkov, N. K., Chizhov, O. S., and Molodtsov, N. V., 1968, Mass spectrometry of oligosaccharides, *Tetrahedron* **24**:5587.

Laine, R. A., and Sweeley, C. C., 1971, Analysis of trimethylsilyl-*O*-methyloximes of carbohydrates by combined gas–liquid chromatography–mass spectrometry, *Anal. Biochem.* **43**:533.

Laine, R. A., and Sweeley, C. C., 1973, *O*-methyl oximes of sugars. Analysis as *O*-trimethylsilyl derivatives by gas–liquid chromatography and mass spectrometry, *Carbohydr. Res.* **27**:199.

Lance, D. G., and Jones, J. K. N., 1967, Gas chromatography of derivatives of the methyl ethers of D-xylose, *Can. J. Chem.* **45**:1995.

Li, B. W., Cochran, T. W., and Vercelotti, J. R., 1977, Chemical ionization mass spectra of per-*O*-acetylated aldononitriles and methylated aldononitrile acetates. *Carbohydr. Res.* **59**:567.

Lindberg, B., 1972, Methylation analysis of polysaccharides, *Methods Enzymol.* **28**:178.

Lindberg, B., and Lönngren, J., 1978, Methylation analysis of complex carbohydrates: General procedure and application for sequence analysis, *Methods Enzymol.* **50**:3.

Lindberg, B., Lönngren, J., and Thompson, J. L., 1973, Methylation studies on levans, *Acta Chem. Scand.* **27**:1819.

Lönngren, J., and Svensson, S., 1974, Mass spectrometry in structural analysis of natural carbohydrates, *Adv. Carbohydr. Chem. Biochem.* **29**:41.

MacLeod, J. K., Summons, R. E., and Letham, D. S., 1976, Mass spectrometry of cytokinin metabolites. Per(trimethylsilyl) and permethyl derivatives of glucosides of zeatin and 6-benzylaminopurine, *J. Org. Chem.* **41**:3959.

McNeil, M. C., and Albersheim, P., 1977, Chemical ionization mass spectrometry of methylated hexitol acetates, *Carbohydr. Res.* **56**:239.

Matsubara, T., and Hayashi, A., 1974, Determination of the structure of partially methylated sugars as *O*-trimethylsilyl ethers by gas chromatography mass spectrometry, *Biomed. Mass Spectrom.* **1**:62.

Moor, J., and Waight, E. S., 1975, The mass spectra of permethylated oligosaccharides, *Biomed. Mass Spectrom.* **2**:36.

Morgenlie, S., 1975, Analysis of mixtures of the common aldoses by gas chromatography of their *O*-isopropylidene derivatives. *Carbohydr. Res.* **41**:285.

Morrison, I. M., 1975, Determination of the degree of polymerization of oligo- and polysaccharides by gas–liquid chromatography, *J. Chromatogr.* **108**:361.

Murata, T., and Takahashi, S., 1978, Characterization of *O*-trimethylysilyl derivatives of D-glucose, D-galactose and D-mannose by gas–liquid chromatography–chemical ionization mass spectrometry with ammonia as reagent gas, *Carbohydr. Res.* **62**:1.

Orme, T. W., Boone, C. W., and Roller, P. P., 1974, The analysis of 2-acetamido-2-deoxy aldose derivatives by gas–liquid chromatography and mass spectrometry, *Carbohydr. Res.* **37**:261.

Pfaffenberger, C. D., Szafranek, J., Horning, M. G., and Horning, E. C., 1975, Gas chromatographic determination of polyols and aldoses in human urine as polyacetates and aldononitrile polyacetates. *Anal. Biochem.* **63**:501.

Pierce, A. E., 1968, Silylation of organic compounds. Pierce Chemical Co., Rockford, Illinois.

Promé, J. C., Lacave, C., Ahibo-Coffy, A., and Savagnac, A., 1976, Séparation et étude structurale des espèces moléculaires de monomycolates et de dimycolate de α-D-Tréhalose presents chez *Mycobacterium phlei*, *Europ. J. Biochem.* **63**:543.

Puzo, G., and Promé, J. C., 1977, Field desorption mass spectrometry of oligosaccharides in the presence of metallic salts. *Org. Mass. Spectrom.* **12**:28.

Puzo, G., and Promé, J. C., 1978, Mass spectrometry of acetylated sugars as trimethylsilyl ether derivatives. A way for location of long chain fatty acyl groups, *Biomed. Mass Spectrom.* **5**:146.

Puzo, G., Tichadou, J. L., and Promé, J. C., 1978a, Aliphatic-chain-sugar interactions under electron impact. The loss of methanol from acylated methyl glucpoyranosides, *Org. Mass Spectrom.* **13**:51.

Puzo, G., Tissié, G., Lacave, C., Aurelle, H., and Promé, J. C. 1978b, Structural determination of cord factor from a *Corynebacterium diphtheriae* strain by a combination of mass spectral ionization methods: Field desorption, cesium cationization and electron impact mass spectrometry studies, *Biomed. Mass Spectrom.* **5**:699.

Puzo, G., Tissié, G., Aurelle, H., Lacave, C., and Promé, J. C., 1979, Occurrence of 3-oxo-acyl group in 6,6′-diesters of α-D-trehalose. New glycolipids related to cord factor from *Corynebacterium diphtheriae*, *Eur. J. Biochem.* **98**:99.

Radford, T., and De Jongh, D. C., 1972, "Carbohydrates," in *Biochemical Applications of Mass Spectrometry* (G. R. Waller, ed.), Wiley-Interscience, New York.

Radjiejewska-Lebrecht, J., Feige, U., Jensen, M., Kotelko, K., Friebolin, H., and Mayer, H., 1980, Structural studies on the glucose–heptose region of the *Proteus mirabilis* R core, *Eur. J. Biochem.* **107**:31.

Radjiejewska-Lebrecht, J., Shaw, D. H., Borowiak, D., Fromme, I., and Mayer, H., 1979, Gas chromatography–mass spectrometry identification of a range of methyl ethers from L-glycero-D-mannoheptose and D-glycero-D-mannoheptose, *J. Chromatogr.* **179**:113.

Reinhold, V. N., Wirtz-Peitz, F., and Biemann, K., 1974, Synthesis, gas–liquid chromatography and mass spectrometry of per-*O*-trimethylsilyl carbohydrate boronates, *Carbohydr. Res.* **37**:203.

Saier, M. H., and Ballou, C. E., 1968, The 6-*O*-methylglucose-containing lipopolysaccharide of *Mycobacterium phlei*. Identification of D-glyceric acid and 3-*O*-methyl-D-glucose in the polysaccharide, *J. Biol. Chem.* **243**:992.

Schoots, A. C., and Leclercq, P. A., 1979, Chemical ionization mass spectrometry of trimethyl-silylated carbohydrates and organic acids retained in uremic serum, *Biomed. Mass Spectrom.* **6**:502.

Semenza, G., Curtius, C. H., Kolinska, J., and Müller, M., 1967, Studies on intestinal sucrase and intestinal sugar transport. VI. Liberation of α-glucose by sucrase and isomaltase from the glycone moiety of the substrates, *Biochim. Biophys. Acta* **146**:196.

Seymour, F. R., Chen, E. C. M., and Bishop, S. H., 1979, Identification of aldoses by use of their peracetylated aldononitrile derivatives. A GLC—MS approach, *Carbohydr. Res.* **73**:19.

Seymour, F. R., Chen, E. C. M., and Stouffer, J. E., 1980, Identification of ketoses by use of their peracetylated oxime derivatives: a GC—MS approach, *Carbohydr. Res.* **83**:201.

Seymour, F. R., Plattner, R. D., and Slodki, M. E., 1975, Gas–liquid chromatography–mass spectrometry of methylated and deutero-methylated per-*O*-acetylaldononitriles from D-mannose, *Carbohydr. Res.* **44**:181.

Seymour, F. R., Slodki, M. E., Plattner, R. D., and Jeanes, A., 1977, Six unusual dextrans: Methylation structural analysis by combined GLC—MS of peracetyl derivatives aldononitriles, *Carbohydr. Res.* **53**:153.

Sikl, D., Maslzr, L., and Bauer, S., 1970, Polysaccharides of *Torulopsis colliculosa* (*Hartman*) *Saccardo*. Isolation and structural features of extra cellular mannan and of cell-wall mannan, *Coll. Czeshovlov. Chem. Commun.* **35**:2965.

Sjostróm, E., Pfister, K., and Seppàlà, E., 1974, Quantitative determination of aldonic and uronic acids in mixtures as deuterium-labelled alditol acetates by gas chromatography-mass spectrometry, *Carbohydr. Res.* **38**:293.

Stewart, T. S., Mendershausen, P. B., and Ballou, C. E., 1968, Preparation of a mannopen-taose, mannohexaose and mannoheptaose from *Saccharomyces Cerevisiae* Mannan, *Biochemistry* **7**:1843.

Størset, P., Stokke, O., and Jellum, E., 1978, Monosaccharides and monosaccharide deriva-tives in human seminal plasma, *J. Chromatogr.* **145**:351.

Sweeley, C. C., Bentley, R., Makita, M., and Wells, W. W. 1963, Gas–liquid chromatography of trimethylsilylderivatives of sugars and related substances, *J. Am. Chem. Soc.* **85**:2497.

Sweet, D. P., Albersheim, P., and Shapiro, R. H., 1975, Partially ethylated alditol acetates as derivatives for elucidation of the glycosyl linkage-composition of polysaccharides, *Carbohydr. Res.* **40**:199.

Sweet, D. P., Shapiro, R. H., and Albersheim, P., 1974, The mass spectral fragmentation of partially ethylated alditol acetates, a derivative used in determining the glycosyl linkage composition of polysaccharides, *Biomed. Mass Spectrom.* **1**:263.

Szafraned, J., Pfaffenberger, C. D., and Horning, E. C., 1973, Separation of aldoses and alditols using thermostable open tubular glass capillary columns, *Anal. Lett.* **6**:479.

Szafranek, J., Pfaffenberger, C. D., and Horning, E. C., 1974, The mass spectra of some per-*O*-acetylaldononitriles, *Carbohydr. Res.* **38**:97.

Tolstikov, G. A., Fal'ko, V. S., Baltina, L. A., and Khvostenko, V. I., 1977, Mass Spectrometry of negative ions and the stereochemistry of organic compounds. 2. Acetates of pentoses and hexoses, *Bull. Acad. Sci. USSR Chem. Sci.* 964.

Valent, B. S., Darvill, A. G., McNeil, M., Robertsen, B. K., and Albersheim, P., 1980, A general and sensitive chemical method for sequencing the glycosyl residues of complex carbohydrates, *Carbohydr. Res.* **79**:165.

Varna, R., and Varna, R. S., 1977, A simple procedure for combined gas chromatography analysis of neutral sugars, hexosamines and alditols. Determination of the degree of polymerization of oligo- and polysaccharides and chain weights of glycosaminoglycans, *J. Chromatogr.* **139**:303.

Vilkas, E., Amar, C., Markovits, J., Vliegenthart, J. F. G., and Kamerling, J. P., 1973, Occurrence of a galactofuranose disaccharide in immunoadjuvant fractions of *Mycobacterium tuberculosis* (cell walls and wax D), *Biochim. Biophys. Acta* **297**:423.

Wiecko, J., and Sherman, W. R., 1975, Mass spectral study of cyclic alkane boronates of sugar phosphates. Evidence of interaction between the phosphate groups and boron under electron impact conditions, *Org. Mass Spectrom.* **10**:1007.

Wiecko, J., and Sherman, W. R., 1976, Boroacetylation of carbohydrates. Correlation between structure and mass spectral behaviour in monoacetyl hexose cylcic boronic esters, *J. Am. Chem. Soc.* **98**:7631.

Wood, P. J., Siddiqui, I. R., and Weisz, J., 1975, The use of butane boronic esters in the gas–liquid chromatography of some carbohydrates, *Carbohydr. Res.* **42**:1.

5

AMINO ACIDS AND PEPTIDES

Samuel L. MacKenzie, Prairie Regional Laboratory,
National Research Council of Canada, 110
Gymnasium Road, Saskatoon, Saskatchewan
S7N 0W9, Canada

1. INTRODUCTION

Developments in the analysis of amino acids and peptides by gas chromatography–mass spectrometry (GC/MS) have occurred in a chemical rather than a microbiological context. The presentation in this chapter, therefore, describes advances in technique and structural determination and is leavened, where possible, with examples drawn from the application of GC/MS to microbial metabolism and biochemistry. This chapter does not purport to be an exhaustive review of the relevant literature. Citations have been omitted not from an inherent lack of merit or relevance but because they are accessible through later citations of the same authors or through the more specialized reviews cited.

Successful GC/MS is highly dependent on a foundation of good derivatization and chromatographic techniques. It is possible to discuss only the most fundamental aspects within the ambit of this chapter and so, for details of reaction and instrumental conditions, the reader is directed to more comprehensive descriptions in other sources (Husek and Macek, 1975; Blau and King, 1978; Knapp, 1979; MacKenzie, 1981).

NRCC No. 21327.

The discussion of the fragmentation patterns of amino acid derivatives emphasizes those derivatives which, because of extensive studies of the quantitative aspects of derivatization and chromatography, form the basis of the preferred methods for analyzing amino acids by gas liquid chromatography (GLC). Such methods are, therefore, most likely to be chosen to address a specific problem in the analysis of amino acids by GC/MS.

The determination of the primary structure of a peptide contains two components: composition and amino acid sequence. Since the former category is usually resolved by analysis of a hydrolysate and then becomes a question of amino acid analysis, the section on peptides deals almost exclusively with sequence analysis by GC/MS.

Most mass spectrometric studies of amino acids and peptides and their derivatives have used electron-impact ionization (EI) and, to a lesser extent, chemical ionization (CI) in low-resolution mass spectrometers. The main discussion of amino acids and peptides is in this context. However, a number of highly specialized, but not generally available, instrumental techniques have also been used. In certain applications, these confer advantages over more standard methods and, therefore, their application to amino acids and peptides is discussed briefly in a separate section.

2. AMINO ACIDS

The identification and assaying of amino acids are important elements in many types of physiological and metabolic studies. Standard protein hydrolysates can be analyzed at the picomole level by GLC alone with mass spectrometry being used only when necessary to confirm structure. Methods have also been developed for the analysis of many nonprotein amino acids and for the resolution of amino acid enantiomers by GLC. These have been described in detail in reviews by Husek and Macek (1975), and MacKenzie (1981). However, unusual amino acids may be encountered as free amino acids, as components of hydrolysates of secondary microbial metabolites such as antibiotics, or as the result of post-translational modification of proteins. Also, increased analytical sensitivity or specificity may be required or stable isotopes may be used in the experimental protocol. Under such circumstances, the role of mass spectrometry becomes central.

The following section emphasizes methodological developments in amino acid analysis by GLC and describes the essential features of the mass spectra of the most commonly used amino acid derivatives. Although these methods have not yet been extensively applied to microbiological problems, it is not unreasonable to suggest that they are logical candidates for future applications.

2.1. Free Amino Acids

Amino acids are inherently nonvolatile and generally degrade at the temperatures required for their volatilization. However, spectra of most amino acids can be obtained by direct sample injection, although some, such as glutamic acid, arginine, and cystine, may be modified in the process. In electron impact ionization, the molecular ion is usually of very low abundance and the predominant fragmentation pathway results from loss of the carboxyl group with charge retention on the residual amine fragment. Fragmentations characteristic of the side chains also occur. Aromatic residues preferentially retain the charge following fission of the C_2–C_3 bond and may undergo ring expansion. Thus the phenylalanyl side chain produces the tropylium ion $(C_7H_7)^+$. Functional groups in the β-position are eliminated and the loss of olefin fragments via a McLafferty rearrangement are characteristic of valine, leucine, and isoleucine.

Cyclic α-amino acids in which the ring size ranges from three to eight carbon atoms also give spectra dominated by the amine fragment ion (Coulter and Fenselau, 1972).

Methane chemical ionization (CI) spectra of the α-amino acids yield an enhanced abundance of the protonated molecular ion MH^+, thus confirming the molecular weight, and also abundant amine fragments (Leclercq and Desiderio, 1973). Arginine and cystine decompose on the heated sample inlet. The methane CI spectra of a series of α,ω-diamino acids, ω-amino acids, and cyclic and acyclic α-amino acids have also been described (Weinkam, 1978).

Isobutane CI spectra of α-amino acids contain mostly the protonated molecular ion, MH^+, and exhibit very little fragmentation (Meot-Ner and Field, 1973).

Chemical ionization spectra of α-amino acids have also been obtained using hydrogen as the reagent gas (Tsang and Harrison, 1976). Compared with methane CI spectra the MH^+ abundance was reduced and fragment ion abundance increased. Using both CD_4 and D_2 as reagent gases, it was concluded that there is "extensive intramolecular proton transfer in MH^+ prior to fragmentation." Earlier mechanistic studies had concluded that protonation occurred at a specific site and that subsequent fragmentation was largely determined by the site of protonation.

A special technique, rapid sample evaporation from a heated Teflon surface coupled with ammonia as a proton source, produced proton transfer spectra in which the protonated molecular ion was observed for 19 α-amino acids and was the base peak for all except arginine and glutamic acid. Only 5–8 nmol of sample was required (Gaffney et al., 1977).

Field desorption (FD) MS of ten α-amino acids representing all the amino acid functional groups and including arginine and cystine, gave spectra in all of which the protonated molecular ion MH^+ or the molecular ion M^+ was detected (Winkler and Beckey, 1972).

The above special techniques have been used to circumvent the lack of volatility of natural amino acids or to obtain a better understanding of their fragmentation. For most practical purposes, however, the barrier of volatility may be more effectively surmounted by appropriate derivatization.

2.2. Derivatized Amino Acids

The structures of the common amino acids include the functional carboxy, sulfhydryl, hydroxy, and amino groups. The consequent polarity and potential for hydrogen bonding results in low volatility. Most mass spectrometric studies have, therefore, been conducted on derivatives which reduce the polarity of the functional groups. Although the volatility is significantly improved by esterification alone, and a number of mass spectrometric studies of amino acid esters have been reported, derivatization of all functional groups is required for GLC of amino acids and, therefore, for GC/MS. The strategies which have proven most effective for quantitative derivatization of a mixture of amino acids require two reactions. The first reaction, esterification, is usually achieved by heating the sample in an acidified solution of the appropriate alcohol. Both the catalyst (HCl) and the alcohol are easily removed by evaporation thus avoiding extractions and transfers. The remaining functional groups are then acylated by heating in the presence of a carboxylic acid anhydride or its perfluorinated analog such as trifluoroacetic anhydride. Because the imidazole ring is a good acyl donor, formation of a chromatographically stable diacyl derivative of histidine requires further treatment. On-column acylation by coinjection of the sample with an acid anhydride is the most simple procedure (MacKenzie, 1981).

2.2.1. Sample Preparation

The following discussion of sample preparation is restricted to salient practical features common to most protocols and should apply equally to derivatization of amino acids and peptides.

Proper cleaning of glassware is often overlooked. Cancalon and Klingman (1974) washed all glassware with hot 1 N NaOH, rinsed thoroughly with distilled water, washed with hot 50% HNO_3, rinsed with double distilled water, and dried at 100°C. Siezen and Mague (1977) removed contaminating organic material by heating at 450 to 500°C for two to six hours.

The literature, particularly dealing with peptides, is replete with procedures which feature relatively large-scale filtration or evaporation *in vacuo* using a rotary evaporator. These manipulations are not compatible with the demands of microanalysis. A number of screw-cap reaction vials having a Teflon-lined seal are commercially available in sizes from 200 μl to 5 ml. These may be centrifuged in a bench-top centrifuge and reagents may be removed using appropriate heating and a gentle stream of clean, dry nitrogen. Alternatively, the vials may be connected directly to a vacuum source. Multiple derivatizations can thus be performed with the minimum of transfers and loss of sample.

All reagents should be analytical grade, should be checked before use, and, if necessary, further purified. The quality of reagents from commercial sources should not be assumed. Since reagents are used in excess relative to the sample, a minor reagent impurity can contribute to undesirable secondary reactions and also can be concentrated in the final sample resulting in GLC or GC/MS interference. Control experiments to monitor reagent purity are mandatory, but it is not sufficient to analyze each reagent or solvent on its own. In multireaction derivatizations a contaminant in a reagent or solvent may not, *per se*, interfere but may react with reagents used later to form volatile, interfering products. Gases used for evaporation are also potential sources of contamination and should be used in conjunction with appropriate filters. Water is probably the single most harmful contaminant since acid catalysed esterification reactions are reversible and many *O*- acyl derivatives are sensitive to the presence of water. Therefore, care should be taken to dry *all* reagents and gases.

Rash *et al.* (1972) thoroughly studied potential sources of contamination. Human sources, including fingerprints, skin fragments, hair, dandruff, and saliva, were the most significant. Other sources were laboratory dust, cigarette smoke, and fibers from cellulosic towels. The problems caused by contamination when derivatizing low picomole quantities of oligopeptide hydrolysates were addressed by Frick *et al.* (1977a) and microtechniques using reaction volumes of 5 μl or less were developed. Although GC–MS is a particularly powerful technique for the analysis of crude samples, biological samples are often partially purified by standard ion-exchange chromatography before analysis (Kaiser *et al.*, 1974; Adams, 1974). Control experiments to assess amino acid recovery must accompany all ion-exchange clean-up procedures and appropriate corrections must be applied in quantitative analyses.

The above items may appear too simple and fundamental for inclusion in the present discussion. However, in the writer's experience the vast majority of derivatization problems can be attributed to contaminated reagents or solvents.

2.2.2. Chromatography

Because derivatives of some amino acids decompose on hot metal surfaces, glass columns configured to permit on-column injection are recommended for the chromatography of all amino acid and peptide derivatives. Packed columns containing nonpolar liquid phases supported on the best quality supports are adequate for separating the derivatives of amino acids derived from protein hydrolysates and for quantitative analysis by GLC (MacKenzie, 1981). However, when analyzing a complex physiological sample, the superior resolution of a capillary column is advantageous and may be necessary. Furthermore, the better peak height:area ratio obtainable with capillary columns increases the instantaneous concentration of the sample components and thus improves the probability of obtaining good spectra in GC/MS. A further advantage is that the entire effluent from a capillary column may be introduced to the mass spectrometer and, if the column is operated in the splitless mode, the quantity of sample may routinely be in the picomole range. Capillary columns have been used to separate amino acid derivatives by Jonsson *et al.* (1973), Adams *et al.* (1977), Pearce (1977), Poole and Verzele (1977), Degres *et al.* (1979), and Bengtsson and Odham (1979).

The recently developed fused silica capillary columns offer distinct advantages over glass columns in inertness, durability, and ease of handling. They are also sufficiently flexible to be routed from the gas chromatograph directly into the mass spectrometer, thus minimizing connections and leaks (Jennings, 1980). Improved surface deactivation procedures have extended the operating range of capillary columns to 350°C (Grob *et al.*, 1979). These developments, perhaps combined with direct sample injection on wide-bore capillary columns, will no doubt contribute substantially to future GC/MS analysis of amino acids and peptides.

2.2.3. N-Acyl Amino Acids Esters

By far the most commonly used derivatives for the analysis of amino acids by GLC are various *N*-acyl esters and, in particular, *N*-perfluoroacyl alkyl esters. The foundations for quantitative amino acid analysis by GLC using derivatives of this general type were laid by Gehrke and his associates, who developed a two-column separation of the *N*-trifluoroacetyl (TFA) *n*-butyl esters (see, for example, Kaiser *et al.*, 1974). Single column separations were developed for the *N*-heptafluorobutyryl (HFB) *n*-propyl esters (Moss *et al.*, 1971; March, 1975), the *N*-HFB isoamyl esters (Zanetta and Vincendon, 1973) the *N*-acetyl *n*-propyl esters (Adams, 1974), and the *N*-HFB isobutyl esters (MacKenzie and Tenaschuk, 1974, 1979a,b). The specific reaction conditions for formation of these derivatives are too varied to be presented succinctly so the reader is directed to the original literature.

 2.2.3a. Electron Impact Ionization. The electron impact (EI) mass spectra of *N*-acyl amino acid esters have been more thoroughly documented than those of any other category (Gelpi *et al.*, 1969; Manhas *et al.*, 1970; Lawless and Chadha, 1971; Felker and Bandurski, 1975; MacKenzie and Hogge, 1977; Leimer *et al.*, 1977a). It is, therefore, appropriate to describe the spectra in some detail. Since mass spectra can depend on the specific instrument used and on operating conditions such as the pressure in the ion source and the method of sample injection, numerical values for ion abundances will not be given. The mass spectra of the various ester groups studied (except methyl esters) represent essentially the same fragmentation processes. These are described using the C_4 esters and appropriate adjustment in masses will, in most cases, provide the corresponding ions for C_3 and C_5 alkyl esters. The most probable EI fragmentation pathways of the *N*-perfluoroacyl alkyl (C_3 to C_5) amino acids have been elucidated by Gelpi *et al.*, (1969) and are supported by metastable transitions.

 The molecular ions in EI spectra of aliphatic amino acids are generally of insignificant abundance. In the absence of the molecular ion, the largest ions observed represent cleavage of bonds in the ester moiety with charge being retained on the amine fragment. Typically, fragments M-55, M-73, and M-101, of which the M-101 ion is most abundant, are observed (Figure 1). Even if the molecular ion is absent, the presence of all three fragments allows the molecular weight to be inferred with some confidence. Ester cleavage is further manifested by frequently abundant $C_4H_7^+$, $C_4H_8^+$, or $C_4H_9^+$ ions. Depending on the derivative, ions representing the perfluoroalkane series CF_3, C_2F_5, and C_3F_7 are also observed.

 Ions characteristic of the amino acid side chain are also observed and are essential for distinguishing leucine and isoleucine. The side chain is lost as the corresponding olefin (Gelpi *et al.*, 1969), but the abundances of identical fragment ions differ sufficiently to distinguish these amino acids. Since the stability of neutral radicals progresses in the order primary < secondary < tertiary, cleavage is favored between carbons 3 and 4 for leucine (loss of m/z 43) and between carbons 2 and 3 for isoleucine (loss of m/z 57). These losses are usually observed in conjunction with the loss of other fragments such as the alkoxy and a perfluoroalkane moiety.

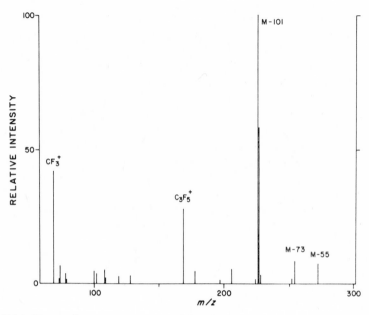

FIGURE 1. Electron impact mass spectrum of N-HFB glycine isobutyl ester. Ions of relative abundance <1% have been omitted.

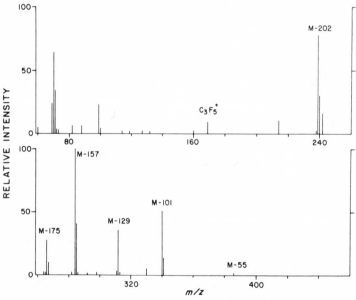

FIGURE 2. Electron impact mass spectrum of N-HFB aspartic acid isobutyl ester. Ions of relative abundance <1% have been omitted.

The valine spectrum contains significant amounts of two ions, M-166 and M-155, which may be useful for diagnostic purposes.

Dicarboxylic amino acids are characterized by losses which reflect the presence of two carboxylic acid ester groups, for example, $(M - C_4H_8 - C_4H_9)^+$, $(M - C_4H_8 - OC_4H_9)^+$, $(M - C_4H_8 - COOC_4H_9)^+$, $(M - COOC_4H_9 - OC_4H_9)^+$, and $(M - 2COOC_4H_9)^+$ (Figure 2). The ion m/z 116 was reported to be characteristic for N-TFA n-butyl glutamic acid (Gelpi et al., 1969) but, perhaps due to different experimental conditions, was not observed by Leimer et al. (1977a). Diagnostic ions at m/z 99 (aspartic acid) and 85 (glutamic acid) were observed for the n-butyl (Leimer et al., 1977a) and isobutyl esters (MacKenzie and Hogge, 1977).

The hydroxy amino acids, serine and threonine, also show the characteristic ester fragment losses but at a much lower abundance than for the aliphatic amino acids. The imminium ion so formed is further stabilized by the loss of the O-acyl group (Gelpi et al., 1969) to give abundant ions at m/z 139 and 153, respectively, for serine and threonine. Alternatively, since the O-acyl group is extremely labile, it is possible that the loss of this moiety could result in a neutral fragment which subsequently loses the ester series of fragments. The observation of ions corresponding to $(M-CF_3CONH_2-55)^+$ would tend to support this hypothesis. The spectra are somewhat more complicated when the homologues of TFA are used, but allow further distinctions to be made. Thus, for N-HFB isobutyl derivatives, an abundant ion is observed at m/z 210 for serine but not for threonine.

Proline and pipecolic acid spectra are dominated by the cyclic imminium ion formed on the loss of the carboxylic ester, but for hydroxyproline, stable ion formation is enhanced by the further loss of the O-acyl moiety. The ions thus formed are frequently the most abundant. In addition, the molecular ion may be observed for proline and pipecolic acid and to a lesser extent for hydroxyproline.

The mass spectra of aromatic and heterocyclic amino acid derivatives are strongly influenced by their cyclic moieties. The stabilizing influence of the aromatic ring often results in a significant molecular ion and in abundant ions representing the aromatic moiety after cleavage of ester and acyl groups. Thus for phenylalanine the ion m/z 91 representing $C_6H_5CH_2^+$ has diagnostic value as does m/z 81 for histidine. The actual molecular structures of these ions result from ring expansion (Gelpi et al., 1969). A similar process occurs for tyrosine and tryptophan, but, because of their active groups, the stable ion retains the acyl group. Thus, characteristic ions are observed at m/z 203 and 226, respectively, for N-TFA butyl tyrosine and tryptophan. When the HFB derivative is used in conjunction with the isobutyl or isoamyl ester, this relationship holds true for tryptophan but not for tyrosine since the base peaks observed would suggest preferential

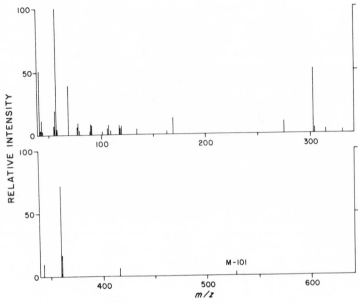

FIGURE 3. Electron impact mass spectrum of N-HFB tyrosine isobutyl ester. Ions of relative abundance <2% have been omitted.

charge retention on the nonaromatic portion of the molecule (Felker and Bandurski, 1975; MacKenzie and Hogge, 1977). However, ions derived from the aromatic moiety are sufficiently abundant to be distinctive (Figure 3).

The sulfur amino acids methionine, cysteine, and cystine give quite characteristic spectra. The methionine spectrum contains fragment ions at m/z 61 $(CH_3SCH_2)^+$ and m/z 75 $(CH_3SCH_2CH_2)^2$, the latter frequently being the base peak. Ions corresponding to loss of these fragments both alone and in combination with ester fragments are also present. Depending on the conditions, the molecular ion may be moderately abundant. An ion at M-74 $(M-CH_2CHSCH_3)^+$, resulting from a McLafferty rearrangement, was the base peak for 20 eV spectra of N-TFA n-butyl methionine (Gelpi et al., 1969), was considerably less abundant for 70 eV spectra (Leimer et al., 1977a) and for N-HFB isoamyl methionine (Felker and Bandurski, 1975), but was not observed for N-HFB isobutyl methionine (MacKenzie and Hogge, 1977).

In most studies to date, the cysteine mass spectrum contains a relatively abundant amine fragment ion $(M-COOC_4H_9)^+$ and an ion at m/z 140 which results from the elimination of trifluorothioacetic acid and $OCOC_4H_9$.

The molecular ion is observed for cystine even for the N-HFB isobutyl ester (m/z 744) although only in low abundance. Except at low ionization potential, the base peak is generally an alkyl fragment derived from the ester. Ions representing M/2 may be observed but larger ions are not usually of significant abundance.

Under standard operating conditions, the most prominent ions in the spectra of N-acyl alkyl esters of lysine are generated by loss of the carboxylic ester to form an imminium fragment followed by elimination of trifluoroacetamide or its homologues. The latter ion is most often the base peak although m/z 67 is the base peak for N-HFB isobutyl esters.

In addition to the molecular ion, spectra of arginine N-acyl esters are characterized by loss of an alkoxycarbonyl group and trifluoroacetamide or its homologues. Although not in themselves of diagnostic value, the perfluoroalkane ions and ester fragment ions may also be abundant. An ion representing $(CF_3CONHCNNCH_2)^+$ (m/z 166) or the corresponding perfluorobutyryl homologue has been noted (Leimer et al., 1977a; Felker and Bandurski, 1975).

Leimer et al. (1977a), in addition to the foregoing spectra of protein amino acids, described the EI mass spectra of a number of other N-TFA n-butyl amino acid esters. These included β-amino acids which gave more complex spectra than the α-amino acids. The ion M-102 was more important than M-101, and M-73 was often the base peak. The latter ion was not usually observed for the n-butyl esters of α-amino acids (Leimer et al., 1977a) but was routinely observed for the isobutyl esters (MacKenzie and Hogge, 1977).

Spectra of N-methyl amino acids contained a characteristic ion m/z 110 attributed to the molecular composition $C_3H_3NF_3$ an ion also noted earlier by Lawless and Chadha (1971). In neither study was the mass verified by high-resolution mass spectrometry. The origin of this ion remains unexplained although Lawless and Chadha verified that the alkyl moiety of the ester was not involved. Spectra of N-TFA n-butyl sarcosine obtained in the writer's laboratory contained an insignificant m/z 110 ion but a more prominent ion at m/z 112 as noted by Lawless and Chadha (1971), while being virtually identical to the spectrum presented by Leimer et al. (1977a) in every other respect. Thus it appears that spectra of N-TFA N-methyl amino acids do not necessarily contain a significant abundance of m/z 110 and care should be taken in interpreting the absence of this ion. The spectrum of N-HFB sarcosine isobutyl ester does not contain the homologous ion m/z 210 (MacKenzie, unpublished results).

2.2.3b. Chemical Ionization Spectra. In chemical ionization (CI), the energy of the electron beam is primarily imparted to a reagent gas present in the ionizing chamber in amounts much greater than the sample.

The ions formed from the reagent then react with the sample in ion/molecule reactions to form characteristic adduct ions. These are for methane $(M + H)^+$, $(M + C_2H_5)^+$, and $(M + C_3H_5)^+$, for isobutane $(M + H)^+$ and $(M + C_4H_9)^+$, and for ammonia $(M + H)^+$ and $(M + NH_4)^+$. Much less energy is transmitted to the sample than in the EI mode and there is less fragmentation. As a consequence, there is a greater probability of observing the molecular ion although, depending on the sample and the conditions, diagnostically useful fragment ions may also be observed.

Documentation of the methane CI spectra of N-acyl amino acid esters is relatively sparse. Leclercq and Desiderio (1973) presented data on methyl esters and N-acetyl amino acids and discussed the fragmentation of the adduct ions but did not examine the combined derivatives. Meot-Ner and Field (1973) examined the isobutane CI spectra of proline, valine, and leucine derivatives. The most extensive studies of CI spectra of N-acyl amino acid esters have been reported by Padieu et al. (1978a,b) and were conducted to "establish conditions required for the use of perdeuterated amino acid derivatives in metabolic studies using GC/MS." These studies focused on the N-acetyl n-propyl esters, the spectra of which were extensively described.

The N-TFA n-butyl esters are probably the most frequently used derivatives for amino acid analysis by GLC. In the methane CI spectra of these derivatives, the protonated molecular ion is the base peak for all except glycine and methionine and the fragment $(M + H - C_4H_8)^+$ is the next most abundant ion. All other fragmentations are weak (Padieu et al., 1978a,b).

Padieu et al. (1978a,b) also described the methane CI spectra of the N-HFB amino acid isobutyl esters but less comprehensively than the above derivatives: arginine, methionine, histidine, and cystine were not examined. Chemical ionization spectra are not as reproducible as EI spectra; methane CI spectra of N-HFB amino acid isobutyl esters obtained in the writer's laboratory, although generally similar, differed significantly in some respects from those described by Padieu et al. (1978a,b). Because of these differences and because the N-HFB amino acid isobutyl esters offer the best GLC separation, the salient features of their methane CI spectra, as obtained by the writer, are described.

The methane CI spectra of the N-HFB isobutyl esters of the protein amino acids all contain the characteristic $(M + H)^+$, $(M + C_2H_5)^+$, and $(M + C_3H_5)^+$ ions in sufficient abundance to render identification of the molecular weight unambiguous. The complete series of adduct ions was infrequently observed by Padieu et al. (1978b). With the exception of glycine, methionine, cystine, phenylalanine, and tyrosine, the $(M + H)^+$ ion constitutes the base peak. In addition, significant amounts of ions $(M -$

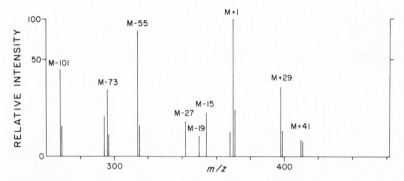

FIGURE 4. Methane chemical ionization mass spectrum of *N*-HFB valine isobutyl ester. Ions of relative abundance <1% have been omitted.

$CH_3)^+$ [including the isobaric ion $(M + C_3H_5 - C_4H_8)^+$], $(M + H - HF)^+$, and $(M - C_2H_3)^+$ [or more probably the isobaric ions $(M + C_2H_5 - C_4H_8)^+$] may also be observed. The presence of the ester is represented by the same series of ions as in EI spectra except that an $(M + H - 76)^+$ ion is often observed (Figure 4). Because of the decreased fragmentation relative to EI, the perfluoroalkane ions are not usually prominent but traces of m/z 169 $(C_3F_7)^+$ are observed.

The methane CI spectra of the aliphatic amino acid derivatives contain few distinctive ions in addition to those mentioned above. The base ion for glycine is $(M + H - 56)^+$ and the valine spectrum contains small amounts of m/z 234 $(M - 135)^+$ not observed for the other aliphatic amino acids. Leucine and isoleucine are difficult to distinguish.

The hydroxy amino acids serine and threonine lose m/z 214, presumably the *O*-acyl moiety, to form a neutral fragment for which a separate series of adduct ions and ester fragment elimination ions are observed. These constitute a false molecular ion pattern which can lead to misinterpretation if the true molecular ion is very weak. Significant amounts of $(M - 185)^+$ are noted only for serine and threonine (Figure 5).

Relatively few useful diagnostic ions are observed in the spectra of the other amino acids. Little or no evidence of the aromatic moiety is observed for phenylalanine and histidine. The ion m/z 298 $(M - SCH_3 - C_4H_8)^+$ is observed in significant abundance only for methionine (Figure 6). The cystine spectrum contains several diagnostic ions. The base peak m/z 284 probably results from cleavage of the disulfide bond followed by elimination of SH to form a neutral fragment of m/z 339 which subsequently is protonated and loses C_4H_8. This is supported by the presence of the corresponding adduct ions and ester fragment elimination ions. Other

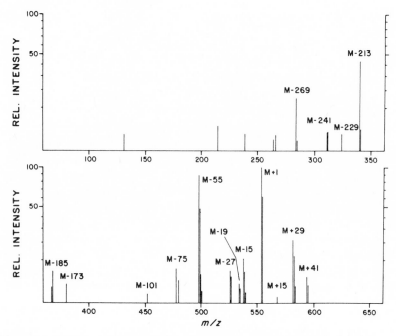

FIGURE 5. Methane chemical ionization mass spectrum of *N*-HFB serine isobutyl ester. Ions of relative abundance <1% have been omitted.

useful ions observed in significant abundance are m/z 623 (structure not rationalized) and m/z 240 $(C_3F_7CONHCHCH_2)^+$.

In contrast to the observations of Padieu *et al.* (1978a,b), the hydroxy tropylium ion was absent from the spectrum of the tyrosine derivative while

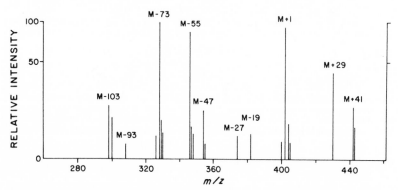

FIGURE 6. Methane chemical ionization mass spectrum of *N*-HFB methionine isobutyl ester. Ions of relative abundance <1% have been omitted.

the acylated hydroxytropylium ion, m/z 303, was present although in low abundance.

Ammonia CI of amino acid derivatives gives a clear indication of molecular weight but virtually no useful fragmentation. When methane CI spectra do not contain the entire series of adduct ions, the addition of a *trace* of ammonia to the methane can be useful for confirming the molecular weight (MacKenzie, unpublished observations).

Most mass spectrometry is concerned with the detection and structure of positive ions, and only in recent years have accessories for negative ion (NI) detection been commercially available. No studies of the NI spectra of *N*-acyl amino acid esters have been reported. Since NI mass spectrometry holds promise of increased sensitivity, especially with perfluorinated derivatives, the NI/CI spectra of the *N*-HFB isobutyl amino acids will be briefly discussed (MacKenzie, unpublished results).

The molecular weight is generally indicated by the ion $(M-1)^-$ generated by loss of a hydrogen radical from M^- (Figure 7). The perfluoroalkane group is represented by elimination of fluorine and a series of HF molecules so that ions $(M-19)^-$ and $(M-20)^-$ are invariably observed. Depending on

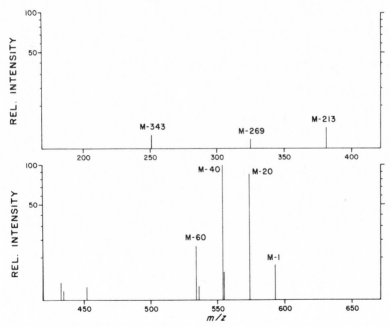

FIGURE 7. Negative ion methane chemical ionization mass spectrum of *N*-HFB lysine isobutyl ester. Ions of relative abundance <0.5% have been omitted.

the number of perfluoroacyl groups, $(M-40)^-$, $(M-60)^-$, and $(M-80)^-$ ions are also observed. Frequently the base ion is a member of this series or a related ion such as $(M-H_2F)^-$ for aspartic acid. Elimination of fragments from the ester group is represented by $(M-C_4H_9)^-$ (aliphatic amino acids), $(M - C_4H_{10})^-$ (aspartic acid, glutamic acid, lysine, and phenylalanine), $(M - OC_4H_9)^-$ (alanine and aspartic acid), $(M - OC_4H_{10})^-$, and $(M - OCOC_4H_{10})^-$ (aliphatic amino acids and phenylalanine). Dicarboxylic esters are characterized by an ion of about 50% relative abundance at M-158 explainable by the elimination of both C_4H_9 and $OCOC_4H_9$. All the spectra of aliphatic amino acids include an ion M-94 attributable to the elimination of both HF and C_4H_9OH. A number of ions resulting from the combined loss of HF or 2HF and an ester fragment are evident. Loss of a series of neutral HF molecules from fragments other than the molecular ion are also noted. Thus, for arginine, the ion m/z 606 (M − $C_3F_7CONH)^-$ is accompanied by ions corresponding to m/z 606-HF, 606-2HF and 606-3HF.

Diagnostically useful ions are also observed. Leucine and isoleucine can be distinguished by the ions $(M-50)^-$ (leucine) and $(M-34)^-$ (isoleucine). The base peak for tyrosine is m/z 325, an ion not detected elsewhere in the total ion record using a single ion search. Likewise, the arginine spectrum contains an $(M-80)^-$ ion in addition to the normal HF series and a diagnostic ion m/z 237 not detected for any other amino acid. Similarly, m/z 178, although detected in trace amounts for several amino acids, is the base ion for proline. The ions m/z 194 and 213 or, in some cases, 193 and 212 are often present with a moderate abundance, but for serine and threonine, these ions carry 60% to 90% of the total ion current and produce the base peak for serine at m/z 213. Threonine, however, has a base peak at m/z 351 (M − C_3F_7 $COOH_3$ or M − C_3F_7CO − F).

The cystine spectrum contains no ion of mass greater than m/z 403, $(M-341)^-$, representing cleavage of the S −CH_2 bond in the $(M-H)^-$ ion with charge retention on the larger fragment. The ion m/z 372, due to S–S cleavage, is observed. Elimination of HS from M/2 produces m/z 339 and, by further elimination of HF and 2HF, abundant ions at m/z 319 and 299, the former being the base peak. The presence of the ester is verified by the loss of 57, 73, and 101 from m/z 339.

2.2.4. Trimethylsilylated Amino Acids

The pertrimethylsilyl (TMS) amino acids are rarely used for quantitative analysis of all the protein amino acids because of multiple derivative formation. However, this does not negate their value for the analysis of specific amino acids. A variety of reagents and conditions for preparing

TMS amino acids are described by Blau and King (1978) and Knapp (1979).

A number of mass spectrometric studies have been made of various TMS amino acids, and knowledge of their spectral characteristics is likely to find application in a qualitative context. The single most comprehensive study of TMS amino acids was reported by Leimer *et al.* (1977b), who presented the EI spectra of 46 TMS amino acids. In general the spectra of TMS amino acids do not provide as much structurally useful fragmentation as those of *N*-acyl esters. The molecular ion is weak and the most characteristic fragments of the TMS α-amino acids are $(M - CH_3)^+$, $(M - CH_3 - CO)^+$, $(M - COOTMS)^+$, and $(M - R)^+$, where R represents the amino acid side chain. The presence of other functional groups produces more complex spectra. For the sulfur amino acids, for example, "additional fragments were observed for C–S and S–S bond cleavage with charge retention on both fragments." Either R^+ or $(M - R)^+$ are the most useful diagnostic ions for the aromatic amino acids. The most characteristic ion for glutamic and aspartic acids is $(M - COOTMS)^+$, but, since this ion is present in the spectra of most TMS carboxylic esters, its presence is not sufficient, *per se*, to indicate a dicarboxylic acid. Arginine is converted to ornithine during silylation.

The β-amino acids have spectra similar to the α-amino acids except for a more abundant molecular ion.

Essentially similar results were obtained by Abramson *et al.* (1974).

Marik *et al.* (1976) thoroughly documented the earlier literature and described the EI mass spectra of di- and tri-TMS C_4 to C_{11} ω-amino acids. These are characterized by m/z 102 $(TMS -NH=CH_2)^+$ and m/z 174 $[(TMS)_2=N=CH_2]^+$. Tri-TMS amino acid spectra also contain a prominent m/z 86 ion $(Me_2Si=N=CH_2)^+$ arising from the persilylated amino group. The ion m/z 147 attributed to the fragment $(Me_2Si=O—SiMe_3)^+$ and produced by rearrangement of the TMS group of polysilyl derivatives is common to the spectra of both di- and tri-TMS derivatives.

Isobutane CI spectra of TMS amino acids have been described by Budzikiewicz and Meissner (1978). In general, the $(M + H)^+$ ion is the base peak but other adduct ions are of low abundance. Abundant $(M + H - CH_2=Si(CH_3)_2)^+$ ions are observed but the ions $(M-15)^+$, $(M-43)^+$, and m/z 147, typical of EI spectra are missing or of very low abundance. The spectra of most of the protein amino acids are unremarkable. Tri-TMS cysteine loses HSTMS from both the $(M + H)^+$ and the $(M + H - 72)^+$ ions. Tetra-TMS cystine rearranges to give the protonated cysteinyl di-TMS ion, which, in turn, loses H_2S or HSTMS. The ion m/z 102, present in EI spectra and characteristic of a terminal TMS-aminomethanyl group, is missing in the spectrum of tri-TMS lysine. Tri-TMS glutamic acid forms a pyrrolidone derivative, and glutamine and asparagine yield abundant $(M + H - 90)^+$ ions. The spectrum of tri-TMS tryptophan is dominated by the ion m/z 202.

2.2.5. Miscellaneous Derivatives

Husek (1974) proposed the formation of volatile amino acid derivatives by reaction with 1,3-dichlorotetrafluoroacetone which condenses with the α-amino and carboxy groups to form 2-bis-(chlorodifluoromethyl)-4-substituted-1,3-oxazolidinones. The remaining polar groups are modified by esterification with methanol and isobutylchloroformate or acylation with heptafluorobutyric anhydride (HFBA). The EI mass spectra of 26 substituted 1,3-oxazolidin-5-ones were reported by Liardon et al. (1979) and the successful chromatography of all the protein amino acid derivatives was demonstrated. A list of diagnostically useful ions was compiled and it was shown that each amino acid could be characterized using only two ions which, with few exceptions, had a $m/z > 150$.

The mass spectra of the protein α-amino acid oxazolidinones show a common fragmentation pathway which is a reflection of the structure of the oxazolidinone ring. Two fragmentation pathways, (1) the loss of chlorine followed by CO and subsequently CF_2 and (2) elimination of CF_2Cl followed by CO, combine to produce a characteristic series M-63, M-85, and M-113 in the spectra of the aliphatic amino acids. Depending on the other functional groups in the molecule, members of this series, but not necessarily all, are usually observed in the spectra of other amino acids and are of diagnostic value when the molecular ion is not detected. As for the perfluorinated amino acid alkyl esters, O- or N-acylation is represented by the perfluoroalkane series of ions and by elimination of an O- or N-acyl fragment from the molecular ion. The aromatic amino acids are best characterized by the elimination of the side chain which retains the charge. The methionine spectrum is dominated by the fragment m/z 61 $(CH_3 -S=CH_2)^+$ and the cystine spectrum by m/z 107, an ion the structure of which remains to be rationalized. Asparagine and glutamine are converted to the corresponding nitriles, spectra of which give the characteristic series of ions described above.

Electron impact spectra of the products of reaction of fluorescamine with 17 naturally occurring amino acids were described by Shieh et al. (1976). Each amino acid could be unequivocally identified in low-resolution spectra.

Although there are numerous descriptions of the gas chromatographic and mass spectral characteristics of other derivatives of amino acids, these have dealt with only a few protein amino acids or amino acids of limited natural distribution.

2.2.6. Stable Isotope Techniques

Apart from their obvious potential for the elucidation of fragmentation pathways, stable isotopes can be usefully employed in a number of quantitative low-resolution GC/MS analyses having applications in biological studies. These fall into two categories, firstly, methods directed to ultramicro quantitative analysis of amino acids and, secondly, methods for determining isotopic enrichment in studies of amino acid metabolism, determination of amino acid pools, and protein turnover.

The addition of known amounts of highly enriched perdeuterated amino acids to a sample before derivatization and GC analysis, in effect, provides a separate internal standard for each amino acid. With certain types of samples, this eliminates the need for ion-exchange purification and, in simplifying sample manipulation, minimizes sample loss. Schulman and Abramson (1975) described the use of this approach for the analysis of free amino acids as their N-TFA n-butyl esters in combination with EI mass spectrometry. The relative labeled/unlabeled ion ratios were determined by integrating structurally significant ions throughout the appropriate GC peak envelope. Integration, rather than using individual scan ratios, was necessitated by the partial GC separation of labeled and unlabeled amino acid derivatives.

Pereira et al. (1973) and later Summons et al. (1974) used deuterated amino acids as internal standards in the analysis of up to 12 amino acids as the N-TFA n-butyl esters in soil and physiological fluids. As little as 10 ng of an amino acid could be determined.

Mee et al. (1977) used ^{2}H or ^{15}N, the N-acetyl methyl esters and direct inlet probe sample injection to assay free amino acids. The sensitivity was enhanced using isobutane CI mass spectrometry. Direct ion ratios were usable without peak integration and derivatives of arginine, histidine, and cystine, which require a separate analysis on a second column as the N-TFA n-butyl esters, could be analyzed along with the other amino acids.

Rafter et al. (1979) added perdeuterated amino acid standards to proteins before hydrolysis "to increase the precision of the amino acid analysis when working with microquantities of proteins." A mean coefficient of variation of 4.3% was achieved with hydrolyses of 15 to 150 µg of sample analyzed as the N-TFA n-butyl esters by EI GC/MS. Labeled glutamic and aspartic acids had to be added after hydrolysis because of loss of deuterium atoms during hydrolysis. It was also assumed that the added serine and threonine were destroyed at the same rate as when in a protein, an assumption which may not be entirely valid.

The problems of isotopic integrity may be avoided by using ^{13}C isotopes. Kingston and Duffield (1978) used ^{13}C amino acids, the N-TFA n-butyl esters and methane CI GC/MS to assay 16 free α-amino acids at the 1 to 5 ng level of detection.

Measurements of isotope ratios may also be used in an entirely different manner, viz., to determine the amount of biological enrichment in metabolic studies. McReynolds and Anbar (1977) assayed nanomole amounts of ^{15}N labeled alanine and leucine by field ionization (FI) and collision-induced dissociation (CID) mass spectrometry. The determination was interference-free because a secondary ion, formed by CID from a proton-ated molecular ion produced by FI, was monitored. The precision and accuracy were "within 10% of the isotope ratio down to the natural abundance level."

The mass spectrometric technique of FI-CID is not generally available, but similar results may be obtained using EI and CI mass spectrometry. Studies by EI (Fairwell *et al.*, 1970) were followed by a more sensitive isobutane-CI approach in which discrimination between mean levels separated by 0.1 at. % ^{15}N was demonstrated with 68% confidence at the 0.1 nmol level using low-resolution mass spectrometry (Robinson *et al.*, 1978). However, the method is not specific for one isotope, and the authors cautioned that "a positive ratio change could be due to an increase in an isotope other than ^{15}N."

Deuterium labeled tryosine and α-methyl tyrosine have also been used in metabolic studies (Sjoquist, 1979).

The foregoing applications of stable isotope ratio measurements have mostly been drawn from studies of mammalian metabolism but there is, in principle, no barrier to their application in studies of microorganisms.

2.2.7. Selected Ion Monitoring

In GC/MS, the mass range of interest is usually scanned repeatedly and complete spectra are obtained for segments of the GC elution profile, the primary purpose being to obtain spectra for compounds as they are eluted from the GC column. Both in qualitative and quantitative terms, most of the scan time is allocated to the collection of either redundant or nonspecific information. If, instead, the mass spectrometer is tuned or controlled to repetitively integrate only a few ions known to be characteristic for the compounds under study, the sensitivity increases roughly in proportion to the increase in scan time per ion. The ultimate in sensitivity is achieved by monitoring only a single ion for each component of interest.

Since only characteristic ions are detected and measured, the technique is much less sensitive to interference than simple GLC analysis. This technique, variously described as selected ion monitoring (SID), multiple ion detection, and mass fragmentography has been applied to a number of amino acid studies. In that context, it is frequently advantageous to incorporate as internal standards amino acids containing stable isotopes such as deuterium and to monitor the ratios of the protium- and deuterium-containing ions.

Pereira *et al.* (1973) used SID to assay ten amino acids in soil extracts as their *N*-TFA *n*-butyl esters. A mixture of deuterated amino acids was used as an internal standard and specific deuterated and nondeuterated ions were monitored for each amino acid. Quantitation of samples containing as little as 10 ng of an amino acid was achieved with a coefficient of variation of less than 1% for most of the amino acids studied. To avoid deuterium–protium exchange, the standard was added after hydrolysis but no special precautions were required in analyzing free amino acids. The method was later extended to the analysis of 12 amino acids in biological fluids and a precision of better than 10% was obtained at a lower limit of about 1 ng of an amino acid (Summons *et al.*, 1974).

Frick *et al.* (1977a) further extended the analysis to 17 amino acids and developed microderivatization techniques for the hydrolysis and analysis of only 50 pmol of a decapeptide. No deuterated internal standards were used, and, because of instrumental factors limiting data collection to only three channels, "two injections were necessary to analyze all the selected amino acids."

The same principles have also been applied to the analysis of low picomole amounts of individual amino acids, specifically, 5-hydroxy tryptophan (Martinez and Gelpi, 1978), tryptophan (Wegmann *et al.*, 1978), and glutamine (Wolfensberger *et al.*, 1979).

Abramson *et al.* (1974) introduced the concept of functional group mass spectrometry based on the strategy of synthesizing derivatives of functional groups which fragment "to produce a substantial ionization due to a common chemical moiety." This allows multiple compound surveys without repeatedly retuning the mass spectrometer to different values throughout the analysis. Like all GC/MS methods based on only a few ions, the method is strongly complemented by use of the GC retention times for compound identification. Maximum sensitivity is attained by minimizing the number of ions monitored; preferably only one ion at a time is detected and measured. Using this strategy, Abramson *et al.* (1974) were able to assay a substantial number of TMS amino acids at a detection limit ranging from 10 to 100 fmol using only the ions m/z 174 and 218. The former ion, due to the fragment $(TMS_2=N=CH_2)^+$, is abundant in the spectra of TMS ω-amino acids while the latter fragment (TMS $-NH{=}CHCOOTMS)^+$ is abundant in the spectra of α-amino acids.

Iwase and Murai (1977) examined a variety of derivatives of alanine, valine, leucine, proline, and glutamic acid and concluded that the *N*-TFA L-prolyl *n*-butyl, *N*-benzoyl *n*-butyl, *N*-pentafluoro benzoyl *n*-butyl, and *N*-TFA L-menthyl derivatives are suitable for ultramicroanalysis of amino acids. These derivatives produce abundant ions at m/z 166, 105, 195, and 83, respectively. However, the study was limited to mixtures of model compounds and was not applied to a sample of unknown composition.

Functional group mass spectrometry has also been applied to analysis of tryptophan and some of its metabolites in physiological fluids with detection limits of less than 100 pg (Segura *et al.*, 1976). Cattabeni *et al.* (1976) analyzed TMS γ-aminobutyric acid using the ratio of m/z 174:304 to confirm identity and the latter ion for quantitation with a detection limit of 60 fmol in physiological samples.

Amino acid enantiomers can be resolved by GLC without mass spectrometry either on chiral columns or as diastereomers on optically inactive columns after reaction with a chiral reagent. To allow meaningful measurements, there must be no kinetic resolution in the conversion reactions, which, ideally, should go to completion. To avoid this problem, Wiecek *et al.* (1979) reacted amino acids with a mixture of *N*-TFA-*R*-prolyl chloride and *N*-TFA-*S*-(1-^{2}H)-prolyl chloride. The four diastereomers were resolved into two peaks by GLC. The composition of each peak was determined using methane CI mass spectrometry to measure the protonated molecular ions of the labeled and unlabeled compounds.

The determination of D-amino acids in proteins and peptides is complicated by racemization of the amino acids during acid hydrolysis. Liardon *et al.* (1981), therefore, used ^{2}HCl to distinguish D-amino acids formed during hydrolysis from the original constituents of the sample. Any molecule inverted during hydrolysis became labeled with deuterium. The amino acids were separated as their $N(O, S)$-TFA isopropyl esters on a chiral stationary phase and assayed by selected ion monitoring.

2.3. Applications

Applications of GC/MS to the identification of free amino acids in a microbiological context are relatively rare since most studies have relied mainly on GC retention times for identification. However, Bengtsson and Odham (1979) developed a GC method applicable to the study of nanogram amounts of free amino acids in aquatic environments and identified amino acids by GC/MS. The procedure allowed determination of less than 1 ppb of individual amino acids in lake and river water samples and could "also be applied to experimental studies of the transport, distribution and metabolism of free amino acids in laboratory models at natural concentrations." Thus, a laboratory population of the nematophagous fungus *Arthrobotrys oligospora* was found to excrete 0.1 to 4 μg amino acids/h. In addition, valine was easily identified in a sample from an unpolluted watercourse by monitoring only the characteristic ion of m/z 268. Subsequently, Bengtsson *et al.* (1981) detected an enrichment of total amino acids in the surface microlayer of natural waters. They also applied GC/MS and specifically single ion monitoring (m/z 266) to assay ornithine production by *Mycoplasma hominis* as a method to detect mycoplasma contamination of mammalian cell cultures.

Aerobic microorganisms are well known to release amino acids but much less is known about amino acid release by anaerobic microorganisms. Studies indicating that strains of the genus *Bifidobacterium* released significant amounts of amino acids but did not accumulate tryptophan caused Aragozzini *et al.* (1979) to consider the possibility of tryptophan transformation. By using EI mass spectrometry and the *N*-acetyl methyl ester derivatives, they were able to demonstrate that tryptophan was metabolized to indole-3-lactic acid by 51 strains of *Bifidobacterium* spp. This transformation was considered significant in the "pattern of amino acid transformation by intestinal microorganisms, particularly regarding the tryptophan metabolic products involved in various pathogeneses."

Mass spectrometry and GC/MS have played a vital role in the identification of amino acids in microbial products, particularly the antibiotics. Such applications are so numerous that only a few arbitrary examples can be cited. For instance, Katz *et al.* (1974) identified α-amino-β, γ-dihydroxybutyric acid in the hydrolysate of actinomycin Z_1 produced by *Streptomyces fradiae*. The EI spectrum of the persilylated compound contained the characteristic ions $(M - CH_3)^+$, $(M - CH_3 - CO)^+$, and $(M - COOTMS)^+$ and an abundant ion at m/z 218 representing the loss of the amino acid side chain. Later, the same group identified a new amino acid, 3-hydroxy-5-methylproline from the same source as its *N*-TFA methyl ester (Katz *et al.*, 1975). The choice of derivatives produced base-line separation of the four isomers, the mass spectra of which were distinguishable. The specific isomer present in the antibiotic hydrolysate was, therefore, unambiguously identified. These studies of actinomycin Z_1 also identified *N*-methylalanine at a site originally thought to be occupied by *N*-methylthreonine.

Even when the structure of a microbial metabolite is known, classical methodology is not always adequate for modern biosynthetic investigations. Kamal *et al.* (1978), therefore, developed procedures incorporating GC/MS for the rapid separation and identification of the amino acid constituents of etamycin, a peptidolactone antibiotic produced by Streptomyces species. L-Alanine, sarcosine, D-leucine, L-threonine, *allo*-hydroxy-D-proline, phenylsarcosine, and 3-hydroxypicolinic acid were identified as their *N*-TFA methyl esters. Because of steric hindrance *N*-β-dimethylleucine was identified as its *N*-formyl methyl ester. The use of GC/MS in combination with stable isotopes presented opportunities in biosynthetic studies or in combination with amino acid analogs in studies of directed biosynthesis. In either type of study, GC/MS would be an indispensable tool for purposes of identification.

3. PEPTIDES

Satisfactory mass spectra have been obtained for a limited number of underivatized peptides by fairly conventional methods but elaborate instrumentation and highly specialized techniques are required to overcome the lack of volatility of most peptides. Developments in this category are described in Section 4, while this section deals only with derivatized peptides.

In parallel with developments in classical Edman and similar methods for determining an amino acid sequence, there have been, during the past two decades, extensive developments in methods for sequencing proteins by mass spectrometry. Starting with the first studies on small synthetic peptides (Andersson, 1958; Biemann *et al.*, 1959) and progressing through the use of GC/MS in combination with classical techniques (Priddle, 1974), these studies culminated in the first protein sequence determination entirely by mass spectrometry (Batley and Morris, 1977). That peptides and, by extension, proteins could be amenable to sequencing by a technique best suited to the analysis of small molecules was first indicated by the observation of fragmentation at the peptide bond in the spectra of acyl peptide esters (Andersson, 1958). Since then, derivatization strategies have focused on minimizing hydrogen bonding by the free proton in the peptide bond and on creating derivatives in which cleavage is directed towards the peptide bond thus simplifying the spectra and enhancing the abundance of the ions providing sequence information. To refer only to the sequence ions is an oversimplification since many other ions in the spectra, particularly those corresponding to the amino acid side chains or their elimination from the molecular ion, may also be interpreted and are often essential to prevent ambiguous assignments (Dell and Morris, 1974).

The principle of deducing the amino acid sequence of a peptide from sequence ions is illustrated in the generalized tripeptide formula shown in Figure 8, in which, for the sake of simplicity, the peptide bond is not

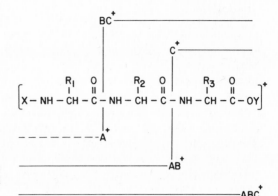

FIGURE 8. Tripeptide structure illustrating formation of *N*- and *C*-terminal sequencing ions.

modified. Since a mass spectrum represents the sum of the fragmentation pathways of a population of sample molecules, the spectrum will contain sequence ions corresponding to A^+, AB^+, and the molecular ion ABC^+. Since the nature of the substituent X and the masses of the amino acids are known, a search of the spectrum in a predefined mass range should yield a prominent ion A^+, the *N*-terminal sequence ion. As each sequence ion is identified, the spectrum is also searched for the corresponding molecular ion, since the numbers of residues cannot be assumed. Depending on the nature of the derivative, charge may be retained on the *C*-terminal end and fragments C^+ and BC^+ may also be observed. Such ions, if present, may allow assembly of the sequence from the *C*-terminal end or, if the molecular ion is absent, may be useful in completing the *N*-terminal sequence. In most cases, the masses of the ions unambiguously identify the amino acids. When mass ambiguity does occur, such as for leucine and isoleucine, other ions and their abundances must be used to resolve the problem.

The GC/MS approach to sequencing peptides presents several advantages over classical methods. For example, the rate-limiting step in classical sequencing techniques is the purification of the peptides produced by chemical or enzymatic cleavage of the original protein. This requirement is largely eliminated in the GC/MS techniques and, furthermore, the criteria for purity of the original sample are not as stringent. Because the GC/MS technique provides specific information such as a precise mass if high resolution mass spectrometry is used, unusual amino acids, branched peptides, blocked groups, and nonprotein components are easier to identify than by classical methods. If methods based on partial acid hydrolysis are used, GC/MS is ideally suited to the determination of the sequence of cyclic peptides. Finally, the sensitivity of the method is such that a complete peptide sequence can be determined using only a few nanomoles of sample (Mudgett *et al.*, 1977). Special variants of the classical methods are more sensitive, but as mass spectrometric techniques such as negative-ion chemical ionization and GLC techniques such as on-column capillary column chromatography become more generally available, it is not unreasonable to foresee an improvement in the sensitivity of the GC/MS method by at least two orders of magnitude. However, it must be emphasized that, to date, GC/MS methods of sequencing peptides have largely been used to complement classical methods.

3.1. Identification of Classical Sequencing Products

The products produced by stepwise cleavage of a single residue either from the *N*- or *C*-terminal of a peptide may be identified by a number of chromatographic techniques including GC (see MacKenzie, 1981) and so mass spectrometry is used mainly for confirmation of structure. Nevertheless, because mass spectrometry may be essential in identifying unusual amino acids, because mass spectrometric data provide specific proof of structure in cases of ambiguity, and because the detection levels may be lower if appropriate techniques are used, a brief discussion is in order.

In what was essentially a GC identification, Rangarajan *et al.* (1973) found that the molecular ion was recognizable in the EI spectra of a large number of TMS-thiohydantoins. Later, chemical ionization studies using either methane (Suzuki *et al.*, 1976), or ammonia or isobutane (Okada and Sakuno, 1978) as the reagent gas and sample injection by the direct inlet probe produced simpler spectra which were readily interpretable using 1 μmol of sample. The 2-anilino-5-thiazolinone, and the *N*-methyl and *N*-phenylthiourea intermediates of the Edman degradation procedure may be thermally induced to form the corresponding thiohydantoin derivatives directly in the mass spectrometer (Fairwell *et al.*, 1973).

The mass spectral properties of amino acid methylthiohydantoins have also been described (Fairwell *et al.*, 1970; Sun and Lovins, 1972).

Both EI (Fairwell *et al.*, 1970) and isobutane CI spectra (Fales *et al.*, 1971) permit rapid identification of amino acid phenylthiohydantoin (PTH) derivatives. The latter method was regarded as 10–100 times more sensitive than GC but, using direct inlet injection, precise temperature control was required. Conversion of PTH amino acids to volatile derivatives was unnecessary and by adding all the pentadeutero PTH amino acids to the unknown, identification and quantitation could be performed simultaneously.

Schulten and Wittmann-Liebold (1976) noted that high-resolution field desorption yielded abundant molecular ions for samples such as

PTH-arginine for which conventional electron impact mass spectrometry gave only characteristic fragments. Individual components of mixtures containing up to 15 PTH amino acids were easily discriminated.

3.2. Sequencing Based on Peptide Bond Modification

3.2.1. N-Acyl Permethylated Peptides

The most widely used mass spectrometric method for peptide sequencing is founded on permethylation of hydroxyl and carboxy groups, and peptide bond nitrogen using a methyl sulfinyl methanyl carbanion to abstract the reactive protons. To avoid the formation of nonvolatile quaternary salts, the primary amino groups must first be masked; acetylation is most commonly used. The application of permethylation to peptides appears to have originated with Das *et al.* (1967), but most of the developments rendering the technique useful for all amino acids can be attributed to Morris and his associates (Morris, 1980).

As in all methods for amino acid sequence determination, proteins must first be reduced to oligopeptides of an appropriate size. For this purpose, the whole range of enzymic and chemical agents used in classical sequencing methods is available. However, cleavage by nonspecific proteases such as elastin and subtilisin and, when necessary, specific chemical agents, is the preferred strategy for producing oligopeptides of about 2–15 residues. Depending on its complexity, the peptide mixture is partially purified by ion-exchange chromatography or by molecular exclusion chromatography to provide samples containing ideally one to five components. These are then derivatized to form *N*-acetyl permethylated peptides; a typical structure is illustrated in Figure 9.

The derivatized peptide mixture is most often introduced into the mass spectrometer via the solid sample inlet. Temperature programming of the

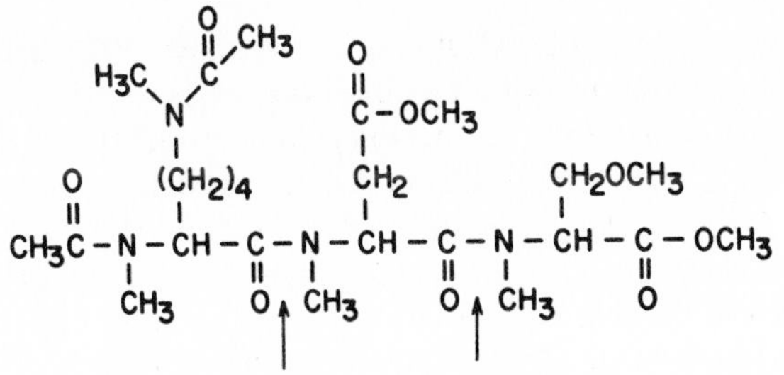

FIGURE 9. Structure of *N*-acetyl permethylated Lys-Asp-Ser. Arrows indicate bonds cleaved to form sequence ions.

inlet results in fractional distillation of the sample. Even if separation of the components is incomplete, analysis of the ion abundances allows ions in areas of peak overlap to be unambiguously attributed to specific components in the mixture. This procedure of mixture analysis obviates the need for most of the time-consuming purification of individual peptides.

The EI mass spectra of N-acetyl permethylated peptides consist largely of ions resulting from cleavage of the peptide bond. Low-resolution spectra are adequate for most purposes, but the specificity of high-resolution spectra has been found to be especially valuable for complex mixtures (Wipf et $al.$, 1973). Because ion abundances in EI spectra decrease rapidly with increasing mass, often preventing the completion of a sequence, CI spectra have also been studied. Gray $et al.$ (1970) obtained good spectra with much less material than for EI spectra and, because the fragmentation occurred almost exclusively at the peptide bond, the spectra were easy to interpret. Furthermore, with methane as a reagent gas, the molecular ion was easy to identify by the presence of the adduct ions $(M + 1)^+$, $(M + 29)^+$, and $(M + 41)^+$. Isobutane CI mass spectra contained three principal types of N-terminal sequence ion (N-imine, N-acylium, and N-amide) and one type of C-terminal sequence ion. The redundant information increased the reliability of the sequence determination (Mudgett $et al.$, 1977, 1978). For most amino acids, but not those containing heterocyclic residues, the N-terminal acylium was the most abundant of the sequence ions. The protonated molecular ion was observed but not the adduct ion $(M + C_4H_9)^+$. Comparison with methane CI spectra showed that, with the particular equipment used, the isobutane spectra had more of the total ion current in sequence ions. Mudgett $et al.$ (1977) concluded that in general isobutane CI spectra "will offer a better opportunity for obtaining both a suitable number and a suitable intensity of sequence ions than will methane CI." However, given that unlike earlier work (Gray $et al.$, 1970), the molecular ion adduct series was not observed in the methane CI spectra, it must be emphasised that the exact form of the spectra depends on the conditions used, especially the source conditions.

A number of variations of the technique have also been introduced. For example, acetylation using an equimolar mixture of acetic anhydride and perdeuteroacetic anhydride produces spectra in which the N-terminal fragments are identified by doublets separated by three mass units (Gray $et al.$, 1970; Hunt $et al.$, 1979) while C-terminal fragments are singlets.

DeJongh $et al.$ (1976) used the N-succinyl derivatives and concluded that the peaks in the high mass region were more abundant. Furthermore, the added mass moved sequencing ions away from the spectral region containing ions characteristic of amino acid side chains.

A sophisticated, but not generally available, mass spectrometric technique was used by Hunt *et al.* (1979). Isobutane CI spectra of a mixture of acetyl and perdeuteroacetyl permethylated peptides yielded abundant *N*-terminal sequence ions as doublets separated by three mass units, a derivatization technique referred to earlier. By using a triple-stage quadropole mass spectrometer, a molecular ion or sequence ion could be selected and made to undergo collision-induced dissociation (CID). Further sequence ions were thus generated. The CID of *N*- and *C*-terminal sequence ions led to complementary sequencing of an unknown peptide.

Hunt *et al.* (1979) also explored the application of electron capture negative ion (NI) CI to peptide sequencing and studied derivatives which might take advantage of the potential 100–1000-fold increase in sensitivity afforded by this technique. *N*-Pentafluorobenzoyl peptide methyl esters and *N*-acetyl peptide pentafluorobenzyl esters yielded a 100-fold increase in ion current for ions characteristic of peptide molecular weight. The application of NI mass spectrometry to peptide sequencing is still in its infancy but holds promise of further significant improvements in sensitivity.

The potential advantages of analyzing peptide mixtures by GC/MS rather than by mass spectrometry alone include clean-up of the sample, removal of secondary reaction products, and improved sensitivity by providing a higher instantaneous concentration of a given compound in the mass spectrometer. The resolving power of capillary columns would enhance these advantages. However, permethylated peptides have low volatility and earlier efforts to chromatograph these derivatives met with limited success even when the more volatile *N*-TFA permethylated derivatives were used (Lindley and Davis, 1974). Priddle *et al.* (1976a) reduced elution times and temperatures by using short columns (2 mm i.d. × 0.3 m) packed with 3% OV-1 or 1% Dexsil 300GC on Diatomite CQ. Spectra of peptides containing up to five residues were obtained using less than 25 nmol of sample. Capillary columns and CI mass spectrometry were also used (Priddle *et al.*, 1976b, 1978). However, these studies were limited to relatively simple peptide mixtures and have yet to be applied to a sample of unknown composition. If the potential advantages of using the combined GC/MS approach for the analysis of permethylated peptides are to be realised, the hydrolytic strategy must produce relatively short peptides. Inevitably, the peptide mixtures will be more complex and chromatographic resolution and the minimizing of secondary reaction products will be more important. Both of these factors require more study.

3.2.1a. Derivatization Procedure. Typically, 1 to 50 nmol of peptides are *N*-acetylated by dissolving the dry residue in an excess of acetic anhydride-methanol (1:4, v/v). After 3 h at room temperature, the reagents are evaporated. A slight excess of a methylsulfinyl carbanion base is added to the acetylated peptide dissolved in 100 μl of dimethylsulfoxide and followed by an excess of methyl iodide. After 1 h at room temperature in a sealed vial, the reaction is terminated by adding water and the whole is extracted with chloroform. The chloroform is washed twice with water and evaporated to dryness *in vacuo*. When dissolved in an appropriate solvent, the sample is ready for analysis (Morris *et al.*, 1971). Reaction conditions must be carefully controlled to prevent *C*-methylation, an effect which is sequence dependent (Mudgett *et al.*, 1975). It is also important to dry the dimethylsulfoxide before preparing the base.

The above procedure is not appropriate for peptides containing histidine, arginine, and the sulfur amino acids because nonvolatile quaternary nitrogen and sulfonium salts are formed. Therefore, subsequent modifications of the procedure are recommended (Morris, 1972; Morris *et al.*, 1973). Morris determined that quaternization of histidine and methionine could be prevented by a very short reaction time (60–75 sec) and that under these conditions even the relatively weak carboxylate anion was methylated. Other procedures based on using precise amounts of the substrate and reagent appear to be less reproducible and are difficult to apply to peptide mixtures of unknown composition. Cysteinyl peptides are most commonly isolated from proteins as the carboxymethyl derivative which can also be identified as a dehydroalanine residue using the short permethylation procedure.

Arginine-containing peptides, the presence of which can be readily determined by a color reaction, require an additional reaction before acetylation. The most gentle procedure, minimizing peptide chain cleavage, is to convert arginine to ornithine by hydrazinolysis. This is achieved by heating the peptide for 15 min at 75°C in 0.5 ml hydrazine:water (1:1) (Morris *et al.*, 1973). After dilution with water and removal of reagents *in vacuo* with warming, the peptide is ready for acetylation. However, even under these mild conditions, peptide bond cleavage can occur (Priddle *et al.*, 1978).

3.2.1b. Applications. There have already been sufficient applications of permethylation to peptide sequencing that a somewhat arbitrary selection must be used to illustrate the range and versatility of the technique. Dell and Morris (1977) developed a mass spectrometric strategy for the rapid screening of homologous proteins. The need for obtaining complete overlap information was eliminated by aligning the peptide sequences with those of a homologous protein. A nonspecific enzyme cleavage of azurin, a respiratory protein from *Pseudomonas fluorescens*, produced a series of peptides most of which were easily aligned with the known sequence of azurin from *P. aeruginosa*.

Most of the applications of permethylation have been to the determination of the sequence of enzymes, such as dihydrofolate reductase from *Lactobacillus casei* (Batley and Morris, 1977), ribitol dehydrogenase from *Klebsiella aerogenes* (Morris *et al.*, 1974), and the type-1 variant of chloramphenicol acetyl transferase specified by most F-like plasmids conferring chloramphenicol resistance (Shaw *et al.*, 1979).

The suitability of GC/MS for the study of unusual structural features was demonstrated by the determination of the sequence of a peptide from *Neurospora crassa* glutamate dehydrogenase which had the blocked *N*-terminal residue *N*-acetyl serine (Morris and Dell, 1975).

3.2.2. Polyamino Alcohols

The first mass spectrometric method proposed for determining amino acid sequences was based on the reduction of the keto group of the peptide chain to a methylene unit by lithium aluminum hydride (Biemann *et al.*, 1959). With reduction as the key feature and in combination with masking of functional groups, Nau (1974) later developed the *O*-TMS dideutero perfluoroalkyl polyamino alcohol derivatives of oligopeptides, a typical structure of which is illustrated in Figure 10. The reduction was preceded by methylation and perfluoroacylation, and followed by trimethylsilylation. These derivatives are particularly volatile and produce mass spectra containing abundant sequence determining ions resulting from cleavage not of the peptide bond but the carbon–carbon bond in the peptide chain. The high volatility suggested a gas chromatographic approach to the separation of derivatized peptide mixtures. In turn, the use of GLC dictated a protein hydrolyzing strategy producing oligopeptides containing ideally 2–5 residues. A GC/MS method for sequencing peptides based on these principles has been described in detail (Nau and Biemann, 1976a,b). In summary, the overall strategy is to hydrolyze a polypeptide to generate a large number of oligopeptides, derivatize the entire mixture, separate the components by GLC, and identify the components by mass spectrometry.

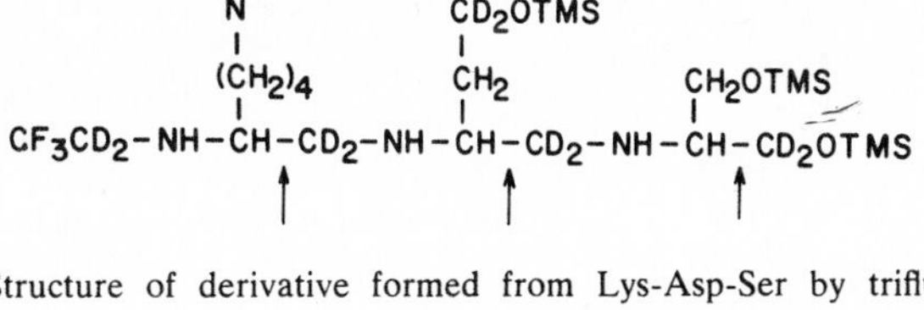

FIGURE 10. Structure of derivative formed from Lys-Asp-Ser by trifluoroacetylation, reduction with LiAlD$_4$, and trimethylsilylation. Arrows indicate bonds cleaved to form sequence ions.

Ideally, the oligopeptide mixture should represent all the peptide bonds present in the original molecule to permit reassembly of the sequence. Since a single hydrolytic method is unlikely to be effective for all proteins, the general hydrolytic strategy is to start with partial acid hydrolysis in $6 N$ HCl at 110°C. Linkages missing due to the acid lability of specific peptide bonds are then determined using another set of oligopeptides generated using a set of enzymes, an enzyme with broad specificity, or a dipeptidyl aminopeptidase.

Asparagine and glutamine are converted to aspartic and glutamic acids during acid hydrolysis and may also be transformed during the derivatization procedure. Nau (1978) circumvented this problem by diborane reduction of the polypeptides before hydrolysis and derivatization.

Even more volatile derivatives have been prepared by combining the permethylation and reduction derivatization procedures (Nau, 1976, 1978). The acetylated or trifluoroacetylated peptides were permethylated, then reduced and finally trimethylsilylated. This procedure opened up yet another sequencing strategy, that is sequencing of large oligopeptides at the permethylation stage followed by analysis of the small oligopeptides in another portion of the sample by GC/MS at the end of the reaction sequence. Good chromatographic peaks and spectra were obtained for the normally troublesome glutamine and asparagine-containing peptides but only a limited number of peptides were examined and the technique has yet to be applied to peptides of unknown composition. Nau (1978) also increased the volatility by using dimethylsilyl instead of TMS derivatives.

Similarly, Mahajan and Desiderio (1978) reduced acetylated permethylated peptides to increase the volatility of large oligopeptides and thus reduce the temperature required for solid probe sample injection.

Chemical ionization mass spectrometry does not appear to have been applied to the analysis of polyamino alcohols. Studies in the writer's laboratory indicate that these derivatives are well suited to the CI technique and that, in addition to defining the molecular weight, sequence ions are formed in sufficient abundance to be useful (Figure 11).

3.2.2a. Derivatization Procedures and Chromatography. Typically, the peptide mixture (50–500 μg) is esterified with 1.5 ml $1.25 N$ methanolic HCl for 45 min at room temperature in a 3 ml vial having a Teflon-lined screw cap (Reactivial or equivalent). Diazomethane should be used if cleavage of labile peptide bonds is to be avoided. After evaporation of the reagents, the methyl esters are dissolved in 300 μl methanol, 100 μl methyltrifluoroacetate, and 50 μl triethylamine. After 2 h at 20°C, the reagents are evaporated and the residue partitioned between water and ethyl acetate. The organic layer is transferred to another vial, evaporated, and 2 ml $1 N$ LiAlD$_4$ in dimethoxyethane added with ice cooling.

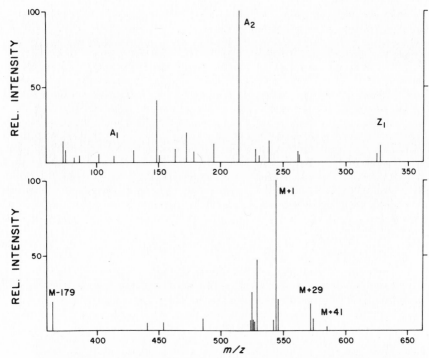

FIGURE 11. Methane chemical ionization mass spectrum of trifluoracetylated, reduced, and trimethylsilylated Gly-Leu-Tyr. Ions of relative abundance <3% have been omitted.

The reaction is stirred continuously while being maintained at 90°C over-night. Excess LiAlD$_4$ is quenched by the slow addition of methanol. One drop of water is added with vigorous stirring. After centrifugation (about 5000 × g) the residue is extracted twice with 0.5 ml hot methanol and the extracts and supernatants are combined and evaporated. The dry residue is extracted twice with 0.5 ml chloroform. The solvents are evaporated and the residue trimethylsilylated by a mixture of 400 μl pyridine and 200 μl TMS-diethylamine (55°C, 1 h). After evaporation of the reagents the sample is dissolved in an appropriate volume of benzene.

The reduction reaction is accompanied by the formation of undesirable secondary reaction products which unnecessarily complicate the chromato-graphic separation. However, reduction with borane or trideuteroborane in tetrahydrofuran substantially reduces the formation of side products (Frank and Desiderio, 1978; Frank et al., 1978). This procedure also reduces the derivatization time from the beginning of reduction to sample injection to less than 3 h and permits the derivatization and detection of as little as 0.2 nmol even of labile peptides such as aspartylphenylalanine.

The *O*-TMS perfluoro polyamino alcohol derivatives have typically been separated on 3-ft glass columns packed with 3% or 10% OV-17 on 100–120 mesh Gas Chrom Q temperature programmed from 80 to 330°C at 8 or 12°C/min (Nau, 1974; Nau and Biemann, 1976a). Frank and Desiderio (1978) used a 12-ft column containing 1% SE30 on Gas Chrom Q. Although the longer column presumably improved the resolution, the chromatographic aspect of the report was of secondary importance and no comment was made. In general, little effort has been made to optimize the chromatography of *O*-TMS polyamino alcohol derivatives. Frank *et al.* (1978) applied a capillary column to the separation for the first time, but only to a simple mixture. In analyzing a test mixture of peptides obtained from glucagon by partial acid hydrolysis, the writer has been able to almost completely resolve the derivatives of threonylphenylalanine and phenyl-alanylthreonine using a 0.3 mm × 25 m SE30 capillary column. Interpretation of the mass spectra was much easier than when these peptides were coeluted from a packed column (MacKenzie, unpublished observations).

3.2.2b. Applications. Peptide sequencing based on reduction to poly-amino alcohols and GLC separation of derivatized peptide mixtures has not yet been widely applied to proteins of unknown sequence despite the thorough studies of reaction conditions and model compounds, and the suitability of the technique to spectral interpretation and sequence assembly by computer.

The probability of obtaining insoluble fragments makes the sequencing of hydrophobic proteins by classical methods difficult. However, the combination of classical and GC/MS methods has been effective in obtaining a partial sequence for rhodopsin from the extreme hapophile *Halobacterium halobium* (Gerber *et al.*, 1979).

Small cyclic peptides are not likely to contain peptide bonds amenable to specific chemical cleavage. Furthermore, the frequent occurrence of D-amino acids prevents cleavage by proteolytic enzymes. Partial acid hydrolysis, however, should yield a mixture of small peptides capable of being reassembled into the original sequence but not requiring identification of a specific *N*- or *C*-terminus. A striking example is the determination of the structure of cyclochlorotine, a cyclic pentapeptide isolated from *Penicillium islandicum* (Anderegg *et al.*, 1979).

3.3. Sequencing Using Dipeptidyl Aminopeptidases

Proteins can be hydrolyzed by dipeptidyl aminopeptidases to produce dipeptides which can be derivatized and analyzed by GLC. Repeating the procedure on another portion of the sample, from which one amino acid has been removed by Edman degradation or to which one amino acid has been added, produces another series of dipeptides. The sequence of the original protein is reassembled from the two sets of analyses and a knowledge of the *N*-terminus and amino acid composition (Caprioli and Seifert, 1975; Krutzsch and Pisano, 1977). This technique is limited by the need for special dipeptidyl aminopeptidases (DAPs) to completely degrade proline-containing polypeptides and by the difficulty of sequence assembly when structural features are repeated. Thus a complete sequence assembly by the DAP method alone is seldom possible.

The interest in sequencing by the DAP method has provoked significant developments in the GC/MS of dipeptide derivatives. Krutzsch and Pisano (1978) examined the EI spectra of about 200 per TMS dipeptides in which the 20 common amino acids were represented in both the amino and carboxy terminals. The same structural criteria were followed in a later study using CI mass spectrometry (Krutzsch and Kindt, 1979). The CI spectra contained two abundant ions useful for molecular weight determination and exhibited a 10-fold increase in sensitivity.

The pentafluoropropionyl (PFP) methyl esters of dipeptides have been studied using both complex mixtures of dipeptides generated by DAP I (Caprioli and Seifert, 1975) and synthetic dipeptides (Seifert *et al.*, 1978). The objective of developing a "generally applicable system of dipeptide analysis" was aided by computerized pattern recognition analysis of EI spectra (Zeimer *et al.*, 1979).

Acetyl acetonyl dipeptide methyl esters have been studied by GC/MS (Frank *et al.*, 1977) and by direct inlet injection (Schier *et al.*, 1974). In the latter case, the methyl esters were generated in the direct inlet probe by pyrolytic conversion of trimethylanilinium salts. A similar method was used for *N*-ethoxycarbonyl propenyl dipeptide methyl esters (Schier *et al.*, 1976).

3.4. Miscellaneous Peptide Derivatives

In addition to the aforementioned procedures, a number of other derivatives have been used either for peptide sequencing in general or for detecting specific amino acid residues. The chromatographic and mass spectral properties of a substantial number of *N*-terminal masking groups have been described (Okada *et al.*, 1974). Patil *et al.* (1973) demonstrated that *C*-terminal fragmentations are more abundant in Schiff base derivatives of peptide esters than in acyl derivatives. In addition to *N*- and *C*-terminal fragments, internal fragments, arising from loss of fragments from both ends, provided sequence information. Peptides containing *N*-terminal tryptophan and histidine produce substituted *β*-carbolines by cyclization followed by dehydrogenation (Jayasimhulu and Day, 1980b).

Leucine and isoleucine residues are difficult to distinguish in the EI mass spectra of N-acylated peptide esters. To resolve this problem, Waern and Falter (1978) studied the mass spectra of model peptide esters containing TFA-perdeuteroleucine as the N-terminus and containing leucine or isoleucine in other positions. The ratio of peak abundances arising from the loss of isobutene from leucine or isoleucine in the peptide chain and perdeuterated leucine from the N-terminus was always significantly smaller when the peptide contained isoleucine. In a more direct method, Jayasimhulu and Day (1980a) used two novel reactions and random or sequential degradation to convert leucine and isoleucine into oxazinones which gave characteristic spectra allowing unambiguous distinction of leucine and isoleucine.

Peptide amides can be differentiated from peptides by reaction with trifluoroacetic anhydride to produce cyclic imidazolinones, the mass spectra of which contain abundant molecular ions (Engelfried and Koenig, 1976).

3.5. General Applications

Some applications of GC/MS to the analysis of peptide and protein sequencing have already been described under the categories of the methods used. These methods were designed to analyze proteins which are gene products. However, microorganisms also produce peptides, lipopeptides, and glycopeptides by a direct enzymic mechanism. This group of microbial products, which may reside as structural elements or as secondary metabolic products, often manifests unusual structural features amenable to analysis by GC/MS. For example, an intracellular peptide isolated from *Cephalosporum* sp. was shown by the EI mass spectrum of its $N(S)$-ethoxycarbonyl methyl ester derivative to be δ-(L-α-aminoadipyl)-L-cysteinyl-D-valine (Loder and Abraham, 1971). Kamiya *et al.* (1978) isolated from *Rhodosporidium toruloides* a pheromonelike substance, named rhodotorucine A, which induces mating tube formation. This compound, probably the first regulatory lipopeptide isolated from nature, was shown by mass spectrometry and nuclear magnetic resonance spectroscopy (NMR) to contain a lipophilic amino acid, S-farnesyl cysteine, at the C-terminus.

The bacterial cell wall peptidoglycan is cross-linked through bridges between the peptide units. Unusual *meso*-diaminopimelyl-*meso*-diaminopimelic acid and D-alanyl-(D)-*meso*-diaminopimelic acid cross links were discovered in the cell wall peptidoglycan of *Mycobacterium smegmatis* (Wietzerbin *et al.*, 1974). The structure of the former cross-link was derived primarily from EI mass spectrometry of the N-acetyl methyl ester of the corresponding dipeptide following partial acid hydrolysis.

Cyclic peptides are not uncommon products of microbial metabolism. Hirota *et al.* (1974) relied exclusively on high-resolution mass spectrometry to determine the structure of a cyclic pentapeptide plant-growth inhibitor isolated from *Cylindrocladium scoparium*. The compound, named, Cyl-2, was determined to have the structure cyclo-D-*O*-methyltyrosyl-L-isoleucyl-L-pipecolyl-2-amino-δ-oxo-9, 10-epoxydecanoyl. Similarly, the structures of host-specific cyclodepsipeptides from *Alternaria mali* were established by study of their mass spectral fragmentation patterns (Ueno *et al.*, 1977).

Mass spectrometry or GC/MS has played a key role in elucidating the structures of peptide antibiotics and in unambiguously interpreting unusual structural features. Some typical examples follow. The presence of one residue of α-amino dehydrobutyric acid was detected in TL-119, an antibiotic isolated from *Bacillus subtilis* (Nakagawa *et al.*, 1975). The *N*-terminal moiety of amastatin was shown to be 3-amino-2-hydroxy-5-methyl hexanoic acid (Tobe *et al.*, 1979). The antibiotic, hypelcin A, isolated from *Hypocrea peltata* contained 10 residues of α-aminobutyric acid and a *C*-terminal leucinol out of a total of 20 residues (Fujita *et al.*, 1979).

Apart from contributing to determining the structures of new antibiotics, mass spectrometry has also been useful in revising incorrect structures derived using classical chemical methods. Largely on the basis of chemical degradation experiments and a molecular formula derived from microanalysis, echinomycin was considered to contain a dithian cross-link. However, the accepted structure was inconsistent with the NMR spectrum of echinomycin, leading Dell *et al.* (1975) to conduct a detailed study including EI and field desorption mass spectrometry. As a result, the cross-link was modified to a thioacetal structure.

4. SPECIALIZED MASS SPECTROMETRIC TECHNIQUES

Most GC/MS studies of amino acids and peptides have involved either EI or CI positive ion spectra. However, although negative ion (NI) spectra have been thoroughly studied for other types of compounds (Dillard, 1973), there have been very few studies of the NI spectra of amino acids and peptides or their derivatives. Stapleton and Bowie (1976) described the EI NI spectra of a number of simple derivatives of glycine and α- and β-alanine. To facilitate capture of secondary electrons and enhance ion currents, electron-attracting moieties were used for acylation with nitrobenzoyl derivatives being favored. Abundant molecular anions were obtained, but apart from confirming the molecular weight, the spectra were not regarded as a viable alternative to positive ion mass spectra. Voigt and

Schmidt (1978) investigated the low-voltage (2–4 eV) NI mass spectra of 20 free α-amino acids and discussed the fragmentation processes.

Because of the generally low ion currents in EI NI mass spectrometry, no significant advantage appears to be gained over the corresponding positive ion spectra. In the CI mode, however, a large population of near thermal electrons is generated, and under these conditions the probability of electron capture without inducing extensive fragmentation and consequently of generating negative ion spectra is enhanced. Hunt and his colleagues examined the analytical potential of CI NI mass spectrometry and described the simultaneous recording of positive and negative CI spectra (Hunt *et al.*, 1976). They also explored the use of various reagent gases to form stable negative ions. A 100- to 1000-fold increase in sensitivity was obtained for the range of compounds studied. The application of pulsed positive–negative CI mass spectrometry to oligopeptide sequencing was demonstrated using a mixture of methane and methyl nitrate as the reagent gas. The two spectra contained complementary information, and, in particular, the *N*-terminal sequence ions were easily recognized by doublets separated by 29 u in the two spectra. Later developments by the same group concentrated on derivatives containing strong electron withdrawing substituents attached to a site of unsaturation (Hunt *et al.*, 1979). Although these studies were limited to only a few model peptides, the potential sensitivity of NI mass spectrometry is certain to induce further developments especially in relation to oligopeptide sequencing.

The lack of volatility of amino acids and peptides and the problems encountered in generating volatile derivatives especially at the ultramicro level have encouraged investigations of other methods of obtaining spectra while minimizing thermal degradation. One relatively simple approach is to desorb the sample from probes which introduce the sample into the ion source close to the electron beam. Applications of this technique, called in-beam mass spectrometry, have been reviewed by Cotter (1980).

In a variant of the in-beam method, the process of desorption has been further enhanced by rapid sample heating, by introducing the sample on an inert surface such as Teflon, and by lining the collision chamber with Teflon (Beuhler *et al.*, 1974a; Hansen and Munson, 1978). Informative CI spectra of underivatized amino acids and peptides were obtained using only a few nanomoles of sample. Beuhler *et al.* (1974b) also studied the ammonia CI spectra of underivatized arginine-containing peptides including the nonapeptide bradykinin in which arginine residues occupy both the *N*- and *C*-terminus. A unique sequence consistent with the known structure could be constructed from the spectrum.

Field desorption (FD) mass spectrometry, in which ionization and desorption occur in the ion source as a consequence of the application of an intense electrostatic field, is probably the most widely used "soft" ionization method for obtaining mass spectra of underivatized oligopeptides. The subject has been thoroughly covered in a review by Schulten (1977). Typical examples are the studies of Pandey et al. (1977) and Calas et al. (1980). Although in some studies abundant molecular weight determining ions have been obtained in concert with abundant sequencing ions, the spectra are generally simpler than EI or CI spectra. Interpretation of FD mass spectra is hampered by the distinct possibility of either M^+ or $(M + H)^+$ or both being present. Frick et al. (1977b) devised a procedure based on the analysis of peptide methylation–methanolysis product mixtures to distinguish these two ions.

Other variants of the FD method include ^{252}Cf plasma desorption (MacFarlane and Torgerson, 1976) and collision-induced dissociation of field desorbed peptides (Weber and Levsen, 1980). Field desorption mass spectra of leucine, isoleucine, and norleucine were compared with more conventional spectra by Fales et al. (1975).

Although applied only to a limited range of amino acids or peptides, a number of other sophisticated mass spectral methods deserve mention. These include secondary ion mass spectrometry (Benninghoven and Sichtermann, 1978), electrohydrodynamic ionization mass spectrometry (Stimpson and Evans, 1978), Curie point pyrolysis in combination with low voltage EI mass spectrometry (Meuzelaar et al., 1973), mass analyzed ion kinetic energy spectrometry (Kondrat and Cooks, 1978), and tandem mass spectrometry (Maugh, 1980). However, these techniques are incompatible with the needs of most laboratories although entirely appropriate for large centers specializing in the study of mass spectrometry. Furthermore, these methods have yet to contribute significantly to the analysis of peptides of unknown structure, and much study remains before they supplant the more conventional methods described earlier.

5. CONCLUSIONS

The last decade has witnessed significant advances in the technique of GLC as applied to amino acids and peptides. There have been improvements in derivatization procedures, packed and capillary columns, detectors, enantiomer resolution, and peptide sequencing methods. When these are combined with improvements in mass spectrometry, the combination is the single most powerful tool available for determining amino acid and oligopeptide structures. However, the opportunities presented by advances in GC/MS have not yet been extensively exploited in a microbiological context. Mass spectrometry has contributed greatly to the determination of the structures of peptide antibiotics but largely in the solid sample-injection mode.

In considering the future, it is appropriate to paraphrase the words of Michael Polanyi and to conclude that it is difficult "to forecast the further progress of science except for routine extensions of the existing system." Nevertheless, it is safe to predict that even routine extensions of existing GC/MS techniques will contribute to the identification of new amino acids either free or as constituents of microbial metabolites. Furthermore, the use of ultrasensitive detection techniques such as SIM and NI–CI will enable problems of microbial biosynthesis to be routinely addressed at a quantitative level now rarely achieved. Whether the problem is related to microbial ecology, metabolic regulation, plant tumorigenesis, or microbial evolution, wherever amino acids or peptides are involved, GC/MS will be a cornerstone of structural studies.

REFERENCES

Abramson, F. P., McCaman, M. W., and McCaman, R. E., 1974, Femtomole level of analysis of biogenic amines and amino acids using functional group mass spectrometry, *Anal. Biochem.* **57**:482.

Adams, R. F., 1974, Determination of amino acid profiles in biological samples by gas chromatography, *J. Chromatogr.* **95**:189.

Adams, R. F., Vandemark, F. L., and Schmidt, G. J., 1977, Ultramicro GC determination of amino acids using glass open tubular columns and a nitrogen-selective detector, *J. Chromatogr. Sci.* **15**:63.

Anderegg, R. J., Biemann, K., Manmade, A., and Ghosh, A. C., 1979, Mass spectrometric peptide sequencing: Cyclochlorotine, *Biomed. Mass Spectrom.* **6**:129.

Andersson, C.-O., 1958, Mass spectrometric studies on amino acid and peptide derivatives, *Acta Chem. Scand.* **12**:1353.

Aragozzini, F., Ferrari, A., Pacini, N., and Gualandris, R., 1979, Indole-3-lactic acid as a tryptophan metabolite produced by *Bifidobacterium* spp., *Appl. Environ. Microbiol.* **38**:544.

Batley, K. E., and Morris, H. R., 1977, Dihydrofolate reductase: Partial sequence of the *Lactobacillus casei* enzyme and homology with other dihydrofolate reductases, *Biochem. Soc. Trans.* **5**:1097.

Bengtsson, G., and Odham, G., 1979, A micromethod for the analysis of free amino acids by gas chromatography and its application to biological systems, *Anal. Biochem.* **92**:426.

Bengtsson, G., Odham, G., and Westerdahl, G., 1981, Glass capillary gas chromatographic analysis of free amino acids in biological microenvironments using electron capture or selected ion-monitoring detection, *Anal. Biochem.* **111**:163.

Benninghoven, A., and Sichtermann, W. K., 1978, Detection, identification and structural investigation of biologically important compounds by secondary ion mass spectrometry, *Anal. Chem.* **50**:1180.

Beuhler, R. J., Flanigan, E., Greene, L. J., and Friedman, L., 1974a, Proton transfer mass spectrometry of underivatized peptides, *Biochemistry* **13**:5060.

Beuhler, R. J., Flanigan, E., Greene, L. J., and Friedman, L., 1974b, Proton transfer mass spectrometry of peptides. A rapid heating technique for underivatized peptides containing arginine, *J. Am. Chem. Soc.* **96**:3990.

Biemann, K., Gapp, F., and Seibl, J., 1959, Application of mass spectrometry to structure problems. I. Amino acid sequence in peptides, *J. Am. Chem. Soc.* **81**:2274.

Blau, K., and King, G., 1978, *Handbook of Derivatives for Chromatography*, Heyden and Son, Ltd., London.

Budzikiewicz, H., and Meissner, G., 1978, Chemical ionization spectra of trimethylsilylated amino acids and oligopeptides, *Org. Mass. Spectrom.* **13**:608.

Calas, B., Mery, J., Parello, J., Prome, J. C., Roussel, J., and Patouraux, D., 1980, Field desorption mass spectrometry of basic peptides. An approach to the sequencing of arginine-containing peptides, *Biomed. Mass Spectrom.* **7**:288.

Cancalon, P., and Klingman, J. D., 1974, An improved procedure for preparing the *n*-butyl-trifluoroacetyl amino acid derivatives and its application in the study of radioactive amino acids from biological sources, *J. Chromatogr. Sci.* **12**:349.

Caprioli, R. M., and Seifert, W. E., 1975, Hydrolysis of polypeptides and proteins utilizing a mixture of dipeptidylaminopeptidases with analysis by GC/MS, *Biochem. Biophys. Res. Commun.* **64**:295.

Cattabeni, F., Galli, C. L., and Eros, T., 1976, A simple and highly sensitive mass fragmentographic procedure for gamma amino butyric acid determinations, *Anal. Biochem.* **72**:1.

Cotter, R. J., 1980, Mass spectrometry of nonvolatile compounds; Desorption from extended probes, *Anal. Chem.* **52**:1589A.

Coulter, A. W., and Fenselau, C., 1972, Variation of fragmentation with ring size in cyclic α-amino acids, *Org. Mass Spectrom.* **6**:105.

Das, B. C., Gero, S. D., and Lederer, E., 1967, *N*-Methylation of *N*-acyl oligopeptides, *Biochem. Biophys. Res. Commun.* **29**:211.

Dell, A., and Morris, H. R., 1974, New observations on the fragmentation properties of peptides under electron impact mass spectrometry, *Biochem. Biophys. Res. Commun.* **61**:1125.

Dell, A., and Morris, H. R., 1977, A mass spectrometric strategy for the rapid screening of homologous proteins: Studies on a *Pseudomonas* azurin, *Biochem. Biophys. Res. Commun.* **78**:874.

Dell, A., Williams, D. H., Morris, H. R., Smith, G. A., Feeney, J., and Roberts, G. C. K., 1975, Structure revision of the antibiotic echinomycin, *J. Am. Chem. Soc.* **97**:2497.

DeJongh, D. C., Faus, F., Nayar, M. S. B., Boileau, G., and Brakier-Gingras, L., 1976. The use of *N*-succinyl derivatives in the study of amino acids and peptides by mass spectrometry, *Biomed. Mass Spectrom.* **3**:191.

Desgres, J., Boisson, D., and Padieu, P., 1979, Gas–Liquid chromatography of isobutyl ester, *N*(*O*)-heptafluorobutyrate derivatives of amino acids on a glass capillary column for quantitative separation in clinical biology, *J. Chromatogr.* **162**:133.

Dillard, J. G., 1973, Negative ion mass spectrometry, *Chem. Rev.* **73**:589.

Engelfried, C., Koenig, W. A., and Voelter, W., 1976, Mass spectrometric investigation of peptide amides, *Biomed. Mass Spectrom.* **3**:241.

Fairwell, T., Barnes, W. T., Richards, F. F., and Lovins, R. E., 1970, Sequence analysis of complex protein mixtures by isotope dilution and mass spectrometry, *Biochemistry* **9**:2260.

Fairwell, T., Ellis, S., and Lovins, R. E., 1973, Quantitative protein sequencing using mass spectrometry: Thermally induced formation of thiohydantoin amino acid derivatives from *N*-methyl and *N*-phenylthiourea amino acids and peptides in the mass spectrometer, *Anal. Biochem.* **53**:115.

Fales, H. M., Nagai, Y., Milne, G. W. A., Brewer, H. B., Jr., Bronzert, T. J., and Pisano, J. J., 1971, Use of chemical ionization mass spectrometry in analysis of amino acid phenylthiohydantoin derivatives formed during Edman degradation of proteins, *Anal. Biochem.* **43**:288.

Fales, H. M., Milne, G. W. A., Winkler, H. U., Beckey, H. D., Damico, J. N., and Barron, R., 1975, Comparison of mass spectra of some biologically important compounds as obtained by various ionization techniques, *Anal. Chem.* **47**:207.

Felker, P., and Bandurski, R. S., 1975, Quantitative gas–liquid chromatography and mass spectrometry of the $N(O)$-perfluorobutyryl-O-isoamyl derivatives of amino acids, *Anal. Biochem.* **67**:245.

Frank, H., and Desiderio, D. M., 1978, Reduction of oligopeptides to amino alcohols with borane, *Anal. Biochem.* **90**:413.

Frank, H., Haegele, K. D., and Desiderio, D. M., 1977, Gas chromatography–mass spectrometry study of acetylacetonyl dipeptide methyl esters, *Anal. Chem.* **49**:287.

Frank, H., das Neves, H. J. C., and Bayer, E., 1978, Use of borane as reducing agent in sequence analysis of peptides by gas chromatography–mass spectrometry, *J. Chromatogr.* **152**:357.

Frick, W., Chang, D., Folkers, K., and Daves, G. D., Jr., 1977a, Critical experimental parameters in gas chromatographic–mass spectrometric analysis of oligopeptide hydrolysates at the picomole level, *Anal. Chem.* **49**:1241.

Frick, W., Daves, G. D., Jr., Barofsky, D. F., Barofsky, E., Fisher, G. H., Chang, D., and Folkers, K., 1977b, Sample derivatization and structure analysis by field desorption mass spectrometry. Peptide methylation-methanolysis, *Biomed. Mass Spectrom.* **4**:152.

Fujita, T., Takaishi, Y., and Shiromoto, T., 1979, New peptide antibiotic, Hypelcin A, from *Hypocrea peltata*, *J. Chem. Soc., Chem. Commun.* 413.

Gaffney, J. S., Pierce, R. C., and Friedman, L., 1977, Mass spectrometer study of evaporation of α-amino acids, *J. Am. Chem. Soc.* **99**:4293.

Gelpi, E., Koenig, W. A., Gibert, J., and Oro, J., 1969, Combined gas chromatography–mass spectrometry of amino acid derivatives, *J. Chromatogr.* **7**:604.

Gerber, G. E., Anderegg, R. J., Herlihy, W. C., Gray, C. P., Biemann, K., and Khorana, H. G., 1979, Partial primary structure of bacteriorhodopsin: Sequencing methods for membrane proteins, *Proc. Nat. Acad. Sci. USA* **76**:227.

Gray, W. R., Wojcik, L. H., and Futrell, J. H., 1970, Application of mass spectrometry to protein chemistry II. Chemical ionization studies on acetylated permethylated peptides, *Biochem. Biophys. Res. Commun.* **41**:1111.

Grob, K., Grob, G., and Grob, K., Jr., 1979, Deactivation of glass capillary columns by silylation. Part I: Principles and basic technique, *HRC CC J. High Resolut. Chromatogr. Chromatogr. Commun.* **1**:31.

Hansen, G., and Munson, B., 1978, Surface chemical ionization mass spectrometry, *Anal. Chem.* **50**:1130.

Hirota, A., Suzuki, A., Aizawa, K., and Tamura, S., 1974, Mass spectrometric determination of amino acid sequence in Cyl-2, a novel cyclotetrapeptide from *Cylindrocladium scoparium*, *Biomed. Mass Spectrom.* **1**:15.

Hunt, D. F., Stafford, G. C., Jr., Crow, F. W., and Russell, J. W., 1976, Pulsed positive negative ion chemical ionization mass spectrometry, *Anal. Chem.* **48**:2098.

Hunt, D. F., Buko, A. M., Ballard, J., and Shabanowitz, J., 1979, Polypeptide sequencing by mass spectrometry: New methodology, *Proc. 27th Annual Conference on Mass Spectrometry and Allied Topics*, Seattle, Washington, p. 608.

Husek, P., 1974, Derivation of amino acids with 1,3-dichlorotetrafluoroacetone and its use in gas chromatography, *J. Chromatogr.* **91**:475.

Husek, P., and Macek, K., 1975, Gas chromatography of amino acids, *J. Chromatogr.* **113**:139.

Iwase, H., and Murai, A., 1977, On the derivatives for the ultramicro determination of amino acids by mass fragmentography, *Anal. Biochem.* **78**:340.

Jayasimhulu, K., and Day, R. A., 1980a, Electron impact fragmentation of oxazoles and oxazinones derived from leucine and isoleucine, *Biomed. Mass Spectrom.* **7**:7.

Jayasimhulu, K., and Day, R. A., 1980b, Mass spectrometric determination of N-terminal tryptophan and N-terminal histidine in peptides, *Biomed. Mass Spectrom.* **7**:321.

Jennings, W., 1980, Evolution and application of the fused silica column *HRC CC, J. High Resolut. Chromatogr. Chromatogr. Commun.* **3**:601.

Jönsson, J., Eyem, J., and Sjöquist, J., 1973, Quantitative gas chromatographic analysis of amino acids on a short glass capillary column, *Anal. Biochem.* **51**:204.

Kaiser, F. E., Gehrke, C. W., Zumwalt, R. W., and Kuo, K. C., 1974, Amino acid analysis. Hydrolysis, ion-exchange clean-up, derivatization, and quantitation by gas liquid chromatography, *J. Chromatogr.* **94**:113.

Kamal, F., Katz, E., and Mauger, A. B., 1978, Electrophoretic, chromatographic and mass spectrometric procedures for the identification and isotopic assay of amino acid constituents in etamycin, *J. Chromatogr.* **151**:245.

Kamiya, Y., Sakurai, A., Tamura, S., Takahashi, N., Abe, K., Tsuchiya, E., Fukui, S., Kitada, C., and Fujino, M., 1978, Structure of rhodotorucine A, a novel lipopeptide, inducing mating tube formation in *Rhodosporidium toruloides, Biochem. Biophys. Res. Commun.* **83**:1077.

Katz, E., Mason, K. T., and Mauger, A. B., 1974, The presence of α-amino-β,γ-dihydroxy-butyric acid in hydrolysates of actinomycin Z_1, *J. Antiobiot.* **XXVII**:952.

Katz, E., Mason, K. T., and Mauger, A. B., 1975, 3-Hydroxy-5-methylproline, a new amino acid identified as a component of actinomycin Z_1, *Biochem. Biophys. Res. Commun.* **63**:502.

Kingston, E. E., and Duffield, A. M., 1978, Plasma amino acid quantitation using gas chromatography chemical ionization mass spectrometry and ^{13}C amino acids as internal standards, *Biomed. Mass Spectrom.* **5**:621.

Knapp, D. R., 1979, *Handbook of Analytical Derivatization Reactions*, Wiley-Interscience, New York.

Kondrat, R. W., and Cooks, R. G., 1978, Direct analysis of mixtures by mass spectrometry, *Anal. Chem.* **50**:81A.

Krutzsch, H. C., and Kindt, T. J., 1979, The identification of trimethylsilylated dipeptides with chemical ionization mass spectrometry, *Anal. Biochem.* **92**:525.

Krutzsch, H. C., and Pisano, J. J., 1977, "Analysis of Dipeptides by Gas Chromatography–Mass Spectrometry and Application to Sequencing with Dipeptidyl Aminopeptidases," *Methods in Enzymology*, (C. W. Hirs and S. N. Timasheff, eds.), Vol XLVII, pp. 391–404, Academic Press, New York.

Krutzsch, H. C., and Pisano, J. J., 1978, Separation and sequence of dipeptides using gas chromatography and mass spectrometry of their trimethylsilylated derivatives, *Biochemistry* **17**:2791.

Lawless, J. G., and Chadha, M. C., 1971, Mass spectrometric analysis of C_3 and C_4 aliphatic amino acid derivatives, *Anal. Biochem.* **44**:473.

Leclercq, P. A., and Desiderio, D. M., 1973, Chemical ionization mass spectra of amino acids and derivatives. Occurrence and fragmentation of ion–molecule reaction products, *Org. Mass. Spectrom.* **7**:515.

Leimer, K. R., Rice, R. H., and Gehrke, C. W., 1977a, Complete mass spectra of N-trifluoroacetyl-n-butyl esters of amino acids, *J. Chromatogr.* **141**:121.

Leimer, K. R., Rice, R. H., and Gehrke, C. W., 1977b, Complete mass spectra of the per-trimethylsilylated amino acids, *J. Chromatogr.* **141**:355.

Liardon, R., Ott-Kuhn, U., and Husek, P., 1979, Mass spectra of α-amino acid oxazolidinones, *Biomed. Mass Spectrom.* 6:381.

Liardon, R., Ledermann, S., and Ott, U., 1981, Determination of D-amino acids by deuterium labelling and selected ion monitoring, *J. Chromatogr.* 203:385.

Lindley, H., and Davis, P. C., Gas chromatography of some dipeptide derivatives, *J. Chromatogr.* 100:117.

Loder, P. B., and Abraham, E. P., 1971, Isolation and nature of intra cellular peptides from a Cephalosporin C-producing *Cephalosporium* sp., *Biochem. J.* 123:471.

Macfarlane, R. D., and Torgerson, D. F., 1976, Californium-252 plasma desorption mass spectrometry, *Science* 191:920.

MacKenzie, S. L., 1981, "Recent Developments in Amino Acid Analysis by Gas–Liquid Chromatography," in *Methods of Biochemical Analysis* (D. Glick, ed.), Vol. 27, pp. 1–88, Wiley-Interscience, New York.

MacKenzie, S. L., and Hogge, L. R., 1977, Gas chromatography–mass spectrometry of the N(O)-heptafluorobutyryl isobutyl esters of the protein amino-acids using electron impact ionization, *J. Chromatogr.* 132:485.

MacKenzie, S. L., and Tenaschuk, D., 1974, Gas liquid chromatography of N-heptafluorobutyryl isobutyl esters of amino acids, *J. Chromatogr.* 97:19.

MacKenzie, S. L., and Tenaschuk, D., 1979a, Quantitative formation of N(O,S)-heptafluorobutyryl isobutyl amino acids for gas chromatographic analysis. I. Esterification, *J. Chromatogr.* 171:195.

MacKenzie, S. L., and Tenaschuk, D., 1979b, Quantitative formation of N(O,S)-heptafluorobutyryl isobutyl amino acids for gas chromatographic analysis. II. Acylation, *J. Chromatogr.* 173:53.

Mahajan, V. K., and Desiderio, D. M., 1978, Mass spectrometry of acetylated, permethylated and reduced oligopeptides, *Biochem. Biophys. Res. Commun.* 82:1104.

Manhas, M. S., Hsieh, R. S., and Bose, A. K., 1970, Mass spectral studies. Part VII. Unusual fragmentation of some N-trifluoroacetyl amino-acid methyl esters, *J. Chem. Soc., C.* 116.

March, J. F., 1975, Modified technique for the quantitative analysis of amino acids by gas chromatography using heptafluorobutyric n-propyl derivatives, *Anal. Biochem.* 69:420.

Marik, J., Capek, A., and Kralicek, J., 1976, Gas chromatography–mass spectrometry of trimethylsilyl derivatives of ω-amino acids, *J. Chromatogr.* 128:1.

Martinez, E., and Gelpi, E., 1978, Mixed pentafluoropropionyl trimethylsilyl derivatives of 5-hydroxytryptophan for mass fragmentographic detection. Development of a retention index model for substituted indoles, *J. Chromatogr.* 167:77.

Maugh, T., 1980, Separations by MS speed up, simplify analysis, *Science* 209:675.

McReynolds, J. H., and Anbar, M., 1977, Isotopic assay of nanomole amounts of nitrogen-15 labeled amino acids by collision-induced dissociation mass spectrometry, *Anal. Chem.* 49:1832.

Mee, J. M. L., Korth, J., Halpern, B., and James, L. B., 1977, Rapid and quantitative blood amino acid analysis by chemical ionization mass spectrometry, *Biomed. Mass Spectrom.* 4:178.

Meot-Ner, M., and Field, F. H., 1973, Chemical ionization mass spectrometry. XX. Energy effects and virtual ion temperature in the decomposition kinetics of amino acids and amino acid derivatives, *J. Am. Chem. Soc.* 95:7207.

Meuzelaar, H. L. C., Posthumus, M. A., Kistemaker, P. G., and Kistemaker, J., 1973, Curie point pyrolysis in direct combination with low voltage electron impact ionization mass spectrometry. New method for the analysis of nonvolatile organic materials, *Anal. Chem.* 45:1546.

Morris, H. R., 1972, Studies towards the complete sequence determination of proteins by mass spectrometry; a rapid procedure for the successful permethylation of histidine containing peptides, *FEBS (Fed. Eur. Biochem. Soc.) Lett.* **22**:257.

Morris, H. R., 1980, Biomolecular structure determination by mass spectrometry, *Nature* **286**:447.

Morris, H. R., and Dell, A., 1975, The sequence of the blocked *N*-terminal peptide from *Neurospora* glutamate dehydrogenase, *Biochem. J.* **149**:754.

Morris, H. R., Williams, D. H., and Ambler, R. P., 1971, Determination of the sequences of protein-derived peptides and peptide mixtures by mass spectrometry, *Biochem. J.* **125**:189.

Morris, H. R., Dickinson, R. J., and Williams, D. H., 1973, Studies towards the complete sequence determination of proteins by mass spectrometry: Derivatization of methionine, cysteine and arginine containing peptides, *Biochem. Biophys. Res. Commun.* **51**:247.

Morris, H. R., Batley, K. E., Harding, N. G. L., Bjur, R. A., Dann, J. G., and King, R. W., 1974, Dihydrofolate reductase: Low-resolution mass-spectrometric analysis of an elastase digest as a sequencing tool, *Biochem. J.* **137**:409.

Moss, C. W., Lambert, M. A., and Diaz, F. J., 1971, GLC of twenty protein amino acids on a single column, *J. Chromatogr.* **60**:134.

Mudgett, M., Bowen, D. V., Kindt, T. J., and Field, F. H., 1975, *C*-Methylation: An Artifact in peptides derivatized for sequencing by mass spectrometry, *Biomed. Mass Spectrom.* **2**:254.

Mudgett, M., Bowen, D. V., Field, F. H., and Kindt, T. J., 1977, Peptide sequencing: The utility of chemical ionization mass spectrometry, *Biomed. Mass Spectrom.* **4**:159.

Mudgett, M., Sogn, J. A., Bowen, D. V., and Field, F. H., 1978, Peptide sequencing by chemical ionization mass spectrometry, *Adv. Mass Spectrom.* **7B**:1506.

Nakagawa, Y., Nakazawa, T., and Shoji, J., 1975, On the structure of a new antibiotic TL-119 (Studies on antibiotics from the genus *Bacillus*. VI), *J. Antibiot.* **28**:1004.

Nau, H., 1974, New dideutero-perfluoroalkylated oligopeptide derivatives for protein-sequencing by gas chromatography-mass spectrometry, *Biochem. Biophys. Res. Comm.* **59**:1088.

Nau, H., 1976, Gas chromatography–mass spectrometry of permethylated peptides and their reduced and trimethylsilylated derivatives, *J. Chromatogr.* **121**:376.

Nau, H., 1978, Sequencing of polypeptides containing asparagine and glutamine residues by gas chromatography–mass spectrometry, *Adv. Mass Spectrom.* **7B**:1518.

Nau, H., and Biemann, K., 1976a, Amino acid sequencing by gas chromatography–mass spectrometry using perfluoro-dideuteroalkylated peptide derivatives. A. Gas chromatographic retention indices, *Anal. Biochem.* **73**:139.

Nau, H., and Biemann, K., 1976b, Amino acid sequencing by gas chromatography–mass spectrometry using perfluoro-dideuteroalkylated peptide derivatives. B. Interpretation of the mass spectra, *Anal. Biochem.* **73**:154.

Okada, K., and Sakuno, A., 1978, Identification of amino acid thiohydantoin derivatives by chemical ionization mass spectrometry, *Org. Mass Spectrom.* **13**:535.

Okada, K., Nagai, S., Uyehara, T., and Hiramoto, M., 1974, A study of some new and useful *N*-terminal groups in mass spectrometry of peptides. The use of 3-hydroxyalkanoyl and unsaturated acyl groups, *Tetrahedron* **30**:1175.

Padieu, P., Desgres, J., and Maume, B. F., 1978a, Chemical ionization in capillary column gas chromatography-quadropole mass spectrometry: An ideal combination for structure proof and fragmentation mechanism elucidation as well as for isotopic dilution mass fragmentography, *Finnigan Spectra* **7**:No. 1.

Padieu, P., Desgres, J., Maume, B. F., Van der Velde, G., and Skinner, R. S., 1978b, Investigation of the chemical ionization of amino acid derivatives for biological studies, *Adv. Mass Spectrom.* **7B**:1604.

Pandey, R. C., Meng, H., Cook, J. C., Jr., and Rinehart, K. L., Jr., 1977, Structure of antiamoebin I from high resolution field desorption and gas chromatographic mass spectrometry studies, *J. Am. Chem. Soc.* **99**:5203.

Patil, G. V., Hamilton, R. E., and Day, R. A., 1973, Schiff base derivatives of peptide esters: Relative abundance of N-terminal, C-terminal and "internal" fragments as a function of the blocking group, *Org. Mass Spectrom.* **7**:817.

Pearce, R. J., 1977, Amino acid analysis by gas–liquid chromatography of N-heptafluorobutyryl isobutyl esters. Complete resolution using a support-coated open-tubular capillary column, *J. Chromatogr.* **136**:113.

Pereira, W. E., Hoyano, Y., Reynolds, W. E., Summons, R. E., and Duffield, A. M., 1973, Simultaneous quantitation of ten amino acids in soil extracts by mass fragmentography, *Anal. Biochem.* **55**:236.

Poole, C. F., and Verzele, M., 1978, Separation of protein amino acids as their N(O)-acyl alkyl ester derivatives on glass capillary columns, *J. Chromatogr.* **150**:439.

Priddle, J. D., 1974, The use of mass-spectrometry to complement conventional techniques for protein sequence determination, *Biochem. J.* **139**:23.

Priddle, J. D., Rose, K., and Offord, R. E., 1976a, The separation and sequencing of permethylated peptides by mass spectrometry directly coupled to gas–liquid chromatography, *Biochem. J.* **157**:777.

Priddle, J. D., Rose, K., and Offord, R. E., 1976b, Direct sequencing of permethylated peptides by gas chromatography–mass spectrometry using electron impact and chemical ionization, *Adv. Mass Spectrom. Biochem. Med.* **II**:477.

Priddle, J. D., Rose, K., and Offord, R. E., 1978, Amino acid sequence determination by direct e.i./c.i. mass spectrometry of N-acyl; N,O-permethylated oligopeptides separated by gas chromatography, *Adv. Mass Spectrom.* **7B**:1502.

Rafter, J. J., Ingelman-Sundberg, M., and Gustafsson, J.-A., 1979, Protein amino acid analysis by an isotope ratio gas chromatography mass spectrometry computer technique, *Biomed. Mass Spectrom.* **6**:317.

Rangarajan, M., Ardrey, R. E., and Darbre, A., 1973, Gas–liquid chromatography and mass spectrometry of amino acid thiohydantoins and their use in protein sequencing, *J. Chromatogr.* **87**:499.

Rash, J. J., Gehrke, C. W., Zumwalt, R. W., Kuo, K. C., Kvenvolden, K. A., and Stalling, D. L., 1972, GLC of amino acids: A survey of contamination, *J. Chromatogr. Sci.* **10**:444.

Robinson, J. R., Starratt, A. N., and Schlahetka, E. E., 1978, Estimation of nitrogen-15 levels in derivatized amino acids using gas chromatography–quadropole mass spectrometry with chemical ionization and selected ion monitoring, *Biomed. Mass Spectrom.* **5**:648.

Schier, G. M., Halpern, B., and Milne, G. W. A., 1974, Characterization of dipeptides by electron impact and chemical ionization mass spectrometry, *Biomed. Mass Spectrom.* **1**:212.

Schier, G. M., Bolton, P. D., and Halpern, B., 1976, The mass spectrometric identification of dipeptide mixtures obtained from dipeptidylaminopeptidase I—Hydrolysates, *Biomed. Mass Spectrom.* **3**:32.

Schulman, M. F., and Abramson, F. P., 1975, Plasma amino acid analysis by isotope ratio gas chromatography–mass spectrometry computer techniques, *Biomed. Mass Spectrom.* **2**:9.

Schulten, H.-R., 1977, "Field Desorption Mass Spectrometry and its Application in Biochemical Analysis," in *Methods of Biochemical Analysis* (D. Glick, ed.), Vol. 24, pp. 313–448, Wiley-Interscience, New York.

Schulten, H.-R., and Wittmann-Liebold, B., 1976, High resolution field desorption mass spectrometry V. Mixtures of amino acid phenylthiohydantoins and Edman degradation products, *Anal. Biochem.* **76**:300.

Segura, J., Artigas, F., Martinez, E., and Gelpi, E., 1976, Adsorption of tryptophan metabolites from physiological fluids on XAD-2 and determination by single ion monitoring, *Biomed. Mass Spectrom.* **3**:91.

Seifert, W. E., Jr., McKee, R. E., Beckner, C. F., and Caprioli, R. M., 1978, Characterization of mixtures of dipeptides by gas chromatography–mass spectrometry, *Anal. Biochem.* **88**:149.

Shaw, W. V., Packman, L. C., Burleigh, B. D., Dell, A., Morris, H. R., and Hartley, B. S., 1979, Primary structure of a chloramphenicol acetyltransferase specified by R plasmids, *Nature* **282**:870.

Shieh, J.-J., Leung, K., and Desiderio, D. M., 1976, Basic mass spectrometric investigations of amino acid-fluorescamine derivatives, *Org. Mass Spectrom.* **11**:479.

Siezen, R. J., and Mague, T. H., 1977, Gas–liquid chromatography of the N-heptafluorobutyryl isobutyl esters of fifty biologically interesting amino acids, *J. Chromatogr.* **130**:151.

Sjöquist, B., 1979, Analysis of tyrosine and deuterium labelled tyrosine in tissues and body fluids, *Biomed. Mass Spectrom.* **6**:392.

Stapleton, B. J., and Bowie, J. H., 1976, Electron impact studies. CIII—Negative ion mass spectra of naturally occurring compounds: Nitrobenzoyl derivatives of amino esters, *Org. Mass Spectrom.* **11**:429.

Stimpson, B. P., and Evans, C. A., Jr., 1978, Electrohydrodynamic ionization mass spectrometry of biochemical materials, *Biomed. Mass Spectrom.* **5**:52.

Summons, R. E., Pereira, W. E., Reynolds, W. E., Rindfleisch, T. C., and Duffield, A. M., 1974, Analysis of twelve amino acids in biological fluids by mass fragmentography, *Anal. Chem.* **46**:582.

Sun, T., and Lovins, R. E., 1972, Quantitative protein sequencing using mass spectrometry: Use of low ionizing voltages in mass spectral analysis of methyl- and phenylthiohydantoin amino acid derivatives, *Anal. Biochem.* **45**:176.

Suzuki, T., Song, K.-D., Itagaki, Y., and Tuzimura, K., 1976, Mass spectrometric identication of amino acid thiohydantoins, *Org. Mass Spectrom.* **11**:557.

Tobe, H., Morishima, H., Naganawa, H., Takita, T., Aoyagi, T., and Umezawa, H., 1979, Structure and chemical synthesis of amastatin, *Agr. Biol. Chem.* **43**:591.

Tsang, C. W., and Harrison, A. G., 1976, Chemical ionization of amino acids, *J. Am. Chem. Soc.* **98**:1301.

Ueno, T., Nakashima, T., Uemoto, M., Fukami, H., Lee, S.-N., and Izumiya, N., 1977, Mass spectrometry of *Alternaria mali* toxins and related cyclodepsipeptides, *Biomed. Mass Spectrom.* **4**:134.

Voigt, D., and Schmidt, J., 1978, Negative ion mass spectrometry of natural products. VIII—α-amino acids, *Biomed. Mass Spectrom.* **5**:44.

Waern, R., and Falter, H., 1978, An approach to the differentiation of leucine and isoleucine residues in EI mass spectra of peptides, *Biochem. Biophys. Res. Commun.* **81**:448.

Weber, R., and Levsen, K., 1980, Collision induced dissociation of field desorbed di- and tripeptides, *Biomed. Mass Spectrom.* **7**:314.

Wegmann, H., Curtius, H.-Ch., and Redweik, U., 1978, Selective ion monitoring of tryptophan, N-acetyltryptophan and kynurenine in human serum. Application to the *in vivo* measurement of tryptophan pyrrolase activity, *J. Chromatogr.* **158**:305.

Weinkam, R. J., 1978, Reactions of protonated diamino acids in the gas phase, *J. Org. Chem.* **43**:2581.

Wiecek, C., Halpern, B., Sargeson, A. M., 1979, The determination of steric purity of amines and amino acids by gas chromatography and mass spectrometry, *Org. Mass Spectrom.* **14**:281.

Wietzerbin, J., Das, C. D., Petit, J.-F., Lederer, E., Leyh-Bouille, M., and Ghuysen, J.-M., 1974, Occurrence of D-alanyl-(D)-*meso*-diaminopimelic acid and *meso*-diaminopimelyl-*meso*-diaminopimelic acid linkages in the peptidoglycan of *Mycobacteria, Biochemistry* **13**:3471.

Winkler, H. U., and Beckey, H. D., 1972, Field desorption mass spectrometry of amino acids, *Org. Mass Spectrom.* **6**:655.

Wipf, H. K., Irving, P., McCamish, M., Venkataraghavan, R., and McLafferty, F. W., 1973, Mass spectrometric studies of peptides. V. Determination of amino acid sequences in peptide mixtures by mass spectrometry, *J. Am. Chem. Soc.* **95**:3369.

Wolfensberger, M., Redwik, U., and Curtius, H.-Ch., 1979, Gas chromatography mass spectrometry and selected ion monitoring of the N,N'-dipentafluoropropionyl hexafluoroisopropyl ester of glutamine, *J. Chromatogr.* **172**:471.

Zanetta, Z. P., and Vincendon, G., 1973, GLC of the N(O)-heptafluorobutyryl isoamyl esters of amino acids. 1. Separation and quantitative determination of the constituent amino acids of proteins, *J. Chromatogr.* **76**:91.

Ziemer, J. N., Perone, S. P., Caprioli, R. M., and Seifert, W. E., 1979, Computerized pattern recognition applied to gas chromatrography–mass spectrometry identification of pentafluoropropionyl dipeptide methyl esters, *Anal. Chem.* **51**:1732.

PART III

APPLICATIONS

6

ANALYSIS OF VOLATILE METABOLITES IN IDENTIFICATION OF MICROBES AND DIAGNOSIS OF INFECTIOUS DISEASES

Lennart Larsson and Per-Anders Mårdh,
Department of Medical Microbiology, University of Lund, Sölvegatan 23, S-223 62 Lund, Sweden

Göran Odham, Laboratory of Ecological Chemistry, University of Lund, Ecology Building, Helgonavägen 5, S-223 62 Lund, Sweden

1. INTRODUCTION

The production of volatile compounds by microbes is dependent upon the environmental conditions, e.g., the chemical composition of the growth medium and whether respiration or fermentation is utilized in the microbial metabolism. Many microorganisms produce a number of volatile metabolites. Under controlled conditions of growth, characteristic combinations of volatiles are often formed by given genera, and sometimes also by given species. Quantitative identification of such microbial products should therefore provide information which is valuable both in the identification of microorganisms for diagnostic purposes and for taxonomic studies (Larsson and Mårdh, 1977).

Gas chromatography (GC) is one of the most specific and sensitive analytical techniques for studying volatile products of microbial metabolism. An additional advantage is that the preparatory steps are few and uncomplicated. A single extraction generally suffices for analysis. Extraction procedures, moreover, can be omitted if head-space techniques are used.

Several methods are now available for reliable and sensitive GC registration of organic and inorganic volatiles, the identities of which can be established by combining GC with mass spectrometry (MS). Use of a computerized mass spectrometer (Chapter 2) greatly simplifies the evaluation of molecular structures of such compounds.

This chapter deals with the use of GC/MS for analysis of microbial volatiles and of volatiles associated with infection-induced changes in body fluids. Although the emphasis is on microbes which are relevant in clinical microbiology, the described analytical methods are applicable in the study of microbial volatiles in general.

2. PREPARATION OF SPECIMENS FOR ANALYSIS OF MICROBIAL VOLATILES

Volatile metabolites produced by microorganisms include such diverse classes of compounds as fatty acids, alcohols, aldehydes, esters, ketones, amines, and sulfur-containing compounds. Among the variety of other products are diacetyl, acetoin, short-chain hydrocarbons, and permanent gases, i.e., hydrogen, oxygen, nitrogen, carbon monoxide, carbon dioxide and methane (Stotzky and Schenck, 1976).

GC and GC/MS analysis of volatile compounds is most conveniently done without prior derivatization. However, electron capture (EC) detection of suitable derivatives of the volatiles can be used when very high sensitivity of the analysis is vital (Chapter 1, also Section 3.2 of the present chapter). The following descriptions concern GC/MS analysis of underivatized microbial volatiles. The procedures include injection of aqueous solutions (spent culture medium) and solvent extracts thereof, "trapping" of volatiles followed by chemical or thermal desorption and various headspace injection techniques.

2.1. Direct Injection of Aqueous Solutions

When aqueous solution is directly injected onto a GC column for analysis of volatiles, the sizes of the peaks in the chromatograms reflect the true concentrations of the studied components, provided that correction is made for detector response. If a flame ionization (FI) detector is used, no solvent peak appears in the chromatogram. A disadvantage of the direct injection is that the volatiles are not concentrated prior to the analysis. Furthermore, the water which is simultaneously introduced onto the column may significantly lower the separation efficiency of the stationary phase, although certain solid polymers can be used as phases in such analyses

(Section 3.1). Aqueous solutions such as broth cultures may contain nonvolatile substances, including metabolites of high molecular weight. If such components are not removed beforehand, their injection can cause a relatively rapid degradation of the GC column (Rogosa and Love, 1968; Henkel 1971; Palo and Ilková, 1970). To overcome this disadvantage, the sample can be purified, e.g., by ultrafiltration and ion-exchange chromatography, prior to analysis (Section 4.1).

2.2. Solvent Extraction

Organic volatiles in aqueous medium can be more or less efficiently transferred to an organic solvent, leaving in the aqueous phase material that is non-relevant for the analysis. The distribution of a compound between two immiscible solvents is lawbound. Assuming that a compound A is allowed to distribute itself between water and an organic solvent, the resulting equilibrium may be expressed as

$$A_w \rightleftharpoons A_o$$

where the subscripts w and o refer to the water and the organic phase, respectively. The partition coefficient, K, can generally be expressed as the ratio of the solubility of A in the two solvents, viz.,

$$K = \frac{\text{Solubility of A in the organic phase}}{\text{Solubility of A in the aqueous phase}}$$

Generally speaking, the smaller and more polar the molecule, the lower will be the partition coefficient. For example, this coefficient in the system diethyl ether: water is 7.4 for butanol but only 0.14 for methanol, which means that the latter alcohol is not efficiently extracted with ether. When extracting, for instance, spent bacterial culture medium, such discrimination of certain microbial volatiles must be considered in the quantitative estimation of metabolites.

The partition coefficient can often be increased by adding an organic salt prior to extraction. Such a "salting-out" procedure decreases the solubility of organic compounds in water and consequently leads to more efficient extraction. Another requirement for efficient extraction is that the organic molecules are in a covalent state. Thus the solution must be acidified for "quantitative" extraction of fatty acids, and alkalized for extraction of underivatized amines. Selective extractions of organic acids and amines from broth cultures and body fluids have been reported (Edman *et al.*, 1981). In this technique the samples are first made acidic (pH 2) to allow acids to be extracted by chloroform. After increasing the pH of the aqueous phase to 10, amines can be extracted using the same (chloroform) solvent.

The choice of solvent for extraction must be made to suit the type of GC detector. Thus, halogen-containing solvents should be avoided when using EC detection. For example, a microliter of carbon tetrachloride or methylene chloride might render the EC detector useless for hours, or even days, because of overloading. Also in detection by thermal conductivity (TC) or FI it is advantageous to use a solvent with a low detector response, in order to ensure the registration of rapidly eluting metabolites which may be more or less efficiently concealed by a large solvent peak. FI detectors have low response to carbon tetrachloride, chloroform, and carbon disulfide, and consequently these solvents are frequently used.

2.3. Head-Space Analysis

In head-space gas chromatography (HSGC), the vapor phase above a liquid or a solid medium is analyzed. In the former case, the liquid sample is transferred to a glass vial, which is thereafter sealed with a rubber membrane (Figure 1). A sample of the atmosphere above the medium is collected with a gas-tight syringe via the membrane and is injected into the gas chromatograph. Alternatively, an automated procedure can be used.

In a state of equilibrium between a liquid and a gaseous phase, the area of a chromatographic peak obtained by head-space analysis of a compound, i, is proportional to its partial vapor pressure, viz.,

$$A_i = c_i p_i$$

where A_i is the peak area, c_i is a constant, depending on the type of GC detector, and p_i is the partial vapor pressure at the sampling temperature. By applying Henry's law,

$$p_i = x_i y_i p_i^0$$

where x_i is the mole fraction, y_i is the activity coefficient, and p_i^0 is the vapor pressure of the pure compound at sampling temperature, the following equation is obtained:

$$A_i = c_i x_i y_i p_i^0$$

The sensitivity of head-space analysis, viz., the magnitude of A_i for a given compound, can be improved by obtaining higher y_i or p_i^0. The activity coefficient (y_i) can be considerably increased by adding, for instance, electrolytes (salting-out effect, Table 1), and the vapor pressure (p_i^0) can

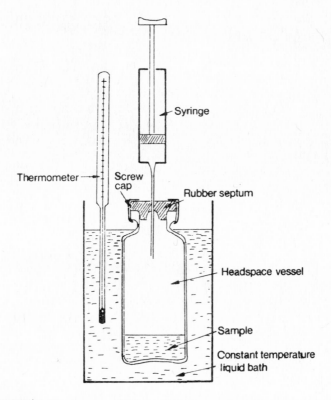

FIGURE 1. Head-space analysis vessel (Hachenberg and Schmidt, 1977).

TABLE 1. Increases in the Concentration of Ethanol in the Head-Space Due to the Addition of Different Salts at a Temperature of 60°C[a]

Salt	Increase in concentration in head-space
Ammonium sulfate	×5
Sodium chloride	×3
Potassium carbonate	×8
Ammonium chloride	×2
Sodium citrate	×5

[a] Hachenberg and Schmidt (1977).

be increased by raising the temperature of the samples according to the Clausius–Clapeyron equation. For further information on theoretical aspects of head-space analysis, the reader is referred to Hachenberg and Schmidt (1977).

Results of head-space analyses when sampling is manually performed, using a gas-tight syringe, are often difficult to reproduce. The syringe should be heated, so as to avoid condensation of volatiles on the cold inner glass surface of the syringe (Kepner *et al.*, 1964; Larsson *et al.*, 1978a). A temperature-stable waterbath may be used for heating the glass vials to the desired temperature.

Automated head-space injection is preferable, however. One such system for direct HSGC (Carlo Erba, Italy), uses a heated syringe which penetrates the membrane of the sealed glass vial and collects 0.1–2.5 ml of the head-space phase, which is then automatically injected onto the GC column. The automatic turntable of the instrument can hold up to 40 vials, the temperature of which can be controlled between 30 and 120°C.

In another type of instrument (Perkin-Elmer, West Germany), the pressure in the vials is increased up to the column head pressure, using preheated carrier gas. Upon injection, which is based on pressure equilibration, a given volume of head-space gas is passed through the sampling capillary into the injector (Kolb, 1976). In this system the heating of the glass vials in the turntable is to a maximum of 190°C. The analyses are made manually (Perkin-Elmer model HS6) or automatically with up to 30 vials (Perkin-Elmer model F45).

Recently, two instruments for automatic sampling from up to 100 (Perkin–Elmer model HS100) and 24 (Dani, Italy) head-space vials were developed. The latter system has the advantage that it can be connected to any gas chromatograph. All of the different systems for automatic head-space injection described above can be used for packed as well as for capillary columns.

For analysis of volatile compounds, HSGC offers several advantages over solvent extraction techniques. The head-space procedure is easy to use, since no extraction procedures are necessary. The absence of a solvent peak when FI detection is used precludes partial or complete concealment of rapidly eluting compounds by such a broad peak. This latter advantage is of major importance in analyses of highly volatile liquids and of gases. HSGC processes do not damage the column, since only volatile compounds are injected. Disadvantages of the technique may include difficulties in making accurate quantitative calculations. In quantitive work true equilibrium between the sample in the liquid and in the gaseous phase is essential. Care must be taken to minimize all risks of absorbance of volatiles into the rubber membrane sealing the vials, and also risk of extraction of organic compounds from the membranes. Teflon-lined membranes should preferably be used.

2.4. Adsorption Followed by Chemical or Thermal Desorption

In order to concentrate volatile organic compounds prior to GC analysis, such compounds can be adsorbed (trapped) and later released by chemical or thermal desorption. This technique can be valuable when trace amounts of volatiles are to be analyzed.

An aqueous solution can be pumped through a bead of adsorbent. In analyses of volatile organic compounds in air, the air can be sucked through a tube which contains the trapping substance. For desorption, a relatively small volume of organic solvent such as diethyl ether or benzyl alcohol can be added to the adsorbent, which is then extracted and analyzed by injection of the extract onto the GC column. When using thermal desorption, the adsorbent is placed directly in a precolumn in the gas chromatograph. By slowly raising the temperature in the injection port, the trapped material is gradually desorbed and transferred by the flow of carrier gas onto the column for chromatographic separation. With this latter technique there will be no solvent peak in the chromatogram, which means that rapidly eluting components can be detected. Thermal desorption can be used only for analysis of volatile compounds. With solvent extraction, also compounds with little or no volatility can be analyzed with GC after derivatization.

Numerous trapping substances are available. Charcoal is frequently preferred (Grob and Grob, 1971), although various polymers have also been employed (Zlatkis *et al.*, 1973; Ryan and Fritz, 1978). The preservation of substances trapped in an adsorbent must be carefully controlled up to the time of analysis, so as to minimize the risk of loss of substance and of further adsorption.

2.5. Other Methods

Distillation of volatiles from fermentation media prior to GC has been used. Doelle and Manderson (1969) reported an appreciable loss of volatile acids in distillation from such solutions, however. This method, moreover, is time-consuming, as are adsorption–desorption techniques, and therefore less suitable in large-scale determinations.

Organic volatiles may be cold-trapped in a precolumn connected to the GC separation column by cooling with, for instance, liquid nitrogen.

The volatiles are thereafter consecutively released in response to temperature programming. Different types of apparatus for cold-trapping are commercially available and can be used with a variety of gas chromatographs.

Another described method utilized a glass ampoule containing the substance (soil) to be studied (Norén and Odham, 1977). The ampoule is placed in a metal container, the temperature of which can be controlled. The ampoule is crushed with a small hammer and the stream of carrier gas is allowed to sweep over the soil sample into the column. This technique gave larger peaks and more complex chromatograms than when the same (soil) sample was analyzed according to the head-space principle, using manual syringe injection.

3. GC ANALYTIC CONDITIONS

In this section, chromatographic columns, stationary phases, and detectors are considered in relation to their ability to give good separation and sensitive registration of microbial volatiles.

3.1. Packing Materials. Columns

Numerous packing materials have been employed for separation of microbial volatiles. In general, solid stationary phases are used for study of volatiles of relatively low molecular weight, while liquid phases are employed to separate somewhat larger molecules. Many solid phases have good thermal stability. In contrast to liquid phases, however, the solid types cannot be used in capillary GC (Chapter 1).

Several solid phases are widely applicable for separating *gases* of microbial origin, including those containing sulfur and nitrogen. Activated alumina and carbon materials, molecular sieves, silica gels and organic porous polymers, e.g., Porapak N, are among these solid phases. *Permanent gases and lighter hydrocarbons* can be separated, for example, with Carbosieve—a high-purity carbon substance—with the aid of temperature programming. One method for isothermal separation of permanent gases requires use of two columns, one packed with a molecular sieve (5A) for separation of H_2, O_2, N_2, CH_4, and CO, in that order, and the other column with, e.g., Chromosorb 102, for separation of CO_2, which is irreversibly adsorbed onto a molecular sieve. Alternatively, the recently introduced "CTR" columns can be employed in which a 1/8 in (outer diameter) tube is surrounded by a 1/4 in tube (Figure 2). These tubes are packed with different materials, e.g., those mentioned for isothermal separation of permanent gases. Thus, they permit analysis of a sample on two different stationary phases simultaneously.

Free fatty acids of low molecular weight can be separated on several porous polymers, e.g., Porapak (Q, P, QS, and PS) and Chromosorb 101. When using a liquid stationary phase, care should be taken to ensure that the solid support has been suitably pretreated to give symmetric peaks, e.g., by incorporating small amounts of terephtalic acid or H_3PO_4 in the liquid phase. Carbowax 20 M (polyethyleneglycol) and modified versions of it such as SP-1000 (Supelco Inc.) and AT-1000 (Alltech Ass.), which may offer better thermal stability, are frequently employed. FFAP (free fatty acid phase) is a terephtalic acid-modified Carbowax 20 M. AT-1200 (Alltech) and SP 1200 (Supelco) are well suited for separation of fatty acids up to about C_6.

Diatomaceous earths are frequently employed as supports. However, graphitized carbons such as Carbopack (Supelco), which in itself can also act as a separating agent, offer excellent thermal stability and separation qualities when supplied with 0.2%–0.4% of the above-mentioned liquid phases. Graphpak (Alltech) is a similar graphitized carbon substance. A disadvantage of these supports is their comparatively high price.

Neutral volatile metabolites, e.g., *alcohols*, *aldehydes* and *ketones*, are readily separated on porous polymers such as the several Porapak resins and Chromosorb 101–103. Alternatives are SP-1000, AT-1000, Carbowax 20 M, and Carbowax 1500, the last-mentioned being most

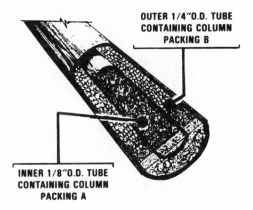

OUTER 1/4″O.D. TUBE
CONTAINING COLUMN
PACKING B

INNER 1/8″O.D. TUBE
CONTAINING COLUMN
PACKING A

FIGURE 2. A CTR column with two different packing materials (Alltech Ass. Catalog, 1980).

frequently employed in study of alcohols. Diatomaceous earths or graphitized carbons are used as supports. *Hydrocarbons* can be chromatographed with a variety of solid phases, for example Carbopack and Graphpak, supplemented with picric acid.

For separating *amines* of low molecular weight, Chromosorb 103 is generally a good choice. Tenax GC, a porous polymer developed for separating small polar compounds, can also be used. The "Pennwalt" packings and Carbowax 20 M, coated on diatomaceous earths or graphitized carbons, give excellent separation of most volatile amines. To avoid peak tailing, it is essential to incorporate KOH into the liquid phase.

Excellent and updated information on packing materials for GC separation of volatiles can be obtained from catalogs issued by the major suppliers of chromatographic equipment.

Use of *capillary columns*, particularly of fused silica type, for the study of microbial volatiles can be expected to expand rapidly in the coming years. Fused silica columns make capillary GC convenient and effective even in routine diagnostic work. Capillary columns have been used in analysis of soil volatiles (Norén and Odham, 1977), and they hold out high diagnostic possibilities as regards, for instance, anaerobes (Larsson and Holst, 1982).

3.2. Detectors

With few exceptions (formic acid, formaldehyde), the FI detector has higher sensitivity than the TC detector in analysis of organic compounds. The FI type therefore is widely used for registration of microbial organic volatiles. For determination of permanent gases, however, a TC detector must be used. The FI detector's lack of sensitivity to water can be advantageous, e.g., in head-space analyses and in direct analyses of aqueous culture media. In neither case will a solvent peak appear in the chromatograms. Sulfur-containing compounds are often registered with a flame photometric detector. Its sensitivity for sulfur compounds is about four times that for most other organic compounds.

Electron capture (EC) detection of microbial volatiles has been accomplished after derivatization with halogen-containing reagents. Metabolites present in amounts of as little as a few picograms can then be detected. Use of this technique for the rapid diagnosis of infectious diseases was reviewed by Edman *et al.*, (1981).

Mass spectrometric detection, i.e., selected ion monitoring (SIM), has so far rarely been used in studies on microbial volatiles. However, SIM is a highly sensitive and selective technique (Labows *et al.*, 1980; Weaver *et al.*, 1980). For details of SIM studies of microbial metabolites, see Chapter 9.

4. ANALYSIS OF MICROBIAL VOLATILES IN SPENT CULTURE MEDIA

Analysis of volatile metabolites by GC is widely used to characterize anaerobic bacteria. Volatiles produced by facultative anaerobes and by aerobes have also been studied, however. Analysis of solvent extracts of spent culture media is usually employed, although direct injection of such media onto the GC column and adsorption/desorption techniques have also been tested. GC head-space analysis for the study of bacterial volatiles in spent culture media has recently attracted interest (Larsson, 1982).

4.1. Direct Injection of Culture Media

Direct injection of spent carbohydrate-containing medium onto the GC column for analysis of volatile microbial alcohols and fatty acids was described by Rogosa and Love (1968) and by Henkel (1971). Cells and cell fragments were first removed by centrifugation or filtration in order to avoid "ghost" peaks and instability of the base line. After such procedures, a satisfactory column stability was reported, although relatively large amounts of nonvolatile material, present in the medium, were simultaneously injected (Rogosa and Love, 1968).

This methodological approach was used in studies on a number of bacterial species cultured in liquid media, usually acidified with sulfuric acid and centrifuged and/or filtrated prior to analysis. Yoshioka *et al.*, (1969), Deacon *et al.*, (1978), and Palo and Ilková (1970) recommended regular repacking of the stationary phase in the column inlet, so as to remove nonvolatile compounds that accumulate after repeated injections.

The genera studied with this technique include *Bacteroides*, *Clostridium*, *Fusobacterium*, *Klebsiella*, *Propionibacterium*, *Pseudomonas*, *Staphylococcus*, and *Streptococcus*. Studies have also been made of *Escherichia coli*. The main volatiles registered by FI detection and identified by comparative GC against the retention times of known compounds were short-chain alcohols, fatty acids, acetoin, and diacetyl. As a rule the GC method enabled differentiation of organisms as to genus, more rarely as to species (Bricknell and Finegold, 1973; Yoshioka *et al.*, 1969; Deacon *et al.*, 1978). A study on the volatiles produced by some species isolated from cows with mastitis was made by Lategan *et al.*, (1978).

A stationary phase consisting solely of a solid porous polymer is frequently employed when aqueous solutions are injected onto a GC column. Such polymers are stable against water. Bricknell and Finegold (1973), however, reported satisfactory results from analyses using a column packed with Porapak Q, coated with 6% FFAP. These authors concluded that the direct injection of acidified supernatant culture was preferable to analysis of ether extracts in the identification of anerobes, provided that the detector was of FI (not TC) type (Bricknell et al., 1975).

In order to avoid rapid deterioration of the column when culture media are injected, the samples can be prepurified. This has been done by applying the supernatants on a packed cation exchange resin, AG 50W-X4, 200–400 mesh, hydrogen form (Sasaki and Takazoe, 1979; Carlsson, 1973). Such purification considerably improved the quality of the chromatograms and also led to a longer lifespan.

4.2. Solvent Extraction

Analysis of underivatized volatile fatty acids by GC after solvent extraction of broth cultures is a standard procedure in identifying anaerobic bacteria in many laboratories. Many anaerobes produce relatively large amounts of short-chain fatty acids, the identification and quantification of which can be of appreciable value in differential diagnosis. GC alone can usually identify anaerobes as to genus (Chatot et al., 1969; Moore et al., 1966; Werner and Hamman, 1980). To establish the species identity of most anaerobic bacteria, however, GC systems with higher differential diagnostic capacity must be developed. Facultative anaerobes and aerobes typically produce relatively small amounts of short-chain organic acids (Parilla et al., 1981) and consequently are less frequently subjected to GC analysis with this method.

Obligate anaerobic bacteria usually are cultured in a peptone-yeast extract medium, sometimes supplemented with glucose (1% w/w) for 2 to 4 days. After acidification to ca. pH 2, using dilute sulfuric acid, the broth cultures are extracted with an organic solvent such as diethyl ether. To increase the efficiency of the extraction, an inorganic salt can be added to give a salting-out effect. Traces of water in the ether solution can be conveniently removed by adding a small amount of, for example, magnesium sulfate to the extract. For the chromatographic analysis an acidic, polar stationary phase is used. As an internal standard, 2-methylpentanoic acid has been suggested (Mayhew and Gorbach, 1977). In samples found to contain no volatile acids, or only acetic and/or formic acid, methylation is done to derivatize nonvolatile fatty acids. Extensive information on GC analysis of solvent extracts from broth cultures of anaerobic bacteria can be obtained from the Anaerobe Laboratory Manual (Holdeman et al., 1977).

Updating of the methodology with respect to preparatory procedures and chromatographic conditions, leading to more rapid and informative results, has been advocated by several authors (Hauser and Zabransky, 1975; Bohannon et al., 1978; Morin and Paquette, 1980; Rizzo, 1980).

Extensive studies with these methods, using large numbers of anaerobic strains, have not yet been published, however. Use of HSGC analysis is another recent and highly promising approach to the study of volatiles produced by anaerobes. Section 4.3 deals with some diagnostic applications.

Analysis of solvent extracts has also been used to detect anaerobes in blood cultures, i.e., in mixtures of broth media and blood from patients with assumed bacteremia or septicemia. The cultures were incubated at 37°C until turbidity of the medium indicated bacterial growth. Samples were then collected, acidified, extracted, and analyzed. On the basis of peaks representing short-chain fatty acids, particularly butyric and isovaleric acids, presence of anaerobes could be established with a high degree of accuracy (Reig *et al.*, 1981). TC (Sondag *et al.*, 1980; Edson *et al.*, 1982) and FI (Wüst, 1977; Reig *et al.*, 1981) detection have been used. The chromatograms sometimes provide information on the genus identity of the causal anaerobic organisms (Wüst, 1977; Sondag *et al.*, 1980). However, facultative anaerobes and aerobes are not usually detectable with these techniques. HSGC (Section 4.3) is superior for this latter purpose.

Phillips and Rogers (1981) used a selective medium supplemented with *p*-hydroxy phenyl acetic acid to study *Clostridium difficile*. GC registration of *p*-cresol in analyses of ether extracts from such broth cultures indicated *C. difficile*. The authors concluded that their technique would be useful for rapid diagnosis of antibiotic-associated colitis; *p*-cresol is responsible for the characteristic odor of *C. difficile*. The study further showed, however, that also *C. scatologenes* produced *p*-cresol.

Aerobes and facultative anaerobes have also been studied. With the aid of chromatograms representing acetoin, diacetyl, and short-chain alcohols and fatty acids, Henis *et al.*, (1966) were able to differentiate strains of *E. coli*, *Aerobacter aerogenes*, *P. aeruginosa*, and some species of the genus *Bacillus*. Both FI and EC were used in detection. For FI analysis a concentrated extract was required, whereas with EC unconcentrated or even diluted extracts were used. Presence of diacetyl and of acetoin could be demonstrated in amounts as small as 5 and 20 pg, respectively, with EC detection. These two components were studied also by Lee and Drucker (1975).

Mitruka and Alexander (1968) analyzed ether extracts of acidified broth media which had been incubated with human pathogenic species. The EC detector demonstrated most metabolites with higher sensitivity than the FI detector, though the reverse was true in some cases. EC permitted identification of some bacterial metabolites in the culture medium after only two hours of incubation.

Cox and Parker (1979) isolated a compound responsible for the typical grape like odor of *P. aeruginosa*. They used preparatory GC of ether extracts from spent culture medium. The isolated compound, which was identified by MS as 2-aminoacetophenone (Figure 3), was suggested as a marker for rapid detection of *P. aeruginosa* in short-term cultures.

A commonly encountered problem in GC analysis of microbial metabolites is that the culture medium as such often contains appreciable amounts of volatiles which can interfere with the evaluation of bacterial products. One way of circumventing this problem was proposed by O'Brien (1967). The bacteria were cultured in trypticase Soy broth for 12 to 16 hours at 25 or 35°C, depending upon their optimum temperature. Thereafter the bacterial cells were thoroughly washed in saline and incubated in a buffer containing 2% glucose for 4 h. Volatile compounds produced by catabolism of glucose were extracted with diethyl ether and analyzed by GC. The chromatographic patterns obtained when culturing the same strain in different media such as brain–heart infusion, trypticase soy broth, or peptone yeast glucose broth were virtually identical. Chromatograms representing the glucose-containing buffer itself contained no or negligible peaks.

Levanon *et al.*, (1977) were able to distinguish between virus-infected tissue culture cells and controls before a virus-induced cytopathogenic

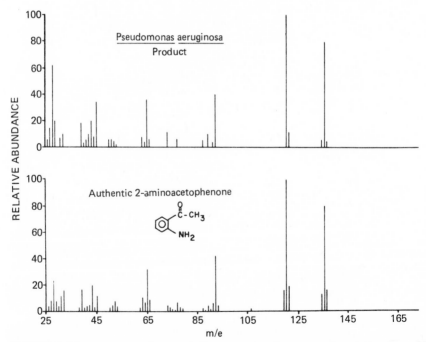

FIGURE 3. Mass spectrum of a metabolite possessing the grape odor of *Pseudomonas aeruginosa*, and of authentic 2-aminoacetophenone (Cox and Parker, 1979).

effect was demonstrable at microscopy (Figure 4). Ethereal extracts of acidified supernatants were analyzed after varying periods of incubation. An FI detector was used. MS identification of the distinguishing peaks was not done. The GC patterns permitted distinction between cells infected with encephalomyocarditis virus, poliomyelitis virus, ECHO and Togavirus.

Some workers have preferred to convert volatile bacterial metabolites into various derivatives prior to GC analysis. Such procedures can improve the chromatographic separation with the available stationary phase. Thus, short-chain fatty acids of *Pseudomonas* (Moss and Dees, 1976), *Peptococcus*, and *Peptostreptococcus* (Lambert and Armfield, 1979) were chromatographed as butyl esters on a relatively nonpolar stationary phase. Amines produced by *Streptococcus* species (Lambert and Armfield, 1979), strains of *Clostridium* (Brooks and Moore, 1969), *Neisseria meningitidis, Klebsiella pneumoniae, Diplococcus pneumoniae,* and *E. coli* (Brooks *et al.,* 1980b) were analyzed as trifluoroacetyl or heptafluorbutyryl derivatives. Such derivatization facilitates the detection of amines of relatively high molecular weight, including phenylethylamine, spermine, spermidine, tryptamine, and

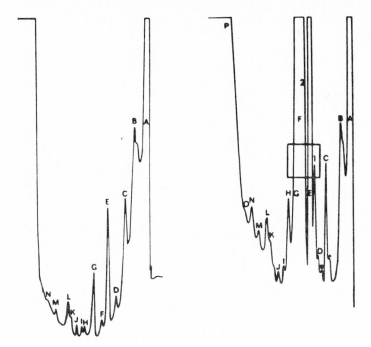

FIGURE 4. Chromatograms representing etheral extracts of uninfected BHK-21 cells (left) and of cells infected by EMC virus, 6 h post-infection (right). Time scale: Right to left (Levanon *et al.,* 1977).

tyramine. These derivatives, moreover, are well suited for EC detection. Fatty acids and alcohols produced by several bacterial genera have also been converted to electron-capturing derivatives in order to increase sensitivity (Brooks *et al.*, 1976, 1980a).

To enhance sensitivity in GC detection of bacterial metabolites in culture media, Mitruka and Alexander (1969, 1972) added a chloro- and bromo-substituted organic acid to the media. After incubation, the media were extracted with diethyl ether and the halogen-containing catabolic products were analyzed, using a gas chromatograph equipped with EC detector. With this technique the sensitivity was increased several thousand-fold as compared to FI detection.

4.3. Direct Head-Space Analysis

HSGC for the detection of microbial metabolites was first applied in dairy research in the mid-1960s. Bawdon and Bassette (1966) studied volatile products of bacteria after 30 h of incubation in milk which previously had been heated and vacuum-distilled to remove volatile compounds. Using a gas-tight syringe, 1 ml of head-space vapor was injected onto the GC column. Characteristic chromatograms were obtained for the various species studied, including *E. coli* and *A. aerogenes*. Pierami and Stevenson (1976) reported similar experience from studies of *Alcaligenes*, *Pseudomonas*, and *Bacillus* microorganisms. Guarino and Kramer (1969) studied various foods infected with *Enterobacter aerogenes* and *Salmonella typhimurium*. Infected material was inoculated into a liquid medium and incubated at 36°C for 18 h, after which HSGC was performed, using manual syringe injection. The chromatographic patterns were readily distinguishable from those representing noninfected foods.

Norrman (1971) analyzed a variety of volatiles—mainly esters and alcohols—produced by a yeast fungus. Harper and Gibbs (1979) found more uncommon metabolites, viz., nitriles and isobutyraldoxime O-methyl ether (Figure 5) in studies on *Aeromonas* and *Moraxella* spp. In both of these studies the technique consisted of manual injection of head-space vapors from above broth cultures, using a gas-tight syringe, and FI detection.

The potentialities of HSGC for study of volatile amines, which are characteristic products of anaerobes (and which account for the often distinctive odor of various species of anaerobic bacteria) have been reported. If acidic and alkaline, as well as neutral, microbial volatiles are to be analyzed, the gas chromatograph should preferably be equipped with facilities for dual column operation, since alkaline and acidic volatiles cannot be efficiently separated on the same stationary phase. Two solid

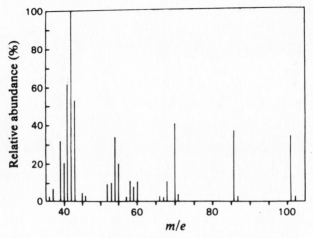

FIGURE 5. Mass spectrum of the O-methyl ether of isobutyraldoxime, produced by *Aeromonas* and *Moraxella* spp (Harper and Gibbs, 1979).

porous polymers, viz., Chromosorb 103 and Porapak Q, were employed for separation of alcohols, fatty acids, and amines produced by *Clostridium septicum*, *Klebsiella pneumoniae*, and *Proteus mirabilis* (Larsson *et al.*, 1978b).

Head-space analysis has been successfully used in the identification of anaerobes. Volatile fatty acids produced by *Clostridium* were studied by Drasar *et al.*, (1976), who used an automated HSGC analyzer. HSGC has greater ability in differential diagnosis than solvent extraction of broth cultures. This superiority is associated with the capacity of HSGC to utilize rapidly eluting compounds such as short-chain alcohols (Larsson *et al.*, 1978a). The advantage of HSGC was further demonstrated in large-scale studies on anaerobes, using an automated HSGC analyzer and a fused silica capillary column (Larsson and Holst, 1982). For example, *C. difficile* could be distinguished from *C. sporogenes* by the latter's production of HSGC-detectable short-chain alcohols (Figure 6). Such differentiation could not be made in analysis of ether extracts, using a packed column (Larsson *et al.*, 1980a). Automated HSGC with FI detection therefore constitutes an excellent alternative to study of solvent extracts in identification of anaerobes.

The differential diagnostic ability of HSGC was further demonstrated (Larsson *et al.*, 1980b) in studies on certain gram-negative rods which are frequently encountered in infections of the urinary tract. Manual, semiautomated and automated techniques of head-space injection were compared. The bacteria were cultured overnight on blood-agar plates and transferred

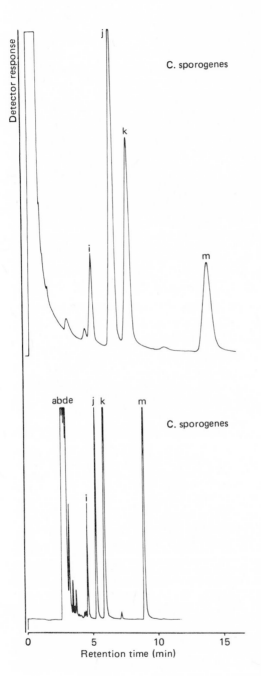

FIGURE 6. Chromatograms representing alcohols (C_2–C_5, a–e) and fatty acids (C_4–C_6, i–m) obtained by analysis of ether extracts using of packed column (upper), and automated headspace analysis of a fused silica capillary column (lower) of broth cultures of *Clostridium sporogenes* (Larsson and Holst, 1982).

to glass ampoules containing buffer solution supplemented with glucose (1% w/w). After incubation for 1 h at 37°C, the head-space vapor was analyzed. *E. coli, K. pneumoniae, Proteus* spp, and other uropathogens were readily differentiated. The study emphasized the advantages of automation in head-space injection and computerized handling of the chromatographic results. Both of these factors are important for application of HSGC in routine diagnostic work. Particularly interesting was the observation that the chromatographic "background," i.e., the sterile buffer, showed no or only very small peaks.

This approach, with study of products from bacterial catabolism of compounds in a defined substrate, has been further developed in studies with automated capillary HSGC, and results evaluated by a small computer (Larsson *et al.*, unpublished). As an example, Figure 7 shows a chromatogram obtained after 1 h incubation of *P. mirabilis* and also its computerized identification. Blood cultures from patients assumed to have septicemia were also studied (Larsson *et al.*, 1982). When occurrence of turbidity indicated bacterial growth, 1 ml portions were transferred to glass ampoules fitted in the analyzer's automated turntable. Of 196 studied specimens, 176 were correctly classified by HSGC as either positive or negative. There were thus 20 false negative results, but no false positives. Of the 20 discordant results, 14 represented staphylococcal species. The method did not detect *P. aeruginosa* or *Campylobacter fetus*—altogether

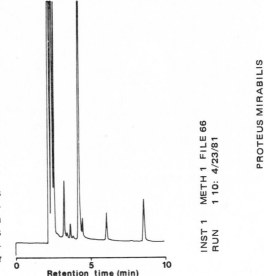

FIGURE 7. Volatile products from amino acid catabolism of *Proteus mirabilis* after 1 h incubation using automated head-space gas chromatography followed by computerized identification (Larsson *et al.*, unpublished results).

three cases. On the other hand, the head-space analysis permitted FI detection of most gram-negative rods such as *E. coli*, various coliforms, *Proteus* and *Salmonella* spp, and of anaerobes (Figure 8).

Rapid screening of urine specimens to detect infections has been done with HSGC (Hayward 1977a, b; Hayward and Jeavons, 1977; Coloe, 1978a, b). The technique involved addition of a carbohydrate and/or amino acid-supplemented growth medium to the urine specimen directly or to urinary sediment obtained by centrifugation. The analyses could be made

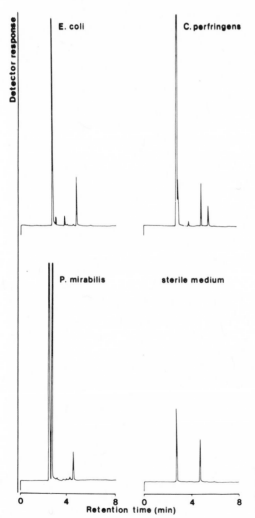

FIGURE 8. Representative chromatograms obtained by head-space analysis of blood cultures (Larsson et al., 1982).

after only a few hours of incubation. Addition of antibiotic to a bacteria-containing urine sample further provided possibilities for studying the antibiotic susceptibility of the infecting bacterium (a). These studies are described in detail in Chapter 7.

4.4. Other Methods

Volatile amine metabolites of *Streptococcus lactis* were studied by Golovnya *et al.* (1969). The volatiles were trapped in a solution of trichloroacetic acid in absolute ethanol, treated with potassium hydroxide and distilled before absorption in hydrogen chloride. After evaporation of the solution, the hydrochloric salts were dried under vacuum and the free bases were analyzed. The chromatograms showed several peaks representing volatile amines.

Volatile metabolites of *Pseudomonas* were concentrated by trapping in tubing containing Tenax GC and then thermally desorbed in the GC injection port (Labows *et al.*, 1980). Series of peaks representing odd-carbon methyl ketones were found in the chromatograms. All of the studied *Pseudomonas* species produced 2-nonanone and 2-undecanone, which were registered by FI and SIM (Figure 9), while 2-aminoacetophenone was

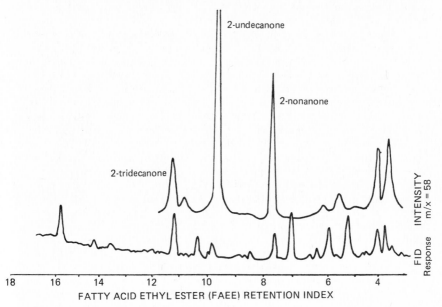

FIGURE 9. Mass spectrometric and flame ionization detection of volatile ketones of *Pseudomonas aeruginosa* (Labows *et al.*, 1980).

produced exclusively by *P. aeruginosa*. All metabolites were identified by MS.

Compounds characteristic of infection with *C. botulinum* in trout were demonstrated at HSGC by Snygg *et al.* (1979). The fish samples were incubated in sealed plastic bags for 7 days at 30°C prior to analysis of head-space vapors, which were cold-trapped in the injection system of the gas chromatograph. The complex chromatographic patterns obtained with a 180-m capillary column were interpreted by a computer and further compared with chromatograms representing trouts experimentally infected with *C. botulinum*. Characteristic combinations of chromatographic peaks were obtained in the tissue infected with *C. botulinum*.

5. DIRECT ANALYSIS OF VOLATILES IN BODY FLUIDS

GC and GC/MS have been used to analyze body fluids from infected hosts, in order to study volatiles associated with the presence of certain microbes. Such analysis may, in some cases, establish the etiology of an infection before results of bacteriological culture are available.

5.1. Direct Injection of Body Fluids

Urine specimens were analyzed after passage through an ultrafiltration membrane for removal of particles and compounds of high molecular weight, followed by shaking with an ion-exchange resin to remove cations (Barrett *et al.*, 1978). The samples were concentrated under a stream of nitrogen and injected onto a 1.5-m GC column. Various porous polymers were used as stationary phase. Acetic acid was detected in urine containing *E. cloacae*, *E. coli*, *Staphylococcus albus*, *Streptococcus faecalis*, *Klebsiella*, or *Proteus* species, but usually only when the bacterial count in the urine exceeded 10^6/ml. *Pseudomonas* and *Candida* species yielded no diagnostically significant peak. The authors concluded that their method for direct analysis of urine was too insensitive for the detection of bacteriuria. Solvent extraction of the specimens did not enhance the sensitivity of the analyses.

Acidified synovial fluids from patients with exudative arthritis were submitted to direct GC analysis (Brook *et al.*, 1980). Two compounds, with retention times corresponding to *n*-valeric and *n*-hexanoic acids, were significantly increased in patients with septic arthritis, but not in those with noninfectious inflammatory or degenerative arthritis. Elevated levels of lactic acid were also observed in septic arthritis, though with the exception of gonococcal arthritis. Using the same injection technique, however, Seifert *et al.* (1978) found high lactic acid concentrations in culture-positive joint fluids and also in two patients with gonorrhea-associated arthritis which were culture-negative for gonococci.

Direct injection of acidified spinal fluid from patients with purulent meningitis was performed by Brook (1979). Four chromatographic peaks of relatively short retention times were significantly increased in these cases as compared with spinal fluid from patients with aseptic meningitis or from controls. As the patients recovered from septic meningitis, the peak areas of the four unidentified components gradually fell to "normal" levels.

Pus specimens containing anaerobes were checked for presence of short-chain fatty acids by direct injection onto GC column packed with Chromosorb 101. Although a small number of false positive and false negative results were obtained, the authors (Phillips *et al.*, 1980) concluded that this simple GC technique is useful for rapid presumptive diagnosis of anaerobic infections.

Gravett *et al.* (1982) recently described that GC can be used as a sensitive (95%) means to detect amniotic fluid infection by registration of acetic, propionic, butyric, and succinic acids in such fluids. In the test system only lactic acid was found in amniotic fluid of healthy women of various gestational length. By other, more sensitive techniques, e.g., GC/MS, approximately 30 different organic acids have been detected in normal amniotic fluid (Nicholls *et al.*, 1976; 1978). These acids stem from the metabolism of the fetus. In all reports mentioned in this section, the detector was of FI type.

5.2. Solvent Extraction

GC analysis of volatile compounds in body fluids has mainly included studies on fatty acids of low molecular weight that are associated with the presence of obligate anaerobes. FI detection is generally used for this purpose. Thus, Gorbach *et al.* (1976), Nord (1977), Watt *et al.* (1982b), Ladas *et al.* (1979), and Phillips *et al.* (1976) extracted acidified pus and/or serous fluid with diethyl ether. Typically, peaks representing short-chain fatty acids indicated infection with anaerobes. More informative chromatograms may be obtained by methylating the fatty acids prior to analysis. In this way, nonvolatile acids, e.g., lactic and succinic acids, also are rendered detectable. Species identity of observed anaerobic organisms cannot as a rule be established with this GC method (Watt *et al.*, 1982a). Similar experience was reported by Thadepalli and Gangopadhyay (1980) in studies of empyema fluids from which anaerobes were recovered, and by Gray *et al.* (1980), who analyzed ether extracts of acidified sputum from patients with anaerobic lung infections. The chromatograms in the last-mentioned study were found to contain peaks of short-chain fatty acids. Sputa from 50 patients with chronic bronchitis did not contain such acids.

Sterile swabs were immersed in pus and then stored at room temperature for 2 h to simulate delay between sample collection and analysis (Reed and Sanderson, 1979). Diethyl ether extracts from the swabs contained short-chain fatty acids in cases of anaerobic infection. The small amounts of fatty acids present in the sterile swabs did not interfere with the analyses.

Volatiles in ether extracts of serum collected from experimentally virus-infected horses were studied by Mitruka *et al.* (1968). They used an EC detector. Changes in the chromatograms could be observed before symptoms of the infection were manifest. The possibility that the method may be useful for early diagnosis of viral infections remains to be explored.

While anaerobic infections have been diagnosed by analysis of volatile metabolites, most of the reports dealing with recognition of infection by GC or GC/MS analysis of body fluids have included studies on derivatized compounds. Elevated levels of lactic acid, shown by GC as methyl ester and indicative of meningitis, septic arthritis, peritonitis, and pleural infection (Brook, 1981) and EC detection of derivatized metabolites such as alcohols, amines, and fatty acids (Edman *et al.*, 1981), represent well-documented GC methods for rapid diagnosis of infectious diseases in body fluid analyses. The last-mentioned technique can not only recognize presence of infection, but also distinguish causal agents.

5.3. Direct Head-Space Analysis

This analytical technique has so far seldom been used to study clinical specimens. In pus from abscesses studied with HSGC and FI detection, Taylor (1980) and Mårdh *et al.* (1981) demonstrated short-chain fatty acids in cases of anaerobic infection. These observations were thus in analogy with results obtained from direct injection or extraction (Sections 5.1 and 5.2).

Using automated capillary HSGC, samples collected from the rectouterine cul-de-sac were studied (Larsson *et al.*, 1980c). Short-chain fatty acids were found in samples from women with peritoneal abscesses caused by anaerobic bacteria, but not in those from women with acute salpingitis of other microbial etiology. The potentiality of HSGC to contribute to the diagnosis of infectious diseases by detecting compounds with short retention times requires further evaluation.

5.4. Other Methods

Volatiles in sera from virus-infected humans were studied by Zlatkis *et al.* (1979). Serum specimens, in aliquots as small as 70 µl, were deposited on a Porasil E microcolumn. The volatiles were stripped from the samples by 2-chloropropane and transferred in vapor phase to a glass-bead collection column. They were then thermally desorbed and analyzed on a capillary column. Computerized interpretation of the chromatographic profiles revealed predictability of 85.7%, i.e., sera correctly classified as normal or virus-infected.

REFERENCES

Alltech Associate Chromatography Products Catalog, Deerfield, Illinois, 1980, Vol. **35**, p. 28.

Barrett, E., Lynam, G., and Trustey, S., 1978, Gas–liquid chromatography for detection of bacteriuria: Examination for volatile acidic and neutral compounds, *J. Clin. Pathol.* **31**:859.

Bawdon, R. E., and Bassette, R., 1966, Differentiation of *Escherichia coli* and *Aerobacter aerogenes* by gas–liquid chromatography, *J. Dairy Sci.* **49**:624.

Bohannon, T. E., Manius, G., Mamaril, F., and Li Wen, L.-F., 1978, Quantitative methods for the gas chromatographic characterization of acidic fermentation by-products of anaerobic bacteria, *J. Chromatogr. Sci.* **16**:28.

Bricknell, K. S., and Finegold, S. M., 1973, A simple, rapid method to process and assay fatty acids and alcohols by gas chromatography, *Anal. Biochem.* **51**:23.

Bricknell, K. S., Sutter, V. L., and Finegold, S. M., 1975, "Detection and Identification of Anaerobic Bacteria," *Gas Chromatographic Applications in Microbiology and Medicine* Mitruka, B. J., ed., John Wiley and Sons, New York.

Brook, I., 1979, Abnormalities in spinal fluid detected by gas–liquid chromatography in meningitis patients, *Chromatographia.* **12**:583.

Brook, I., 1981, The importance of lactic acid levels in body fluids in the detection of bacterial infections, *Rev. Infect. Dis.* **3**:470.

Brook, I., Reza, M. J., Bricknell, K. S., Bluestone, R., and Finegold, S. M., 1980, Abnormalities in synovial fluid of patients with septic arthritis detected by gas–liquid chromatography, *Ann. Rheum. Dis.* **39**:168.

Brooks, J. B., and Moore, W. E. C., 1969, Gas chromatographic analysis of amines and other compounds produced by several species of *Clostridium, Can. J. Microbiol.* **15**:1433.

Brooks, J. B., Selin, M. J., and Alley, C. C., 1976. Electron capture gas chromatography study of the acid and alcohol products of *Clostridium septicum* and *Clostridium chauvoei, J. Clin. Microbiol.* **3**:180.

Brooks, J. B., Kellogg, D. S., Shepherd, M. E., and Alley, C. C., 1980a, Rapid differentiation of the major causative agents of bacterial meningitis by use of frequency-pulsed electron capture gas–liquid chromatography: Analysis of acids, *J. Clin. Microbiol.* **11**:45.

Brooks, J. B., Kellogg, D. S., Shepherd, M. E., and Alley, C. C., 1980b, Rapid differentiation of the major causative agents of bacterial meningitis by use of frequency-pulsed electron capture gas–liquid chromatography: Analysis of amines, *J. Clin. Microbiol.* **11**:52.

Carlsson, J., 1973, Simplified gas chromatographic procedure for identification of bacterial metabolic products, *Appl. Microbiol.* **25**:287.

Chatot, G., Murgat, J. P., and Fontagnes, R., 1969, Étude chimiotaxonomique de 17 souches de bactéries anaérobies par chromatographie en phase gazeuse, *Ann. Inst. Pasteur.* **117**:556.

Coloe, P. J., 1978a, Head-space gas–liquid chromatography for rapid detection of *Escherichia coli* and *Proteus mirabilis* in urine, *J. Clin. Pathol.* **31**:365.

Coloe, P. J., 1978b, Ethanol formed from arabinose: A rapid method for detecting *Escherichia coli*, *J. Clin. Pathol.* **31**:361.

Cox, C. D., and Parker, J., 1979, Use of 2-aminoacetophenone production in identification of *Pseudomonas aeruginosa*, *J. Clin. Microbiol.* **9**:479.

Deacon, A. G., Duerden, B. I., and Holbrook, W. P., 1978, Gas–liquid chromatographic analysis of metabolic products in the identification of *Bacteroidaceae* of clinical interest, *J. Med. Microbiol.* **11**:81.

Doelle, H. W., and Manderson, G. J., 1969, Preparation of extracts of culture liquids for gas-chromatographic determination of acidic fermentation products, *Antonie van Leeuwenhoek* **35**:467.

Drasar, B. S., Goddard, P., Heaton, S., Peach, S., and West, B., 1976, Clostridia isolated from faeces, *J. Med. Microbiol.* **9**:63.

Edman, D. C., Craven, R. B., and Brooks, J. B., 1981, Gas chromatography in the identification of microorganisms and diagnosis of infectious diseases, *CRC Crit. Rev. Clin. Labor. Sci.* 133.

Edson, R. S., Rosenblatt, J. E., Washington II, J. A., and Stewart, J. B., 1982, Gas–liquid chromatography of positive blood cultures for rapid presumptive diagnosis of anaerobic bacteremia, *J. Clin. Microbiol.* **15**:1059.

Golovnya, R. V., Zhuravleva, I. L., and Kharatyan, S. G., 1969, Gas chromatographic analysis of amines in volatile substances of *Streptococcus lactis*, *J. Chromatogr.* **44**:262.

Gorbach, S. L., Mayhew, J. W., Bartlett, J. G., Thadepalli, H., and Onderdonk, A. B., 1976, Rapid diagnosis of anaerobic infections by direct gas–liquid chromatography of clinical specimens, *J. Clin. Invest.* **57**:478.

Gravett, M. G., Eschenbach, D. A., Spiegel-Brown, C. A., and Holmes, K. K., 1982, Rapid diagnosis of amniotic-fluid infection by gas–liquid chromatography, *N. Engl. J. Med.* **306**:725.

Gray, J., Perks, W. H., and Birch, J., 1980, Direct gas–liquid chromatography of clinical material as an aid in the diagnosis of anaerobic lung infections, *Curr. Chemoth. Infect. Dis.* **1**:540.

Grob, K., and Grob, G., 1971, Gas–liquid chromatographic–mass-spectrometric investigation of C_6–C_{20} organic compounds in an urban atmosphere. An application of ultra trace analysis on capillary columns, *J. Chromatogr.* **62**:1.

Guarino, P. A., and Kramer, A., 1969, Gas chromatographic analysis of head-space vapours to identify microorganisms in foods, *J. Food Sci.* **34**:31.

Hachenberg, H., and Schmidt, A. P., 1977, *Gas Chromatographic Head-Space Analysis*, Heyden and Son Ltd, London.

Harper, D. B., and Gibbs, P. A., 1979, Identification of isobutyronitrile and isobutyraldoxime *O*-methyl ether as volatile microbial catabolites of valine, *Biochem. J.* **182**:609.

Hauser, K. J., and Zabransky, R. J., 1975, Modification of the gas–liquid chromatography procedure and evaluation of a new column packing material for the identification of anaerobic bacteria, *J. Clin. Microbiol.* **2**:1.

Hayward, N. J., and Jeavons, T. H., 1977, Assessment of technique for rapid detection of *Escherichia coli* and *Proteus* species in urine by head-space gas–liquid chromatography, *J. Clin. Microbiol.* **6**:202.

Hayward, N. J., Jeavons, T. H., Nicholson, A. J. C., and Thornton, A. G., 1977a, Methyl mercaptan and dimethyl disulfide production from methionine by *Proteus* species detected by head-space gas–liquid chromatography, *J. Clin. Microbiol.* **6**:187.

Hayward, N. J., Jeavons, T. H., Nicholson, A. J. C., and Thornton, A. G., 1977b, Development of specific tests for rapid detection of *Escherichia coli* and all species of *Proteus* in urine, *J. Clin. Microbiol.* **6**:195.

Henis, Y., Gould, J. R., and Alexander, M., 1966, Detection and identification of bacteria by gas chromatography, *Appl. Microbiol.* **14**:513.

Henkel, H. G., 1971, Gas chromatographic analysis of low boiling fatty acids in biological media, *J. Chromatogr.* **58**:201.

Holdeman, L. V., Cato, E. P., and Moore, W. E. C., 1977, *Anaerobe Laboratory Manual*, 4th edn., Blacksburg, V. P. I. and S. U. Anaerobe Laboratory.

Kepner, R. E., Maarse, H., and Strating, J., 1964, Gas chromatographic head-space technique for the quantitative determination of volatile components in multicomponent aqueous solutions, *Anal. Chem.* **36**:77.

Kolb, B., 1976, Application of an automated head-space procedure for trace analysis by gas chromatography, *J. Chromatogr.* **122**:553.

Labows, J. N., McGinley, K. J., Webster, G. F., and Leyden, J. J., 1980, Head-space analysis of volatile metabolites of *Pseudomonas aeruginosa* and related species by gas chromatography–mass spectrometry, *J. Clin. Microbiol.* **12**:521.

Ladas, S., Arapakis, G., Malamou-Ladas, H., Palikaris, G., and Arseni, A., 1979, Rapid diagnosis of anaerobic infections by gas–liquid chromatography, *J. Clin. Pathol.* **32**:1163.

Lambert, M. A., and Armfield, A. Y., 1979, Differentiation of *Peptococcus* and *Peptostreptococcus* by gas–liquid chromatography of cellular fatty acids and metabolic products, *J. Clin. Microbiol.* **10**:464.

Larsson, L., 1982, "Head-Space Gas Chromatography in Clinical Microbiology," in *Rapid Methods and Autom. in Microbiol.* Tilton, R. C., ed., ASM Press, p. 151.

Larsson, L., and Holst, E., 1982, Feasibility of automated head-space gas-chromatography in identification of anaerobic bacteria, *Acta Pathol. Microbiol. Scand., Sect. B.* **90**:125.

Larsson, L., and Mårdh, P.-A., 1977, Application of gas chromatography to diagnosis of microorganisms and infectious diseases, *Acta Pathol. Microbiol. Scand., Sect. B Suppl.* **259**:5.

Larsson, L., Mårdh, P.-A., and Odham, G., 1978a, Detection of alcohols and volatile fatty acids by head-space chromatography in identification of anaerobic bacteria, *J. Clin. Microbiol.* **7**:23.

Larsson, L., Mårdh, P.-A., and Odham, G., 1978b, Analysis of amines and other bacterial products by head-space gas chromatography, *Acta Pathol. Microbiol. Scand. Sect. B* **86**:207.

Larsson, L., Holst, E., Gemmell, C. G., and Mårdh, P.-A., 1980a, Characterization of *Clostridium difficile* and its differentiation from *Clostridium sporogenes* by automatic head-space gas chromatography, *Scand. J. Infect. Dis. Suppl.* **22**:37.

Larsson, L., Mårdh, P.-A., and Odham, G., 1980b, Analysis of glucose catabolites using head-space gas chromatography for differentiation of some selected gram-negative species, *Current Microbiol.* **4**:143.

Larsson, L., Mårdh, P.-A., and Odham, G., 1980c, "Automated Head-Space Gas Chromatography as a Differential Diagnostic Aid in Intra-Abdominal Inflammatory Processes, Including PID," in *Abstracts of 3rd Meeting of I.S.S.T.D.*, Antwerp, Belgium, Oct. 2–3, 1980, p. 58.

Larsson, L., Mårdh, P.-A., Odham, G., and Carlsson, M.-L., 1982, Diagnosis of bacteremia by automated head-space capillary gas chromatography, *J. Clin. Pathol.* **35**:715.

Lategan, P. M., DuPreez, J. C., and Potgieter, H. J., 1978, Preliminary findings on the characterization of bacteria causing mastitis by gas chromatography, *Br. Vet. J.* **134**:342.

Lee, S. M., and Drucker, D. B., 1975, Analysis of acetoin and diacetyl in bacterial culture supernatants by gas–liquid chromatography, *J. Clin. Microbiol.* **2**:162.

Levanon, A., Klibansky, Y., and Kohn, A., 1977, Picorna- and Togavirus infection of cells detected by gas chromatography, *J. Med. Virol.* **1**:227.

Mårdh, P.-A., Larsson, L., and Odham, G., 1981, Head-space gas chromatography as a tool in the identification of anaerobic bacteria and diagnosis of anaerobic infections, *Scand. J. Infect. Dis. Suppl.* **26**:14.

Mayhew, J. W., and Gorbach, S. L., 1977, Internal standards for gas chromatographic analysis of metabolic end products from anaerobic bacteria, *Appl. Environ. Microbiol.* **33**:1002.

Mitruka, B. M., and Alexander, M., 1968, Rapid and sensitive detection of bacteria by gas chromatography, *Appl. Microbiol.* **16**:636.

Mitruka, B. M., and Alexander, M., 1969, Cometabolism and gas chromatography for the sensitive detection of bacteria, *Appl. Microbiol.* **17**:551.

Mitruka, B. M., and Alexander, M., 1972, Halogenated compounds for the sensitive detection of clostridia by gas chromatography, *Can. J. Microbiol.* **18**:1519.

Mitruka, B. M., Norcross, N. L., and Alexander, M., 1968, Gas chromatographic detection of *in vivo* activity of equine infectious anaemia virus, *Appl. Microbiol.* **16**:1093.

Moore, W. E. C., Cato, E. P., and Holdeman, L. V., 1966, Fermentation patterns of some *Clostridium* species, *Int. J. Syst. Bacteriol.* **16**:383.

Morin, A., and Paquette, G., 1980, Rapid temperature programmed gas–liquid chromatography of volatile fatty acids (C_1–C_7) for the identification of anaerobic bacteria, *Experientia* **36**:1380.

Moss, C. W., and Dees, S. B., 1976, Cellular fatty acids and metabolic products of *Pseudomonas* species obtained from clinical specimens, *J. Clin. Microbiol.* **4**:492.

Nicholls, T., Hähnel, R., Wilkinson, S., and Wysochi, S., 1976, Identification of 2-hydroxy-butyric acid in human amniotic fluid, *Clin. Chim. Acta* **69**:127.

Nicholls, T. M., Hähnel, R., and Wilkinson, S., 1978, Organic acids in amniotic fluid, *Clin. Chim. Acta* **84**:11.

Nord, C.-E., 1977, Diagnosis of anaerobic infections by gas–liquid chromatography, *Acta Pathol. Microbiol. Scand., Sect. B Suppl.* **259**:55.

Norén, B., and Odham, G., 1977, Intake system using glass ampoules for gas chromatographic analysis of volatile compounds of biological origin, *Acta Pathol. Microbiol. Scand., Sect. B, Suppl.* **259**:29.

Norrman, J., 1971, Studies on volatile organic compounds produced by fungi with emphasis on *Dipodascus aggregatus*, thesis, Uppsala, Sweden.

O'Brien, R. I., 1967, Differentiation of bacteria by gas chromatographic analysis of products of glucose catabolism, *Food Technol.* **21**:1130.

Palo, V., and Ilková, H., 1970, Direct gas chromatographic estimation of lower alcohols, acetaldehyde, acetone and diacetyl in milk products *J. Chromatogr.* **53**:363.

Parilla, E., Soriano, F., Ponte, M. C., and Ales, J. M., 1981, Aportación de la cromatografia gas–liquido al diagnóstico precoz de las bacteriemias, *Rev. Clin. Esp.* **161**:177.

Phillips, K. D., and Rogers, P. A., 1981, Rapid detection and presumptive identification of *Clostridium difficile* by p-cresol production on a selective medium, *J. Clin. Pathol* **34**:642.

Phillips, K. D., Tearle, P. V., and Willis, A. T., 1976, Rapid diagnosis of anaerobic infections by gas–liquid chromatography of clinical material, *J. Clin. Pathol.* **29**:428.

Phillips, I., Taylor, E., and Eykyn, S., 1980, The rapid laboratory diagnosis of anaerobic infection, *Infection* **8**:S155.

Pierami, R. M., and Stevenson, K. E., 1976, Detection of metabolites produced by psychrotropic bacteria growing in milk, *J. Dairy, Sci.* **59**:1010.

Reed, P. J., and Sanderson, P. J., 1979, Detection of anaerobic wound infection by analysis of pus swabs for volatile fatty acids by gas–liquid chromatography, *J. Clin. Pathol.* **32**:1203.

Reig, M., Molina, D., Loza, E., Ledesma, M. A., Meseguer, M. A., 1981, Gas–liquid chromatography in routine processing of blood cultures for detecting anaerobic bacteraemia, *J. Clin. Pathol.* **34**:189.

Rizzo, A. F., 1980, Rapid gas-chromatographic method for identification of metabolic products of anaerobic bacteria, *J. Clin. Microbiol.* **11**:418.

Rogosa, M., and Love, L. L., 1968, Direct quantitative gas chromatographic separation of C_2–C_6 fatty acids, methanol, and ethyl alcohol in aqueous microbial fermentation media, *Appl. Microbiol.* **16**:285.

Ryan, J. P., and Fritz, J. S., 1978, Determination of trace organic impurities in water using thermal desorption from XAD resin, *J. Chromatogr. Sci.* **16**:488.

Sasaki, N., and Takazoe, I., 1979, A gas chromatographic determination for analysis of bacterial volatile fatty acid and non-volatile fatty acid products, *Bull. Tokyo Dent. Coll.* **20**:31.

Seifert, M. H., Mathews, J. A., Phillips, I., and Gargan, R. A., 1978, Gas–liquid chromatography in diagnosis of pyogenic arthritis, *Br. Med. J.* 1402.

Snygg, B. G., Andersson, J., Krall, C., Stöllman, U., and Åkesson, C.-Å., 1979, Separation of botulinum positive and negative fish samples by means of a pattern recognition method applied to head-space gas chromatograms, *Appl. Environ. Microbiol.* **38**:1081.

Sondag, J. E., Ali, M., and Murray, P. R., 1980, Rapid presumptive identification of anaerobes in blood cultures by gas–liquid chromatography. *J. Clin. Microbiol.* **11**:274.

Stotzky, G., and Schenck, S., 1976, Volatile organic compounds and microorganisms. *CRC Crit. Rev. Microbiol.* 333.

Taylor, A. J., 1980, "The Application of Gas Chromatographic Head-Space Analysis to Medical Microbiology," *Applied Head-Space Gas Chromatography*, Kolb, B., ed., Heyden, London.

Thadepalli, H., and Gangopadhyay, P. K., 1980, Rapid diagnosis of anaerobic empyema by direct gas–liquid chromatography of pleural fluid, *Chest* **77**:507.

Watt, B., Geddes, P. A., Greenan, O. A., Napier, S. K., and Mitchell, A., 1982, Can direct gas–liquid chromatography of clinical samples detect specific organisms? *J. Clin. Pathol.* **35**:706.

Watt, B., Geddes, P. A., Greenan, O. A., Napier, S. K., and Mitchell, A., 1982, Gas–liquid chromatography in the diagnosis of anaerobic infections: a three-year experience, *J. Clin. Pathol.* **35**:709.

Weaver, J. C., Pungor, E., and Cooney, C. L., 1980, Mass-spectrometer monitoring of volatile compounds in the gas and liquid phases of fermentors, *Abs. Pap. Acs.* **180**:24.

Werner, H., and Hamman, R., 1980, Feasibility of exact species identification in routine diagnosis of anaerobic bacteria, *Infection* **8**:S158.

Wüst, J. 1977, Presumptive diagnosis of anaerobic bacteremia by gas–liquid chromatography of blood cultures, *J. Clin. Microbiol.* **6**:586.

Yoshioka, M., Kitamura, M., and Tamura, Z., 1969, Rapid gas chromatographic analysis of microbial volatile metabolites, *Jpn. J. Microbiol.* **13**:87.

Zlatkis, A., Lichtenstein, H. A., and Tishbee, A., 1973, Concentration and analysis of trace volatile organics in gases and biological fluids with a new solid adsorbent, *Chromatographia* **6**:67.

Zlatkis, A., Lee, K. Y., Poole, C. F., and Holzer, G., 1979, Capillary column gas chromatographic profile analsis of volatile compounds in sera of normal and virus-infected patients, *J. Chromatogr.* **163**:125.

7

HEAD-SPACE/GAS–LIQUID CHROMATOGRAPHY IN CLINICAL MICROBIOLOGY WITH SPECIAL REFERENCE TO THE LABORATORY DIAGNOSIS OF URINARY TRACT INFECTIONS

Nancy J. Hayward, Bacteriology Department, Alfred Hospital, Commercial Road, Prahran, Melbourne 3181, Victoria, Australia

1. ANALYSIS BY HEAD-SPACE/GAS–LIQUID CHROMATOGRAPHY

Head-space/gas–liquid chromatography (HS/GLC) has been used in food microbiology since 1965. So far it has not been widely applied in clinical microbiology; in fact, considerably less than has GLC using other methods of sample preparation. This is surprising because HS samples can be easily prepared and quickly analyzed, and HS/GLC yields chromatograms that are readily interpreted, three valuable qualities in a busy clinical laboratory.

There are two widely dissimilar techniques for HS analysis, indirect and direct HS analysis.

1.1. Indirect Analysis

In indirect HS analysis the vapor is swept out in a stream of inert gas to a collector which is later heated to flush volatile compounds into the gas chromatograph for analysis. This technique was developed for the quantitative analysis of more than 200 volatile compounds from urine in an attempt to find differences between the compounds in the vapor above urine specimens from healthy and diseased persons that could be used for laboratory diagnosis (Robinson *et al.*, 1973).

Indirect HS analysis was applied by Labows *et al.* (1980) to species of *Pseudomonas* growing on solid medium in flasks. For 2 h the vapor phase above cultures at 37°C was carried in a stream of nitrogen to a porous polymer collector, which was then heated at 220°C for 10 min to release the volatile compounds for an analysis by GLC/MS lasting about 1 h. The results suggested that a combination of volatile compounds, such as 2-aminoacetophenone, methyl ketones, and sulfides, might be used to identify *Pseudomonas aeruginosa* in clinical specimens.

1.2. Direct Analysis

Direct HS analysis is simpler to perform than is such indirect analysis. A liquid or solid is sealed in a vial. Vials covered by metal caps with central holes over underlying rubber septa are suitable. The sealed vial is heated for a few minutes in a bath to increase the vapor pressure of volatile compounds in the sample. The rubber septum is pierced with the needle of a gas-tight syringe to remove vapor for analysis. Vapor samples can be removed *manually*, using a syringe heated in an oven to the same temperature as the bath containing the sample vials; thereby condensation in the syringe is prevented. Since the sample is invisible, it should be removed slowly and steadily. Care must be taken to ensure that there is no partial blockage of the syringe needle impeding the passage of vapor from the HS to the barrel of the syringe. Duplicate analyses are essential. The temperature of vials cannot be too hot to handle, 60°C and 75°C having been used.

A gas chromatograph with an *automatic* HS injector is a marked advantage over manual injection. The sealed vials are heated in a turntable. At intervals selected by the operator the rubber septum of each vial in succession is pierced automatically, and carrier gas from the supply line flows into the vial and pressurizes it. When the valve from the supply is closed, a mixture of carrier gas and volatile compounds flows into the gas chromatograph for analysis. Results are highly reproducible, and duplicate analyses are not essential. Automatic analysis has four major advantages over manual injection. It requires less operator time, less operator skill, only half the materials, and the temperature of the vials can be as high as 150°C.

The vapor pressures of volatile compounds in HS can often be increased by adding an inorganic salt in excess of the amount needed to produce a saturated solution ("salting out"). The neutral salt, Na_2SO_4, has been used. If H_2SO_4 is added with Na_2SO_4, acidic compounds—or if instead solid NaOH is added with Na_2SO_4, alkaline compounds—are salted out. Other inorganic salts that have been used are the acid salt, $MgSO_4$, to salt out acidic compounds, and the alkaline salt, K_2CO_3, for alkaline compounds. Neutral compounds may be salted out by either acid or alkaline salts.

Drasar *et al.* (1976) record that they used direct HS analysis to detect volatile acids formed by *Clostridium*. They acidified cultures without the addition of an inorganic salt and heated them to 90°C before automatic injection. Their paper does not state which acids were detected.

Larsson *et al.* (1978a) analyzed acidified cultures of *Bacteroides fragilis*, *Clostridium perfringens* and *Propionibacterium acnes* by GLC. They compared direct injection of spent sterile culture medium with injection of ether extracts and with HS analysis after salting out with Na_2SO_4 at 75°C. Fatty acids were detected by all three methods, but ethanol, propanol, and *n*-butanol could be identified only in the chromatograms from HS analysis. They found that more information, both in number of compounds and in their resolution, can be obtained from HS than by other techniques.

The effectiveness of HS analysis in disclosing compounds that are not detected by other methods of sample preparation was further shown with cultures of *C. difficile* and *C. sporogenes* (Larsson *et al.*, 1980a) and with *Peptococcus asaccharolyticus* (Mårdh *et al.*, 1981).

HS analysis after salting out with acidified Na_2SO_4 has also been used for the detection of *Bacteroides* in pus (Mårdh *et al.*, 1981), for the detection of growth in blood cultures (Larsson *et al.*, 1982), and for detecting metabolic products after incubating various gram-negative aerobes in phosphate buffer containing glucose (Larsson *et al.*, 1980b).

The same authors (Larsson *et al.*, 1978b) extended their technique for HS analysis to include samples salted out with alkaline Na_2SO_4 in addition to acidified Na_2SO_4 at 75°C. They showed that alcohols up to octanol, acids up to octanoic, and amines up to decylamine could be detected. Chromatograms from *C. septicum*, *Klebsiella pneumoniae*, and *Proteus mirabilis* in chopped-meat glucose medium led them to conclude that direct HS analysis was a potential tool for the characterization of both aerobic and anaerobic bacteria.

Hayward and Spicer (1983) detected the growth of bacteria in blood cultures by HS analysis after salting out with K_2CO_3.

Larsson *et al.* (1981) inoculated buffered carbohydrates and amino acids with bacteria from urine cultures, incubated 1 h, analyzed the HS, and used a computer to recognize patterns in the chromatograms.

In applications of GLC in microbiology it has been usual to identify a particular microorganism by pattern recognition. Often the metabolic products have not been identified by MS. Even when products have been identified, the substrate in the culture medium from which they were formed has not been determined.

A very different approach was adopted by Hayward *et al.* (1977a,b). Once a product was positively identified—by GLC retention time and MS followed by checking with an authentic sample—they took the further step of determining from which substrate in the culture medium the metabolic product arose. A high concentration of this substrate added to the medium gave a product signal that stood out above the background products from other components of the culture medium.

This approach led to the development of a rapid test for the diagnosis of urinary tract infections (Hayward and Jeavons, 1977; Coloe, 1978). The test uses direct HS analysis of volatile compounds salted out with K_2CO_3 at 60°C from short-term cultures of urine in medium enriched with known substrates. In 4 h it is possible to report provisionally on the identity of the infecting organism, and the antibiotic susceptibility of Enterobacteria, in 90% of urine specimens from infected patients, by screening with quantitative microscopy followed by rapid diagnosis with HS/GLC.

2. SUITABILITY OF DIRECT HEAD-SPACE/ GAS–LIQUID CHROMATOGRAPHY FOR RAPID TESTS

Direct HS analysis is well suited to rapid GLC tests. Samples are easy to prepare, by merely mixing a few ml of liquid with a salt in a glass vial, sealing with a rubber septum and metal cap, and heating for a few minutes. If the HS sample is prepared at a relatively low temperature, the compounds in the vapor will be few and highly volatile. Under such conditions the GLC oven temperature can be relatively low, and analyses can, without programming, be completed in a few minutes. At a low oven temperature, stationary phases of column packings deteriorate only slowly and the compounds in the HS, being highly volatile, cause minimal soiling of detectors. The analysis has no solvent peak, and compounds with very short retention times can be detected.

3. DISCOVERY OF SPECIFIC PRODUCTS TO ACT AS HEAD-SPACE/GAS–LIQUID CHROMATOGRAPHY MARKERS FOR THE RAPID DETECTION OF BACTERIA AS EXEMPLIFIED BY COMMON URINARY TRACT PATHOGENS

3.1. Ethanol as Marker for Escherichia coli and Related Species

For *Escherichia coli* and species of related genera, viz., *Klebsiella*, *Enterobacter*, and *Citrobacter* (coliforms), ethanol produced from arabinose is a marker (Hayward *et al.*, 1977b; Coloe, 1978). Ethanol appears early in growth at the start of the logarithmic phase. The amount of ethanol at any time early in growth is related to the original number of bacteria in a urine specimen. Therefore, a time of incubation can be chosen such that $\geqslant 10^5$ colony-forming units (CFU)/ml of *E. coli* or a coliform, in urine cultured in a particular medium enriched with arabinose, will have produced sufficient ethanol to be detected in an HS/GLC test. When 1% arabinose is added to a peptone yeast-extract medium, the incubation time necessary for the detection of ethanol from $\geqslant 10^5$ CFU/ml urine of *E. coli*, or a coliform, in unshaken cultures of urine is about 3.5 h. Production of ethanol from arabinose is depressed by aeration and is considerably less in shaken than unshaken cultures. Only one other species that may be found in urine, viz., *Streptococcus faecalis*, produces ethanol from this medium in either shaken or unshaken cultures in small amounts detectable in 3.5 h. A gram stain of a urine specimen shows whether it contains a large number of streptococci and thereby distinguishes a specimen containing a high concentration of *S. faecalis*, viz., 10^7 CFU/ml urine, from a specimen containing a borderline number of *E. coli* or a coliform, viz., 10^5 CFU/ml urine.

Ethanol is salted out by K_2CO_3. If a urine specimen contains $>10^5$ CFU *E. coli* or a coliform/ml, ethanol is found in the HS above the urine culture when added to K_2CO_3 in vials which are then sealed and heated to 60°C. The total time needed from inoculating urine into arabinose medium to obtaining a chromatogram indicating whether or not $\geqslant 10^5$ CFU/ml of *E. coli* or a coliform are present in the urine is less than 4 h.

Unshaken cultures of nearly all strains of *E. coli* and some strains of coliforms also produce *n*-propanol.

Lactose was used as substrate in earlier work on the rapid HS/GLC test (Hayward and Jeavons, 1977) but was later replaced by arabinose because some strains of *E. coli* do not ferment lactose, viz., the Alkalescens–Dispar group. All *Escherichia*, both classical strains of *E. coli* and the

Alkalescens–Dispar group, ferment arabinose, as do all species of *Kleb-siella, Enterobacter*, and *Citrobacter* encountered so far in urine specimens.

3.2. Methyl Mercaptan and Dimethyl Disulfide as Markers for Proteus spp.

For all *Proteus* spp., viz., *P. vulgaris, P. mirabilis, P. morganii, P. rettgeri, P. inconstans A*, and *P. inconstans B*, methyl mercaptan produced from methionine is a marker (Hayward *et al.*, 1977a). In the presence of oxygen, methyl mercaptan oxidizes rapidly and spontaneously to dimethyl disulfide. Methyl mercaptan is detected in addition to dimethyl disulfide only when the oxygen reserve in the medium has been depleted. Like ethanol, methyl mercaptan is produced at the start of the logarithmic phase of growth. The amount of this compound at any time early in growth is related to the original number of bacteria present in a urine specimen. When $0.1 M$ methionine is added to the medium used for the detection of coliforms, the incubation time necessary for the detection of $\geqslant 10^5$ CFU/ml urine of a *Proteus* sp. in shaken cultures of urine is the same as for *E. coli* and coliforms, viz., 3.5 h. Production of methyl mercaptan and dimethyl disulfide by *P. vulgaris, P. mirabilis*, and *P. rettgeri*, but not *P. morganii*, is depressed by restriction of oxygen and is small in unshaken cultures. Therefore, two cultures must be made from each urine, one unshaken for the detection of metabolic products of *E. coli* and coliforms, and the other shaken for the detection of metabolic products of *Proteus* spp. Other organisms that may be found in urine produce at most only traces of dimethyl disulfide in 3.5 h, whether the cultures are shaken or unshaken.

If a urine specimen contains $\geqslant 10^5$ CFU/ml urine of a *Proteus* spp., both methyl mercaptan and dimethyl disulfide or dimethyl disulfide alone are found in HS above the urine culture when it has been mixed with an excess of K_2CO_3 at 60°C. Therefore one set of HS/GLC conditions can be used to detect $\geqslant 10^5$ CFU/ml urine of *E. coli*, coliforms, and *Proteus* spp. within 4 h of the receipt of the urine specimen in the laboratory.

3.3. Production of Markers in Mixed Cultures

If *E. coli*, a coliform or a *Proteus* sp. in numbers of $\geqslant 10^5$ CFU/ml urine is mixed with another microorganism in urine, the metabolic products of *E. coli*, the coliform or the *Proteus* sp. allow their detection. When a species of *Proteus* and either *E. coli* or a coliform are present in numbers of $\geqslant 10^5$ CFU/ml urine, both can usually be detected by their specific metabolic products. However, when one outnumbers the other by 100-fold or more, only the presence of the predominant species may be detected (Hayward and Jeavons, 1977).

3.4. Trimethylamine as Marker for Some Proteus spp.

P. mirabilis, P. vulgaris, P. rettgeri, and *P. inconstans A,* but not *P. morganii* and *P. inconstans B,* form trimethylamine and ethanol from acetylcholine. When acetylcholine is added to the arabinose peptone medium used for the detection of *E. coli* and coliforms, $\geqslant 10^5$ CFU/ml urine of these *Proteus* spp. can be detected in unshaken cultures of urine after 3.5 h. Production of trimethylamine, like production of ethanol, is depressed by aeration. Trimethylamine can be detected by the same sample preparation and analysis method as ethanol. If it were used as the marker for *Proteus* spp., only one culture, incubated unshaken, would need to be made from each urine. The use of acetylcholine instead of methionine as the substrate for the marker of *Proteus* spp. warrants further investigation.

3.5. Necessity for Mass Spectrometry

The work described above was aimed at producing a method that could be introduced into clinical diagnostic laboratories. Although the complexity and cost of a mass spectrometer is today beyond the resources of many hospitals, MS nevertheless plays an essential part in the development of HS/GLC tests due to its power to identify metabolic products. When in the studies cited above it was noted that the chromatogram from a culture of a given organism showed a characteristic peak, the metabolic product responsible for the peak was identified by MS.

The mass spectra of dimethyl disulfide, with distinctive peaks at 94, 45, 79, and 46 in order of intensity, and of methyl mercaptan, with distinctive peaks at 47, 48, 45, 46, 44, and 49, are shown as examples in Figures 1 and 2. The differences between these spectra and those in the library file are not more than would be expected as a result of the different ion intensities from different types of mass spectrometer.

When a metabolic product had been identified a search based on chemical probability was made for the possible precursor of the product. When such a precursor was found, experiments were designed to determine whether the metabolic product was suitable for a marker to be used in a rapid HS/GLC test, in particular whether the marker was formed in the early stages of growth, could be detected in a short HS/GLC analysis, and was specific for a group of organisms known to cause urinary tract infections.

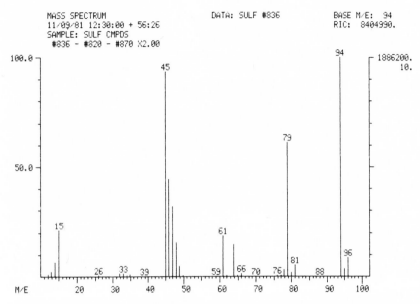

FIGURE 1. Mass spectrum of the metabolic product identified as dimethyl disulfide. The spectrum was kindly provided by Dr. E. Ramshaw of the Division of Food Research, Commonwealth Scientific and Industrial Research Organization, Australia.

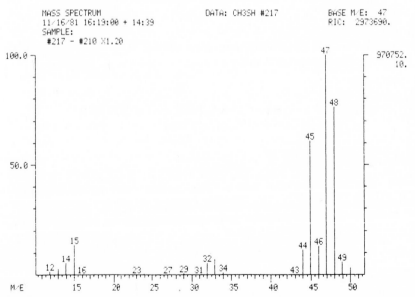

FIGURE 2. Mass spectrum of the metabolic product identified as methyl mercaptan. The spectrum was kindly provided by Dr. E. Ramshaw of the Division of Food Research, Commonwealth Scientific and Industrial Research Organization, Australia.

If the marker was found suitable, the amount of precursor could be adjusted to give an optimum yield of the marker that could be identified on the basis of retention time (RT) by GLC alone.

Among the markers found, ethanol, *n*-propanol, methyl mercaptan, and dimethyl disulfide were easy to identify because their mass spectra are well known. A feature of MS is that a spectrum that is not in any library file often provides enough information for an informed guess as to the identity of the compound(s) that caused it. The accuracy of the guess can be checked by testing a pure sample(s). An unsuspected failure of the gas chromatograph to resolve peaks can often be detected in this way by MS. For example, in the development of the HS/GLC analysis for the detection of pathogens in urine, trimethylamine and methyl mercaptan were not separated on a 1-m column. The mass spectrum of trimethylamine has distinctive peaks, at 58, 42, 59, and 30, which were clearly distinguishable from methyl mercaptan in the composite peak. Its presence was confirmed by separating the two compounds with a 2-m column.

4. SAME-DAY HEAD-SPACE/GAS–LIQUID CHROMATOGRAPHY TEST ON URINE SPECIMENS

4.1. Introduction

The HS/GLC test described above for the early and rapid detection of the majority of organisms that cause urinary tract infections is under trial in a clinical bacteriology laboratory in Melbourne (Hayward, 1983). The gas chromatograph is a Perkin-Elmer F45 with an automatic head-space injector, a stainless steel column, 2 m × 3 mm, packed with 0.4% Carbowax 1500 on 60–80 mesh graphite, and a flame ionization detector. The injector needle temperature is 150°C, the oven temperature 115°C, and the injector and detector temperatures 140°C; the N_2 carrier gas pressure is 180 kPa, H_2 400 kPa, and air 380 kPa. The injection time is 3 s. On this machine the RT of methyl mercaptan is 0.28 min, ethanol 0.33 min, trimethylamine 0.45 min, *n*-propanol 0.55 min, and dimethyl disulfide 1.7 min. An analysis time of 1.8 min is sufficient for the detection of all markers so far shown to be significant in the laboratory diagnosis of urinary tract infections.

Mixtures of equal volumes of each urine under test and arabinose methionine medium are prepared in duplicate and incubated at 37°C, one mixture shaken and the other unshaken. After 3.5 h the cultures are mixed with K_2CO_3 and heated to 60°C in the turntable of the gas chromatograph before analysis. Analyses of a medium blank and of nonincubated urine are made. The medium blank consists of equal parts of water and medium incubated and analyzed with the urine cultures. Equal parts of each urine and water are mixed and analyzed without incubation. About 25 analyses are done in succession in 1 h.

Coloe (1978) centrifuged urine specimens and used the sediment as inoculum for the rapid test. Centrifugation has been omitted from the present test because results from uncentrifuged urine specimens agree with those from routine laboratory cultures. No advantage has been proven to warrant the extra work of centrifugation. On the other hand, it has the disadvantage of requiring ten times the volume of urine that is needed for the present test; patients with urinary tract infections often cannot provide samples of large volume.

The interpretation of chromatograms does not require sophisticated equipment because the markers are present in large amounts relative to the background, having been formed by metabolism of high concentrations of defined substrates in the culture medium. The chromatograms obtained from analyses of cultures of urine specimens are assessed in the light of chromatograms of medium blank and nonincubated urine analyses. The medium blank usually contains a trace of dimethyl disulfide. Some urine samples contain ethanol. The amounts of these are subtracted from the yields of dimethyl disulfide or ethanol in urine cultures. Other volatile compounds with RT that differ from the markers for *E. coli*, coliforms and *Proteus* spp. may be found in urine and in culture medium. These can be neglected in the interpretation of chromatograms.

4.2. Examples of Chromatograms Produced by Urine Specimens

Typical chromatograms are shown in Figures 3, 4, and 5.

The medium blank analyses in Figure 3 show only one peak (d) of any size, with an RT of 1.3 min. This is *n*-butanol, which is in the peptone. The peak is present in chromatograms from all urine cultures.

Urine 1 (Figure 3) yielded some ethanol (a) and dimethyl disulfide (b) in the shaken culture and, most importantly, a large amount of ethanol and some *n*-propanol (c) in the unshaken culture. This chromatogram is typical of a urine specimen containing $\gg 10^5$ CFU/ml, that is, a high cell count of *E. coli* or a coliform, probably *E. coli* because of the *n*-propanol peak. On the following day, routine cultures confirmed the diagnosis of *E. coli*.

Two unidentified substances were present in the nonincubated urine. In the urine cultures, peaks with the same RT as the larger of them were also present. These did not interfere with the interpretation of the chromatograms.

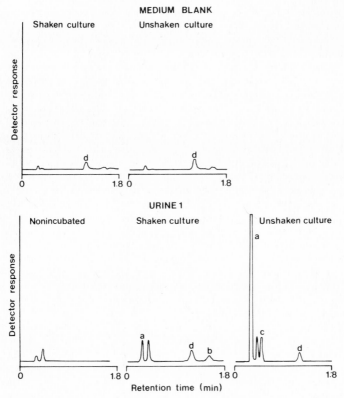

FIGURE 3. Chromatograms of medium blanks, and of nonincubated urine and cultures of Urine 1 in which the infecting organism was *Escherichia coli*. a, ethanol; b, dimethyl disulfide; c, *n*-propanol; d, *n*-butanol.

Urine 2 (Figure 4) yielded only a little ethanol and no dimethyl disulfide from the shaken culture. The most important finding was that the unshaken culture contained a moderate amount of ethanol. This chromatogram is typical of a urine specimen containing $\geqslant 10^5$ CFU/ml, that is, not a very high cell count of *E. coli* or a coliform, probably not *E. coli* because of the absence of an *n*-propanol peak. On the following day, cultures showed that the urine contained approximately 10^5 CFU/ml of *K. pneumoniae*.

An unidentified substance was present in the nonincubated urine and a similar peak, appearing shortly after ethanol, was seen in the chromatogram of the shaken culture. A much larger peak with the same RT was found in the unshaken culture. A peak with this RT, viz., 0.45 min, frequently appears in urine cultures but has not yet been identified by MS. Its significance has not yet been determined.

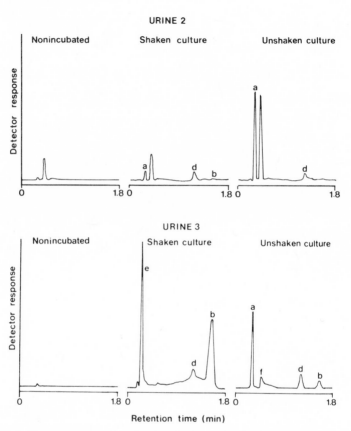

FIGURE 4. Chromatograms of nonincubated urine and cultures of Urine 2 in which the infecting organism was *Klebsiella pneumoniae*, and Urine 3 in which the infecting organism was *Proteus mirabilis*. a, ethanol; b, dimethyl disulfide; d, *n*-butanol; e, methyl mercaptan; f, trimethylamine.

Urine 3 (Figure 4) yielded large amounts of both methyl mercaptan (e) and dimethyl disulfide (b) in the shaken culture. In the unshaken culture there was no methyl mercaptan and only a little dimethyl disulfide. There was also a moderate amount of ethanol and a little trimethylamine (f) which arise from choline compounds in the medium peptone. These chromatograms are typical of urine with a very high count of a *Proteus* sp. On the following day, cultures showed that the urine contained $\gg 10^5$ CFU/ml of *P. mirabilis*.

The nonincubated urine did not contain any significant compounds.

Urine 4 (Figure 5) yielded a large amount of ethanol (a) and a very large amount of dimethyl disulfide (b) in the shaken culture. In the unshaken

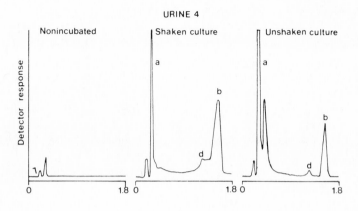

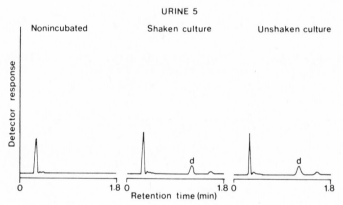

FIGURE 5. Chromatograms of nonincubated urine and cultures of Urine 4 in which there was a mixed infection with *Escherichia coli* and *Proteus mirabilis*, and Urine 5 in which there was not an infection with either a coliform or a *Proteus* sp. a, ethanol; b, dimethyl disulfide; d, *n*-butanol.

culture it yielded a very large amount of ethanol, a large amount of dimethyl disulfide, and a peak with the same RT as the unidentified peak detected in Urine 2 (Figure 4). This unidentified peak possibly masked any trimethyl-amine that was produced. These results indicate that the urine specimen contained $>10^5$ CFU/ml of both a *Proteus* sp. and *E. coli* or a coliform, the latter occurring in large numbers. On the following day, cultures showed that the urine contained $>10^5$ CFU/ml of both *E. coli* and *P. mirabilis*.

Cultures of *Urine 5* (Figure 5) did not produce any compounds to distinguish them from the medium and nonincubated urine. The chromatograms are typical of a urine specimen that does not contain either *E. coli*, a coliform, or a *Proteus* sp. Routine cultures confirmed this finding.

5. ANTIBIOTIC SUSCEPTIBILITY TESTS USING HEAD-SPACE GAS–LIQUID CHROMATOGRAPHY

Investigations in progress suggest that it is possible to determine the antibiotic susceptibility of bacteria by HS/GLC as exemplified by tests of amoxycillin and urine specimens from infected patients. Cultures of urine specimens in arabinose methionine medium containing antibiotic are incubated and analyzed in parallel with cultures in medium without antibiotic. The absence of specific metabolic products from cultures in medium containing antibiotic indicates susceptibility to the antibiotic.

Typical chromatograms from antibiotic susceptibility tests are shown in Figures 6 and 7.

The shaken and unshaken cultures of *Urine 6* (Figure 6) showed that it contained a high count of *E. coli* or a coliform. In the presence of amoxycillin, ethanol production was completely suppressed in the shaken culture and reduced to a small amount in the unshaken culture, indicating that the *E. coli* or coliform was amoxycillin-susceptible. Two days later the susceptibility to amoxycillin of the strain of *E. coli* isolated from the urine sample was confirmed by cultures and conventional susceptibility-testing.

Urine 7 (Figure 7) also contained *E. coli* or a coliform. However, ethanol production was not suppressed by amoxycillin in cultures from this urine. Two days later conventional diagnostic procedures showed that the patient was infected by an amoxycillin-resistant strain of *E. coli*.

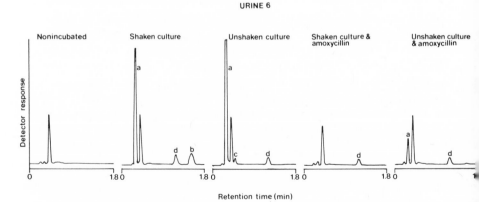

FIGURE 6. Chromatograms of nonincubated urine and cultures of Urine 6 in which the infecting organism was amoxycillin-susceptible *Escherichia coli*. a, ethanol; b, dimethyl disulfide; c, *n*-propanol; d, *n*-butanol.

URINE 7

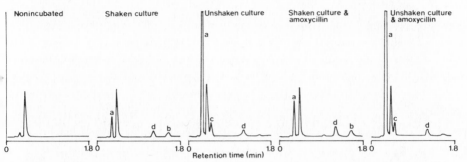

FIGURE 7. Chromatograms of nonincubated urine and cultures of Urine 7 in which the infecting organism was amoxycillin-resistant *Escherichia coli*. a, ethanol; b, dimethyl disulfide; c, *n*-propanol; d, *n*-butanol.

In the case of *Proteus* spp. and amoxycillin, reduced yields of methyl mercaptan and dimethyl disulfide indicate susceptibility as determined by agar-dilution tests. However, somewhat surprisingly, absence of reduction in the yields of the markers does not necessarily mean that the strain of *Proteus* in the urine is amoxycillin-resistant. The nonreduction of the markers of some amoxycillin-susceptible strains is related to the size of the inoculum of bacteria. When the cell count of a *Proteus* sp. in urine is very high, amoxycillin does not suppress the formation of markers even when the strain is amoxycillin-susceptible. When the cell count is close to 10^5 CFU/ml urine, the markers are suppressed by amoxycillin if the strain is susceptible, but not suppressed if it is resistant to the drug. Whether this observation has a clinical counterpart remains to be determined.

Within 4 h of the receipt of a urine specimen in the laboratory, HS/GLC tests can provide the clinician with an answer concerning growth of at least the majority of gram-negative rods encountered in urinary tract infections, viz., species of *Escherichia*, *Klebsiella*, *Enterobacter*, *Citrobacter*, and *Proteus*. During this time period the susceptibility to antibiotics can also be tested.

6. COMPARISON BETWEEN SAME-DAY HEAD-SPACE/GAS–LIQUID CHROMATOGRAPHY TEST AND OTHER RAPID TESTS FOR DETECTION OF URINARY TRACT INFECTIONS

In clinical microbiology laboratories there are two major problems in the analysis of urine specimens, and several rapid tests have been proposed for solving them.

One problem is that urine specimens are usually the most numerous of all types of specimen received. Their analysis is a significant part of the laboratory work load but the majority are from patients with no bacteriuria. The aim of screening tests is to reduce this work load by detecting most *nonpathologic* specimens quickly and easily, so that efforts can be concentrated on the remainder which include all pathologic specimens.

The second problem is that the results of cultures take one day and, in the case of growth, antibiotic susceptibility tests add another day before the laboratory can provide the clinician with an answer. The aim of rapid diagnostic methods, including the same-day HS/GLC test, is to assist therapy by reducing this long time interval. These methods detect most *pathologic* specimens, and give information about the identity, and sometimes also the antibiotic susceptibility, of the infecting organism within one working day.

Rapid tests that detect the presence of microorganisms in a urine sample may provide an answer in a matter of minutes and include bioluminescence to assay bacterial adenosine 5'-triphosphate (Alexander *et al.*, 1976; Johnston *et al.*, 1976), microscopy (Lewis and Alexander, 1976; Barbin *et al.*, 1978), particle size distribution analysis (Dow *et al.*, 1979), and the Limulus assay for endotoxin (Nachum and Shanbrom, 1981).

Rapid tests that detect the growth of microorganisms in short-term urine cultures require hours to provide an answer and include flow microcalorimetry (Beezer *et al.*, 1974), an electrochemical technique (Lamb *et al.*, 1976), impedimetric screening (Cady *et al.*, 1978), an inexpensive manual screening test (Murray and Niles, 1981), the AutoMicrobic System (AMS, Vitek Systems, Inc.), and two methods using light scatter photometry (Autobac, Pfizer Diagnostics, New York).

Tests using the AutoMicrobic System or light scatter photometry are nearest to the same-day HS/GLC test in providing early information to assist therapy and will be described in more detail.

The AutoMicrobic System (Isenberg *et al.*, 1979) detects and identifies more species of infectious agent than light scatter photometry or the HS/GLC test. The time taken to detect, enumerate, and identify a microorganism when present in numbers $>10^5$ CFU/ml urine is sometimes only a few hours, but 13 h are required before a urine specimen can be discarded as negative. The AutoMicrobic System is expensive to buy and to run. The test does not include determination of antibiotic susceptibilities.

The two methods using light scatter photometry (Heinze *et al.*, 1979; Jenkins *et al.*, 1980) include antibiotic susceptibility tests. Only gramnegative rods are identified. The total time to obtain a result is either 4 or 7 h. The Autobac instrument is less expensive than the AutoMicrobic System but involves the recurring expense of Autobac cuvettes.

Rapid tests using the AutoMicrobic System or light scatter photometry detect nonspecific effects of growth and consequently fail when a urine specimen contains more than one organism in numbers $>10^5$ CFU/ml urine, viz., a mixed infection. A method using a gas chromatograph has greater intrinsic potential because it separates the specific products of growth formed by mixtures of organisms and allows their identification. Therefore *E. coli*, coliforms, and *Proteus* spp. can be detected in mixed infections.

In the clinical bacteriology laboratory at Alfred Hospital, Melbourne, where the same-day HS/GLC test is in use, 8-ml aliquots of all urine specimens are first centrifuged and screened by quantitative microscopy including gram stains of deposits. Specimens that are found to be non-pathologic are screened-out, viz., not analyzed by HS/GLC. Analysis of the remaining specimens by HS/GLC detects and gives information about the identity of the majority of infecting organisms and can determine their antibiotic susceptibilies (e.g., *E. coli*, coliforms, and sometimes *Proteus* spp. to amoxycillin). In addition, if the HS/GLC test is negative, the gram stain that was done as part of the initial screening usually gives information that identifies the infecting organism if it is a streptococcus, staphylococcus or yeast. Antibiotic therapy can be guided by this information. Thus microscopy, which yields most information about gram-positive organisms, complements the same-day HS/GLC test for gram-negative rods, and provisional reports to assist therapy can be provided for 90% of infected patients in 4 h.

A gas chromatograph with an automatic head-space injector (Perkin-Elmer Corporation, Norwalk, Connecticut, U.S.A.) is not expensive compared with other instruments recommended for rapid diagnosis. There are no disposable items to constitute a recurring expense.

7. CONCLUSION

Most applications of direct HS/GLC in clinical microbiology have been directed to the analysis of acids. Although the study of acidic metabolic products has been rewarding with GLC samples prepared by other methods, neutral and alkaline compounds are more volatile than acids, and HS analysis is better suited to them than to acids. The ease and speed of direct HS analysis to detect neutral and alkaline products of microorganisms will probably expand the number of applications of this technique in clinical microbiology in the near future.

REFERENCES

Alexander, D. N., Ederer, G. M., and Matsen, J. M., 1976, Evaluation of an adenosine 5'-triphosphate assay as a screening method to detect significant bacteriuria, *J. Clin. Microbiol.* **3**:42.

Barbin, G. K., Thorley, J. D., and Reinarz, J. A., 1978, Simplified microscopy for rapid detection of significant bacteriuria in random urine specimens, *J. Clin. Microbiol.* **7**:286.

Beezer, A. E., Bettelheim, K. A., Newell, R. D., and Stevens J., 1974, Diagnosis of bacteriuria by flow microcalorimetry, *Science Tools* **21**:13.

Cady, P., Dufour, S. W., Lawless, P., Nunke, B., and Kraeger, S. J., 1978, Impedimetric screening for bacteriuria, *J. Clin. Microbiol.* **7**:273.

Coloe, P. J., 1978, Head-space gas liquid chromatography for rapid detection of *Escherichia coli* and *Proteus mirabilis* in urine, *J. Clin. Pathol.* **31**:365.

Dow, C. S., France, A. D., Khan, M. S., and Johnson, T., 1979, Particle size distribution analysis for the rapid detection of microbial infection of urine, *J. Clin. Pathol.* **32**:386.

Drasar, B. S., Goddard, P., Heaton, S., Peach, S., and West, B., 1976, Clostridia isolated from faeces, *J. Med. Microbiol.* **9**:63.

Hayward, N. J., 1983, Head-space gas–liquid chromatography for the rapid laboratory diagnosis of urinary tract infections caused by *Enterobacteria*, *J. Chromatogr.* **274**:27.

Hayward, N. J., and Jeavons, T. H., 1977, Assessment of technique for rapid detection of *Escherichia coli* and *Proteus* species in urine by head-space gas–liquid chromatography, *J. Clin. Microbiol.* **6**:202.

Hayward, N. J., and Spicer, W. J., 1983, Laboratory diagnosis of bacteraemia by head-space gas–liquid chromatography, *Pathology* **15**:161.

Hayward, N. J., Jeavons, T. H., Nicholson, A. J. C., and Thornton, A. G., 1977a Methyl mercaptan and dimethyl disulfide production from methionine by *Proteus* species detected by head-space gas–liquid chromatography, *J. Clin. Microbiol.* **6**:187.

Hayward, N. J., Jeavons, T. H., Nicholson, A. J. C., and Thornton, A. G., 1977b, Development of specific tests for rapid detection of *Escherichia coli* and all species of *Proteus* in urine, *J. Clin. Microbiol.* **6**:195.

Heinze, P. A., Thrupp, L. D., and Anselmo, C. R., 1979, A rapid (4–6 hour) urine-culture system for direct identification and direct antimicrobial susceptibility testing, *Am. J. Clin. Pathol.* **71**:177.

Isenberg, H. D., Gavan, T. L., Sonnenwirth, A., Taylor W. I., and Washington, J. A., 1979, Clinical laboratory evaluation of automated microbial detection/identification system in analysis of clinical urine specimens, *J. Clin. Microbiol.* **10**:226.

Jenkins, R. D., Hale, D. C., and Matsen, J. M., 1980, Rapid semiautomated screening and processing of urine specimens, *J. Clin. Microbiol.* **11**:220.

Johnston, H. H., Mitchell, C. J., and Curtis, G. D. W., 1976, An automated test for the detection of significant bacteriuria, *Lancet* **ii**:400.

Labows, J. N., McGinley, K. J., Webster, G. F., and Leyden, J. J., 1980, Headspace analysis of volatile metabolites of *Pseudomonas aeruginosa* and related species by gas chromatography–mass spectrometry, *J. Clin. Microbiol.* **12**:521.

Lamb, V. A., Dalton, H. P., and Wilkins, J. R., 1976, Electrochemical method for the early detection of urinary tract infections, *Am. J. Clin. Pathol.* **66**:91.

Larsson, L., Mårdh, P.-A., and Odham, G., 1978a, Detection of alcohols and volatile fatty acids by head-space gas chromatography in identification of anaerobic bacteria, *J. Clin. Microbiol.* **7**:23.

Larsson, L., Mårdh, P.-A., and Odham, G., 1978b, Analysis of amines and other bacterial products by head-space gas chromatography, *Acta Pathol. Microbiol. Scand., Sect. B.* **86**:207.

Larsson, L., Holst, E., Gemmell, C. G., and Mårdh, P.-A., 1980a, Characterization of *Clostridium difficile* and its differentiation from *Clostridium sporogenes* by automatic head-space gas chromatography, *Scand. J. Infect. Dis. Suppl.* **22**:37.

Larsson, L., Mårdh, P.-A., and Odham, G., 1980b, Analysis of glucose catabolites by using head-space gas chromatography for differentiation of some selected gram-negative species, *Current Microbiology* **4**:143.

Larsson, L., Mårdh, P.-A., and Odham, G., 1981, Automated head-space gas chromatography in the identification of anaerobes; and of bacteria encountered in urinary tract infections, *3rd International Symposium on Rapid Methods and Automation in Microbiology, Abstracts* No. 54.

Larsson, L., Mårdh, P.-A., Odham, G., and Carlsson, M.-L., 1982, Diagnosis of bacteraemia by automated head-space capillary gas chromatography, *J. Clin. Pathol.* **35**:715.

Lewis, J. F., and Alexander, J., 1976, Microscopy of stained urine smears to determine the need for quantitative culture, *J. Clin. Microbiol.* **4**:372.

Mårdh, P.-A., Larsson, L., and Odham, G., 1981, Head-space gas chromatography as a tool in the identification of anaerobic bacteria and diagnosis of anaerobic infections, *Scand. J. Infect. Dis. Suppl.* **26**:14.

Murray, P. R., and Niles, A. C., 1981, Detection of bacteriuria: Manual screening test and early examination of agar plates, *J. Clin. Microbiol.* **13**:85.

Nachum, R., and Shanbrom, E., 1981, Rapid detection of gram-negative bacteriuria by *Limulus* amoebocyte lysate assay, *J. Clin. Microbiol.* **13**:158.

Robinson, A. B., Partridge, D., Turner, M., Teranishi, R., and Pauling, L., 1973, An apparatus for the quantitative analysis of volatile compounds in urine, *J. Chromatogr.* **85**:19.

8

ANALYSIS OF CELLULAR COMPONENTS IN BACTERIAL CLASSIFICATION AND DIAGNOSIS

Erik Jantzen, National Institute of Public Health, Postuttak, Oslo 1, Norway.

1. INTRODUCTION

1.1. General

Bacterial cell walls have always represented a challenge to the chemist. Substances not found elsewhere in nature are frequently detected and the chemical problems of isolation, purification, and structure elucidation of bacterial compounds are both fascinating and demanding. As a result, several fundamental achievements in organic chemistry, spectroscopy, and chromatography have been carried out by chemists working on bacterial compounds, and a multitude of bacterial substances have been described over the last decades.

Concurrently to the accumulation of chemical data on bacterial cell walls, it has become clear that this kind of chemical evidence is of a high value in a precise definition of a bacterial species. This fact has convincingly been demonstrated by the extensive taxonomic studies of coryneform bacteria and related taxa (cf. Keddie and Cure, 1978; Minnikin *et al.*, 1978; Minnikin and Goodfellow, 1980).

In bacterial diagnostic work, however, chemical analysis of cell wall substances has only played a minor role. This may in part be explained by the late development of convenient chemical techniques. Until the last decade structural elucidation of even a fairly simple organic molecule was a time-consuming and demanding task. The introduction of new and powerful chemical and physical techniques, i.e., gas chromatography (GC) and GC coupled on-line with a mass spectrometer (GC/MS), has changed this situation and made both analysis of complex mixtures of organic compounds and structural elucidations considerably easier. These techniques have made chemical cell wall analysis applicable not only in taxonomic studies, but also in routine diagnosis (Mitruka, 1974; Moss, 1978; Drücker, 1981).

Traditional bacterial classification, as presented in the last edition of *Bergey's Manual of Determinative Bacteriology* (Buchanan and Gibbons, 1974), is based on morphology, nutritional requirements, metabolic activities, and serology. Such noninstrumental techniques afford effective determinative keys which assign the majority of human pathogens as to genus and species.

Why then introduce expensive and complicated instrumental technique in bacterial identification? Firstly, many of the conventionally used tests and techniques are time-consuming, which may lead to a delay in the proper treatment of the patients. Secondly, the use of such techniques may increase the diagnostic accuracy. Thirdly, unusual organisms, which may not be identified without tedious methods, might more easily be recognized by these new chemical techniques.

1.2. Cellular Substances of Diagnostic Potential

A useful chemotaxonomic marker should involve for its biosynthesis a relatively high number of genes to attain a certain stability (Mandel, 1969). The peptidoglycans and lipopolysaccharides in particular seem to fulfill these requirements; both are large and complex molecules of vital importance for the cell and require at least 20 biosynthetic enzymes (Schleifer and Kandler, 1972; Stocker and Mäkelä, 1971). However, compounds of low molecular weight such as the fatty acid constituents of phospholipids have also been shown to be both stable and useful taxonomic markers (see Section 4.1). On the other hand, capsular polysaccharides of certain bacteria have been shown to be markedly influenced by growth conditions, and may in some cases almost disappear completely after a few generations on artificial media (Ellwood and Tempest, 1972). Occurrence of so-called "secondary metabolites," including several pigments, has also

TABLE 1. Cellular Constituents and Their Potential Use in GC Classification and Diagnosis of Bacteria

Cellular constituents	Procedures prior to GC or GC/MS	Monomers of diagnostic potential	Evaluation
"Whole cells"	Hydrolysis or methanolysis (derivatization)	Fatty acids, fatty alcohols, fatty aldehydes, monosaccharides	Rapid and informative in taxonomy and diagnosis
Carbohydrates	Extraction, hydrolysis, derivatization	Rare sugars not found in human cells	Of high potential in taxonomy and diagnosis
Proteins	Extraction, fractionation, purification, hydrolysis, derivatization	Amino acids not found in human cells	Time consuming; probably impractical in diagnosis
Peptidoglycans	Extraction, fractionation, purification, hydrolysis, derivatization	D-amino acids, peptides	Time consuming; probably impractical in diagnosis; valuable in taxonomy
Lipopolysaccharides	Extraction (purification), methanolysis, derivatization	Rare sugars, hydroxy-fatty acids	Valuable in taxonomy and of high potential in diagnosis
Lipids	Extraction, hydrolysis, derivatization	Fatty acids, alcohols, aldehydes	Valuable in taxonomy; in diagnosis, whole cell fatty acids probably preferable
Teichoic acid	Extraction, purification, hydrolysis, derivatization	Sugars, polyalcohols	Valuable in taxonomy; probably impractical in diagnosis

been shown to be biosynthetically unstable and should be used with caution in bacterial classification.

Comparative analyses of commonly occurring cell wall substances have been carried out (Table 1). Thus, structure elucidation of the complex lipids of *Mycobacterium, Nocardia*, and other acid-fast bacteria, and their distribution within the genera, have been given much attention (e.g., Asselineau, 1966; Lechevalier, 1977; Minnikin and Goodfellow, 1980). The peptidoglycan structures of a large number of gram-positive bacteria has also been thoroughly elucidated (Schleifer and Kandler, 1972). This work has markedly influenced the classification of gram-positive bacteria and the peptidoglycan structure is established as a fundamental characteristic in bacterial taxonomy.

The lipopolysaccharides (LPS) of the outer membrane of gram-negative bacteria have also attracted attention (Lüderitz *et al.*, 1971; Wilkinson, 1977). Although most work has been performed on LPS of *Salmonella* and other genera within the family *Enterobacteriaceae*, it has been demonstrated that the LPS constituents represent potentially useful taxonomic markers (Rietschel and Lüderitz, 1980).

1.3. Current GC Approaches in Bacterial Taxonomy

The use of GC in bacterial classification and identification has recently been reviewed (Mitruka, 1974; Drücker *et al.*, 1976; 1981; Larsson and Mårdh, 1977; Moss, 1978), including studies of bacterial interrelationships (chemotaxonomic investigations), and development of techniques for sensitive and rapid identification of infectious agents. Three principally different types of samples have been utilized:

1. Spent growth medium for analysis of metabolic products;
2. Whole cells or simple extracts for analysis of cellular constituents (e.g., fatty acids, alcohols, amino acids, or sugars);
3. Infected material (fluid or tissue) for specific detection of bacterial components.

GC analysis of metabolic products has been adopted as a routine method by numerous laboratories and has proved to be of great assistance in bacterial identification, especially of anaerobic bacteria (see Chapter 6).

Recently GC analyses of cell wall constituents have been applied as a diagnostic tool by many clinical microbiological laboratories. Particularly fatty acid profiles are made use of (see Sections 4 and 5), but the lack of a detailed manual has probably limited their more extensive use in diagnosis.

Analysis of bacterial compounds in clinical samples is a rather ambitious approach due to the often complex host-specific background, but appears an attractive analytical means due to its relative simplicity and rapidity (see Section 5.5).

This chapter will be focused on cellular constituents readily analyzable by GC and GC/MS, without time-consuming work-up procedures. Analysis of cell wall peptidoglycan and proteins will only be dealt with exceptionally. Emphasis will be placed on lipid and carbohydrate constituents, which apparently are easily analyzed in whole cells or simple extracts.

2. INSTRUMENTATION

2.1. Gas Chromatograph

In the diagnostic laboratory, a standard gas chromatograph equipped with a temperature programmer will generally fulfill most requirements for GC analysis of bacteria.

2.2. Detectors

A flame-ionization detector (FID) is the standard "all-purpose" detector in GC. It is robust, sensitive, and linear over a large concentration range and is recommended as the first choice detector also for analyses of microorganisms.

The electron capture detector (ECD) is selective and very sensitive to compounds having affinity for electrons, e.g., substances containing halogen atoms and certain polynuclear hydrocarbons. In microbiology, one has taken advantage of the EC detector's sensitivity in analyses of halogen-containing derivatives of certain metabolites, as well as trifluoroacetylated sugars and heptafluorobutyrated hydroxy-fatty acids (see Section 5.5).

Several other commercially available detectors, both general and selective, are in common use, but should be considered as "special purpose" detectors.

2.3. Columns

Several different types of columns are available: packed, micropacked, and the two main types of capillary, namely, solid coated open tubular (SCOT) and wall coated open tubular (WCOT) columns. Columns of glass are preferable to metal for analysis of biological substances due to higher inertness and therefore less degradation of fragile compounds. The recently introduced fused-silica capillary columns are very useful due to their flexibility (almost unbreakable), high inertness, and resolving power. However, capillary columns require some supplementary equipment and are technically more demanding. Two standard packed glass columns (200 × 0.2 cm) should be available for analyses of bacterial substances; one *nonpolar* where the stationary phase usually is a methylsilicone (SE-30, OV-1, OV-101, SP-2100) and one *polar* where the stationary phase consists of a polyester (diethylene glycol succinate, diethylene glycol adipate) or cyanosilicone (OV-275, SILAR 10C, SP-2340). Normally supports are coated with 1%–10% of stationary phase.

2.4. Integrator–Recorder

Usually GC analysis of bacterial samples provides complex elution profiles. An electronic integrator which prints out retention times and peak areas is therefore quite useful. Computerized combinations of recorder and integrator are now commerciably available. These feasible devices print the retention times of each peak directly on the chromatograms thereby providing a complete analytical report subsequent to each analytical run.

2.5. Ancillary Equipment

Automatic injection systems can be coupled to a gas chromatograph and are of great help in laboratories with large GC routines. Reliable commercial devices are available which make it possible to operate a gas chromatograph on a 24-h basis.

A gas chromatograph may also be used in a preparatory step. Using an automatic repetitive injector, a fraction collector, and a sample valve connected to the detector, the separation process is repeated until sufficient material of a certain number of the substances eluting has been collected.

Spectroscopy and MS (see Section 2.6) can be used as adjuncts to the gas chromatograph and serve as useful supplementary techniques for identification of the individual substances separated by GC.

Computers of various complexity can be linked on-line, or used separately, for handling and interpretation of extensive amounts of data generated by one or more gas chromatographs.

2.6. Gas Chromatograph/Mass Spectrometer

Two main types of instruments are commercially available; magnetic and quadrupole mass spectrometers. Both types have their limitations and advantages, and a proper choice of instrument is dependent on several factors, including the training and professional background of the operators. A modern GC/MS instrument is extensively computerized and the computer is of central importance for interpretation of the vast amount of analytical data produced, especially in connection with the analyses of complex biological samples. A modern instrument is equipped for capillary columns and may have an ion source constructed for both electron impact (EI) and chemical ionization (CI) fragmentation. This combination of high-resolution chromatography, alternative fragmentation (EI and CI), as well as selective ion monitoring (SIM), opens up a multitude of interesting applications in microbiology.

3. METHODS

3.1. Growth and Harvesting of Bacteria

A standardized routine solid medium should be selected as first choice. In our laboratory we use a medium consisting of Tryptose Blood Agar Base supplemented with 0.1% dextrose and 5% horse blood.

Harvesting of bacteria is usually done by washing bacterial colonies off the plate with a right-angled glass rod and a small amount of distilled water. The cells are centrifuged at 2500 rpm in a swing-out rotor for 15 min, washed twice with distilled water, processed further, or freeze-dried for storage.

3.2. Methanolysis, Hydrolysis, and Saponification

Fatty acids of bacterial lipids are most commonly determined after bond-splitting with either dilute alkali in aqueous methanol (saponification) or by acidic methanolysis (Figure 1). The alkaline degradation is rapid and selective but has two main limitations: carbohydrates present in the sample are degraded and the yield of amide-bound hydroxy-fatty acids is low (Jantzen *et al.*, 1978; Fautz *et al.*, 1979) apparently mainly due to production of artifacts (Rietschel *et al.*, 1972).

For determination of monosaccharides in polymeric material, bond-splitting by dilute mineral acid or trifluoroacetic acid is commonly used. Alkaline reagents are avoided due to the instability of sugars at high pH. However, the stability of the various glycosidic linkages differs, and destruction of certain sugars occurs easily, also at low pH. A complete sugar analysis may therefore require several different hydrolytic conditions.

Depolymerization of complex lipids and carbohydrates by acidic methanolysis, e.g., treatment with anhydrous methanolic HCl, seems to be

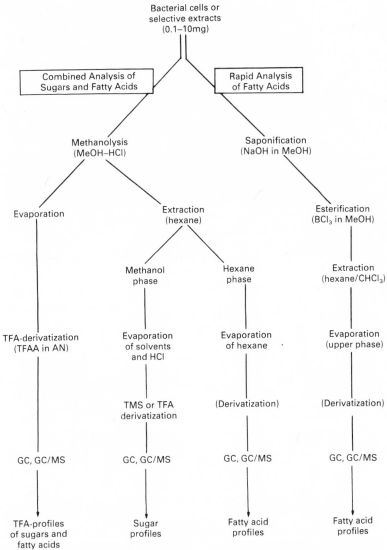

FIGURE 1. Outline of the described assays for cellular fatty acids and sugars (Jantzen *et al.*, 1978).

a reasonably mild method of bond splitting. This reagent liberates the fatty acids directly as methyl esters and the yield of amide-linked constituents are almost quantitative (2 M HCl, 85°C, 18 h). Most glycosidic linkages of glyco-conjugates are also broken at these conditions, and the monosaccharides are liberated, and stabilized as methylglycosides. Thus the method provides a minimum of sugar degradation (Zanetta *et al.*, 1972). Even fragile structures, like *N*-acetylneuraminic acid, are stable in 2 *M* HCl in methanol at 85°C (Clamp *et al.*, 1971). However, certain linkages are not cleaved, e.g., the phosphate ester linkages of teichoic acids and related polymers, and linkages involving unacylated glycosamines. Furthermore, cyclopropane substituted fatty acids are disintegrated. Apparently, both saponification and methanolysis have to be considered depending on the constituents of interest. The experimental details are given in Sections 3.4–3.7.

3.3. Derivatization of Hydroxylated Fatty Acids

In GC analyses of hydroxylated fatty acids which are present in most gram-negative bacteria, a derivatization step prior to GC is required to obtain a reproducible detector signal from these constituents (Jantzen *et al.*, 1978). Trifluoroacetylation and trimethylsilylation are commonly used. The author prefers trifluoroacetylation because a TFA-derivatized hydroxy-acid is well separated from the nonhydroxylated, and because this derivative has a FID detector response similar to a nonhydroxylated fatty acid methylester. Furthermore, the reaction is quantitative and rapid, and the reactants are highly volatile and pure; no extra peaks occur in the chromatograms.

An aliquot of a fatty acid methyl esters in hexane is transferred to a 10-ml glass-stoppered glass centrifuge tube, the solvent is removed by evaporation (nitrogen), and equal volumes of trifluoroacetic anhydride and acetonitrile are added (100 μl). Subsequently, the reaction mixture is heated, for instance by a hair-dryer, to boiling for 2–4 min. After 10–30 min at room temperature, the reaction mixture can be injected on the gas chromatograph. Storage at −20°C in sealed glass capillaries is recommended.

3.4. Combined Analysis of Fatty Acids and Monosaccharides

Bacterial cells or extracts (0.1–10 mg) and 2 *M* HCl in dry methanol (3 ml) are heated under nitrogen in 10-ml glass tubes with Teflon-lined screw caps, at 85°C for 18 h (Jantzen *et al.*, 1978). When cool, the mixture is concentrated to about half its initial volume by a stream of nitrogen (to remove HCl), and then centrifuged for 15 min at 2500 rpm. The supernatant is then transferred by a Pasteur pipette to a 10-ml centrifuge glass tube and concentrated just to dryness by nitrogen (prolonged evaporation time causes loss of methyl esters of chain length up to 14 carbon atoms). Subsequent TFA derivatization is performed as described above (Section 3.3) before injection onto the gas chromatograph.

3.5. Analysis of Fatty Acids Released by Methanolic HCl

The cells (freeze-dried or wet, directly from plate) and $2\,M$ HCl in methanol (3 ml) are heated under nitrogen in 10-ml glass tubes with Teflon-lined screw-caps, at 85°C for 18 h (Jantzen *et al.*, 1978). When cool, the mixture is concentrated to about half its volume by a stream of nitrogen (to remove HCl). Then an aqueous solution of sodium chloride (half saturated) is added (2 volumes), and the fatty acids extracted by 2×3 volumes hexane (centrifugation). The combined hexane phases are concentrated just to dryness by nitrogen and then redissolved in hexane (10–100 µl) before derivatization (see Section 3.3) or injection on GC.

3.6. Rapid Analysis of Fatty Acid Released by Saponification

Several saponification methods for the release of fatty acids have been introduced. Moss *et al.* (1974) compared four commonly used methods of liberation and subsequent methylester formation, and they suggested the following single-tube method.

Bacterial cells are heated for 15 min at 100°C in 5 ml 5% NaOH in 50% aqueous methanol. The saponified material is then cooled and acidified to pH 2 by $6\,M$ aqueous HCl. A 4-ml portion of borontrichloride-methanol reagent is added, and the mixture is heated for 5 min at 100°C and transferred to a tube containing 10 ml of saturated aqueous sodium chloride. The methyl esters are then extracted twice with an equal volume of 1:4 chloroform–hexane. The combined extract is evaporated to 0.1 ml before derivatization (see Section 3.3) or direct injection on the gas chromatograph.

3.7. Analysis of Monosaccharides

Several different methods for GC analysis of monosaccharides have been introduced (cf. Mitruka, 1974; Drücker, 1981). Generally, GC patterns of whole cell carbohydrates are very complex and not easily interpretable. A primary extraction step is therefore probably preferable in most cases (see Section 4.2). Both procedures given below are based on linkage-cleavage by acidic methanolysis. Method A (see below), described by Yamakawa and Ueta (1964), is based on the biomass obtained from one standard agar plate and on a standard GC technique with FID detection. Method B (Pritchard *et al.*, 1981) is a currently developed more sophisticated technique capable of giving carbohydrate fingerprint chromatograms of single bacterial colonies. This procedure, however, requires capillary columns and an ECD detector.

Method A. Bacterial cells or extracts (0.1–10 mg) are methanolyzed as described (Section 3.4). After extraction of the fatty acid methyl esters by 2 × 1 volume of hexane, the methanol layer is deacidified by the addition of a resin (Amberlite IR-4B, OH-form) suspended in methanol. After removal of the resin by filtration, the methanol is evaporated under reduced pressure, and the dried residue is TMS-derivatized with 0.2 ml hexamethyldisilazane and 0.1 ml of trimethylchlorosilane in 0.5 ml of anhydrous pyridine. To complete the reaction, the mixture is heated at 60 to 70°C for a few minutes. After 30 min at room temperature, 3 ml of chloroform is added to the mixture. Subsequently, pyridine and inorganic material are removed by washing with water. The chloroform layer is then concentrated to dryness with a rotary evaporator and redissolved in 100 μl chloroform before injection on the gas chromatograph (see Section 3.8).

Alternatively, the deacidified residual methanolysate may be TFA derivatized as described (Section 3.3) and analyzed on a moderately polar column (Zanetta *et al.*, 1972; Jantzen *et al.*, 1974c).

Method B. Bacteria (amount down to a portion of a colony) grown on blood agar plates are transferred to a capillary tube using an inoculating loop. The sample is rinsed off the loop into a conical plastic micropipette tip attached to the capillary by a 2-mm length of polyethylene tubing. After brief centrifugation, the sample is rapidly dried under vacuum. Approximately 10 μl of 1.4 *N* methanolic HCl is then added by use of a glass micropipette. After sealing, the capillary is placed in a heating block for 24 h at 80°C. The capillary is broken open, and the sample taken to dryness under vacuum. Subsequently, derivatization is accomplished by adding 5 μl of a freshly prepared solution consisting of 100 μl of 10% *N*-methyl-bis-trifluoroacetamide in *N,N*-dimethylformamide and 10 μl pyridine. The reaction mixture is kept at room temperature for at least 2 h.

This procedure has been used for characterization of various Lancefield's streptrococcal groups. Due to the specific detector used (ECD), no interference of proteins, lipids or other cellular components is observed (see Section 4.2).

3.8. GC Conditions

A gas chromatograph equipped with a temperature programmer and a flame ionization is required. Fused silica capillary columns are recommended due to their excellent resolving power, inertness, and flexibility, but they require some supplementary equipment and are more expensive. They are also technically somewhat more demanding. Three standard packed glass columns (200 × 0.2 cm) are recommended for analyses and identification of fatty acids and sugars. The following four columns are currently used by the author.

    a. Combined analysis of fatty acids and sugars: 3%–10% methyl-silicone (SE-30, OV-1, SP-2100) on Gas Chrom Q (or Chemosorb WHP), 100–120 mesh.

    b. Fatty acid methyl esters:
       3%–10% methyl-silicone (see a) and 5%–10% cyano-silicone (Silar 10C) on Gas-Chrom Q, 100–120 mesh.

    c. Sugars: *TFA-methylglycosides*: 5% SP-2401 on Chemosorb WHP, 100–120 mesh and alternatively, 3%–10% methylsilicone. *TMS-methylglycosides*: 3%–10% methyl-silicone.

    d. Both fatty acid methyl esters and methylglycosides: A glass (or fused silica) capillary column (25 m × 0.2 mm) coated with a methyl silicone is a general and high-resolution column.

The initial oven temperature is commonly set to 120°C for fatty acid methyl esters and TMS sugars, and to 90°C when TFA sugars are present. The programming rate is dependent on the complexity of the sample. When analyzing an unknown sample, 2°C/min for a packed column and 8°C/min for a capillary column may be used.

4. APPLICATIONS IN CLASSIFICATION

4.1. Lipid Constituents

Fatty acids are the most generally occurring constituents of bacterial lipids and most interest has been focused on their implication in bacterial taxonomy. However, several additional lipid types, also detectable by GC or GC/MS, are frequently reported. For instance, wax esters and their long-chain alcohol constituents have been reported to be interesting taxonomic markers of species in family *Neisseriaceae* (Bryn et al., 1977) and of genus *Chloroflexus* (Knudsen et al., 1982). As indicated in Figure 2, characteristic wax patterns can be obtained after a simple solvent extraction and high-temperature GC. Methanolysis liberates their alcohol constituents which may be analyzed directly or derivatized. In any case they are easily recognized by retention times and MS fragmentation (Figure 3).

Hydrocarbons, complex mixtures of C_{15}–C_{30} straight-chain and branched isomers, have been found to occur in *Vibrio* and *Micrococcus* (Tornabene et al., 1969; Morrison et al., 1971). Several groups of anaerobic bacteria contain alk-1'-enyl glyceryl ethers (plasmalogens) (Kamio et al., 1969; Lechevalier, 1977) which yield, upon methanolysis, GC-analyzable, but unstable, fatty aldehyde methylacetals.

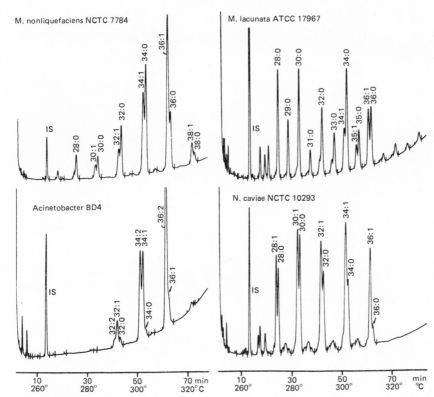

FIGURE 2. GC wax ester profiles of four strains of *Neisseriaceae*. In the wax designation, the first number indicates the number of carbon atoms, and the second, the number of double bonds. IS (internal standard), methyl tricosanoate ($C_{23:0}$). (Bryn *et al.*, 1977.)

Major types of fatty acids and their distribution within bacterial groups are listed in Tables 2 and 3 (see also Goldfine, 1972; Lechevalier, 1977; and Shaw, 1974). Presently, only limited information is available since a restricted number of species and strains of each species have been analyzed. Furthermore, much of the available data on fatty acid compositions are incomplete due to insufficient peak identification or technical shortcomings. Thus the fatty acids of "bound lipids" have often not been included in the studies. Hydroxy fatty acids are easily lost unless precautions are taken, e.g., on-column injection and derivatization. Nevertheless, certain taxonomic features are presently evident:

1. Two large, distinct groups of bacteria can be recognized; those organisms synthesizing methyl branched fatty acids (*iso-* and *anteiso-*) and

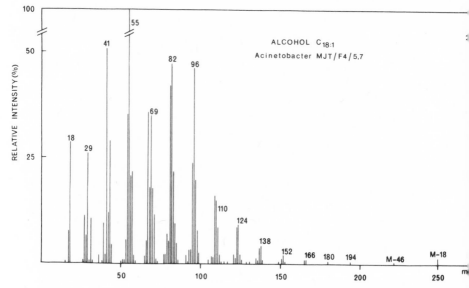

FIGURE 3. Mass spectrum of the characteristic octadecenol wax constituent of a *Acinetobacter* strain. (Bryn and Jantzen, unpublished.)

those making straight-chain, saturated and monounsaturated fatty acids. This distinction was previously believed in principle to coincide with the gram-stain characteristics. However, as evident from Tables 2 and 3, there are several exceptions.

2. Gram-negative bacteria are, with few exceptions, characterized by the presence of hydroxy fatty acids (constituents of lipopolysaccharides). However, hydroxylated fatty acids may also be found in gram-positive bacteria. *Listeria monocygotes*, for instance, has recently been found to contain lipopolysaccharides similar to those found in gram-negative cells (Wexler and Oppenheim, 1980). Furthermore, both gram-positive and gram-negative bacteria may contain hydroxy fatty acid-containing lipids of other kinds (Lechevalier, 1977).

3. Acid-fast bacterial species (e.g., coryneform bacteria, *Nocardia*, and mycobacteria) are characterized by very complex lipids which are entirely different from the lipids of other prokaryotic organisms (see Minnikin and Goodfellow, 1980). Whole cell fatty acid profiles of these bacteria show abundant amounts of 10-methyl-stearic acid (tuberculostearic acid) and certain other methyl-branched acids (Asselineau, 1966; Julak et al., 1980).

TABLE 2. Fatty Acid Composition of Some Genera of Gram-Negative Bacteria[a]

Genus/group	Straight chain	Unsaturated[b]	Hydroxylated[c]	Cyclopropane[d]	Branched chain[e]	References
Achromobacter 1[f]	14, 15, **16**, 17, *18*, 19	16, 18	2-12, 2-14, 3-14, 3-16	**17**, 19	—	Dees and Moss (1978)
Achromobacter 2[f]	16, 17, 18, 19, 20	16, 18	—	17, **19**	—	Dees and Moss (1978)
Acinetobacter	12, 14, 15, **16**, 17, 18	**16**, 17, 18	2-12, 3-*12*, 3-14	—	—	Jantzen et al. (1975)
Actinobacillus	*14*, **16**, 18	**16**, 18	3-14	—	—	Jantzen et al. (1980)
Alcaligenes	12, 14, **16**, 18	16, 18	2-12, 2-14, 3-14, 3-16	*17*, 19	—	Dees et al. (1980)
Bacteroides	14, 15, 16	—	3-b15, 3-15, 3-16, 3-b**17**, 3-17	—	14, **15**	Mayberry (1980); Miyagawa and Suto (1980)
Bordetella	12, 14, 15, **16**, 17, 18	*16*, 18	3-10, 2-12, 3-12, 3-14, 3-16	17, 19	—	Jantzen et al. (1982)
Brucella	16, 17, *18*	16, 17, **18**	t	**19**	—	Dees et al. (1981)
Campylobacter	*14*, **16**	**16**, **18**	3-14	19	—	Blaser et al. (1980)
Cardiobacterium	12, *14*, **16**, 18	**16**, **18**	3-14, 3-16	—	—	Bøvre and Hagen (unpublished)
Cytophaga	14, 15, 16	15, b16, 16	3-b*14*, 3-14, 3-b**16**, 3-*16*	—	14, 16	Fautz et al. (1979)
Eikenella	12, 14, 15, **16**, 18	16, **18**	3-12, 3-14, 3-16	—	—	Bøvre and Hagen (unpublished)
Enterobacteriaceae	12, 13, 14, 15, **16**, 17, 18	**16**, *18*	3-13, 2-14, 3-14	*17*, 19	—	Jantzen and Lassen (1980; unpublished)

(continued overleaf)

TABLE 2. (cont.)

Genus/group	Straight chain	Unsaturated[b]	Hydroxylated[c]	Cyclopropane[d]	Branched chain[e]	References
Flavobacterium	14, 16, 18	16, b17	2-b15, 3-b15, 3-16, 3-b17	—	14, 15, 16	Oyaizu and Komagata (1981); Yano et al. (1976)
Francisella	10, 14, 16, 18, 20, 22, 24, 26	16, 18, 20, 22, 24, 26	2-10, 3-16, 3-18	—	—	Jantzen et al. (1976)
Flexibacter	14, 15, 16, 18	b16, 16	2-b14, 3b-14, 3-14, 2b16, 3-16	—	12, 15, 16	Fautz et al. (1979)
Fusobacterium	12, 14, 15, 16, 18	16, 18	3-12, 3-14, 3-16	—	—	Jantzen and Hofstad (1981)
Gardnerella	14, 16, 18	16, 18	—	—	—	Moss et al. (1969)
Haemophilus	14, 16, 18	16, 18	3-14	—	—	Jantzen et al. (1980)
Kingella	12, 14, 16, 18	16, 18	3-12, 3-14	—	—	Bøvre and Hagen (unpublished)
Legionella	14, 15, 16, 17, 18, 19	14, b16, 16	t	—	14, 15, 16, 17	Moss et al. (1977); Mayberry (1981)

					Reference
Leptotrichia	12, 14, 15, **16**, 18	16, **18**	3-14	—	Hofstad and Jantzen (1981)
Moraxella/ Branhamella	12, 14, *16*, 18	16, **18**	3-12, 3-14,	—	Jantzen et al. (1978)
Neisseria	12, 14, 15, **16**, 18	16, 18	3-12, 3-14, 3-16	—	Jantzen et al. (1978) Lambert et al. (1971)
Pasteurella	*14*, **16**, 18	16, 18	3-14	—	Jantzen et al. (1980)
Pseudomonas (see Table 4)					
Treponema pallidum	14, 15, 16, *17, 18*	16, **18**, 20	—	—	Matthews et al. (1979)
Treponema refringens	14, 15, 16, 18	16, **18**	—	—	Cohen et al. (1970)
Treponema denticola	14, 15, **16**	**18**	—	14, 15, 16	Livermore and Johnson (1974)

[a] Number indicates chain length. The relative amount is indicated as follows: bold, above 20%; italic, between 10% and 20%; normal, below 10%.
[b] Monounsaturated fatty acids; no distinction between isomers has been made.
[c] First number indicates position of OH-group, second number denotes number of carbon atoms in chain. t, trace amounts. The prefix b indicates a methylbranched chain.
[d] Cyclopropane substituted fatty acids.
[e] No distinction between iso and anteiso isomers has been done.
[f] See Table 5.

4. Certain gram-positive and gram-negative bacteria synthesize characteristic cyclopropane substituted fatty acids.

5. Within each of these biosynthetically distinct groups a large number of group- or species-specific fatty acid patterns are occurring. Some examples can be read from Tables 2 to 6.

4.2. Sugars and Amino Acids

Cummins and Harris (1956) introduced cell-wall analysis in classification of bacteria. They examined several gram-positive bacterial species and on the basis of a few key components of cell-wall hydrolysates they were able to divide the examined bacteria into distinct groups.

Following these studies, analysis of cell-wall diaminoacids (e.g., isomers of diaminopimelic acid, diaminobutyric acid, ornithine and lysine) and certain sugars (e.g., arabinose, glucose, galactose, and mannose) has become a well-established and widely used criterion in the classification of gram-positive bacteria (see Keddie and Bousfield, 1980).

The originally employed technique required considerable amounts of purified cell walls and includes several quite laborious methodological steps. Attempts have therefore been made for developing methods more suitable for routine work. However, the production of purified cell wall material is still relatively laborious. Presently therefore the applications are mainly limited to the classification and identification of organisms belonging to taxonomically problematic organisms. Recently, Guerrant et al. (1979) demonstrated the presence of diaminopimelic acid in *Legionella pneumonia* by GC/MS technique (Figure 4). The amount detected was comparable to that of most gram-negative organisms, and the result was helpful for the classification of these taxonomically odd organisms.

Identification of many gram-positive bacteria is guided by the detection of spores in stained smears. This characteristic is particularly important in the identification of members of genus *Clostridium*, but the test is frequently inconclusive due to poor sporulation and heat-labile strains. Dipicolinic acid occurs in substantial amounts in bacterial spores. Tabor et al. (1976) have developed a rapid GC method for the detection of dipicolinic acid in the spores which may be useful clinically for the rapid detection of *Clostridium* species.

Moss et al. (1971) examined the amino acid composition of 14 treponemes by GC technique but no distinct differences among the strains could be observed. Unfortunately, however, such studies of bacterial whole cell amino acids have not been followed up and the potential of this kind of data in bacterial taxonomy is presently uncertain.

TABLE Fatty Acid Composition of Some Genera of Gram Positive Bacteria

Genus/group	Straight chain	Unsaturated	Branched chain	Cyclopropane	References
Actinomyces	10, 12, 14, **16**, **18**, 20	16, 18, 20	—	—	Amdur et al. (1978)
Bacillus	12, 14, 15, **16**, 17, 18	—	12, 13, 14, *15*, 16, 17, 18	—	Kaneda (1977); Niskanen (1978)
Bifidobacterium	12, 13, 14, 15, **16**, 18	14, 16, 17, **18**	15	19	Veerkamp (1971)
Clostridium	12, 14, 15, **16**, 17, *18*, 20	16, *18*	—	—	Ellender et al. (1970); Elsden et al. (1980)
Corynebacterium	11, 12, 14, 15, **16**, 17, 18, 20	16, 18	**15**, *17*	—	Moss et al. (1967); Bowie et al. (1972)
Lactobacillus	12, 14, **16**, 18	16, 17, **18**	14, 15, 16, 18	**19**	Veerkamp (1971)
Listeria	12, 14, 15, 16, 17, 18, 19, 21, 22, 23	16, 18	**15**, **17**	—	Raines et al. (1968); Tadayon and Carrol (1971)
Micrococcus	14, 15, 16, 17, 18, 19	16, 18	14, **15**, *16*, *17*	—	Jantzen et al. (1974a); Brooks et al. (1980)
Mycobacterium	14, 15, **16**, 17, 18, 20, 22, 24, 26	16, **18**	15, 17, *19*	17, 19	Thoen et al. (1971; 1973); Tisdall et al. (1979, 1982); Chomarat et al. (1981)
Pediococcus	12, 14, **16**, 18	12, 14, *16*, **18**, 20	—	17, 19	Uchida and Mogi (1972)
Peptostreptococcus	10, 12, 13, 14, 15, 16, 17, 18	18	10, 11, 12, 13, *14*, 15, *16*, 17, 18	—	Lambert and Armfield, (1979)
Propionibacterium	12, 14, 16, 18, 19, 20, 21, 22	16, 18	**15**, **17**	—	Moss et al. (1969)
Staphylococcus	14, 16, 17, *18*, 19, 20	18	14, **15**, 16, *17*, 18, 19	—	Jantzen et al. (1974a); Dees and Moss (1978)
Streptococcus	14, **16**, *18*, 20	16, 17, **18**, 20	14, **15**, 16, *17*, 18, 19	19[b]	Fischer (1976); Lambert and Moss (1976)

[a] See Table 2 for explanations.
[b] Only in some species.

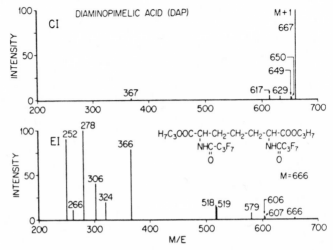

FIGURE 4. Mass spectra (EI and CI) of N-heptafluorobutyryl-n-propyl derivative of diaminopimelic acid obtained from *Legionella pneumonia* (Guerrant et al., 1979).

The analyses of whole cell sugars provide very complex GC profiles but have been demonstrated to be of use in the classification of bacterial isolates. Yamakawa and Ueta (1964) used a technique based on methanolyzed whole cells, removal of the liberated fatty acid methyl esters by extraction, and subsequent trimethylsilylation (TMS-derivatization) of the remaining sugar fraction. They observed that "*Neisseria haemolysans*" differs from *Neisseria* species both in fatty acid and sugar content. This organism is now named *Gemella haemolysans* (Buchanan and Gibbons, 1974).

Farshy and Moss (1970) were able to differentiate ten species of *Clostridium* by a rapid and simple GC technique based on TMS derivatized whole-cell methanolysates. These kinds of complex and mostly unresolved elution profiles include sugars as TMS methylglycosides and fatty acids as methyl esters, and are mainly intended to serve as a kind of easily available chemical "fingerprints" (TMS-profiles) of bacterial isolates.

A similar technique substituting the TMS derivatization with trifluoroacetylation (TFA-derivatization) has also been reported (Jantzen et al., 1972, 1978). TFA-sugars are generally more volatile than the TMS analogs. Pentoses, hexoses, heptoses, and hexosamines all elute before methyl dodecanoate (12:0) on a methyl silicone column and GC profiles of TFA derivatized whole cell methanolysates represent a rough estimate of both fatty acids and monosaccharides present in the sample. Such TFA profiles were tried out on representative strains of *Neisseria*, *Moraxella*, and *Acinetobacter*. The results generally reflected the preconceived classification (Jantzen *et al.*, 1978; Bøvre, 1980). TFA profiles have also been assessed for characterization of mycobacterial isolates (Larsson and Mårdh, 1976; Larsson *et al.*, 1979). The examined species were distinguished by this simple GC technique.

A major improvement of sugar-profiles was recently described by Pritchard *et al.* (1981). They demonstrated a capillary column version

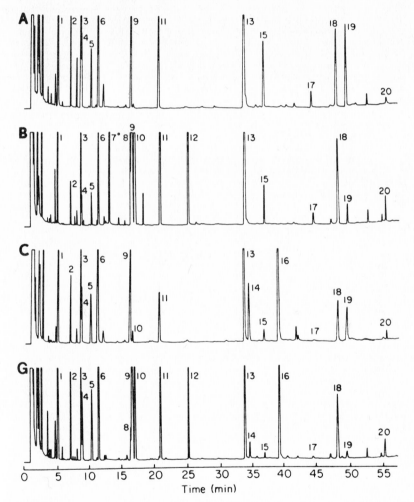

FIGURE 5. GC profiles of 2-μg samples of lyophilized whole cells group A, B, C, and G streptococci run as trifluoroacetylated methylglycosides on a 30-m WCOT (OV-105) glass capillary column. (EC detector) peaks are numbered as follows: rhamnose (1, 3), ribose (2, 4, 5, 6), 1,4-anhydroglucitol (7), galactose (8, 10, 12), glucose (9, 11), *N*-acetylglucosamine (13, 15, 17, 19), *N*-acetylgalactosamine (14, 16), and *N*-acetylmuramic acid (18, 20). (Pritchard *et al.*, 1980.)

TABLE 4. Characteristic LPS Constituents of Some Gram-Negative Bacterial Groups

Bacterial group	Heptose	KDO	Non-OH fatty acid	3-OH fatty acid[a]	2-OH fatty acid[a]	References
Salmonella	+	+	12[b], 14, 16	14	14	Bryn and Rietschel (1978)
Serratia	+	+	12, 14, 16	14	14	Bryn and Rietschel (1978)
Klebsiella	+	+	12, 14, 16	14	14	Bryn and Rietschel (1978)
Enterobacter	+	+	14, 16	14	—	Bryn and Rietschel (1978)
Proteus	+	+	14, 16	14	—	Bryn and Rietschel (1978)
Escherichia	+	+	12, 14, 16	14	—	Bryn and Rietschel (1978)
Yersina	+	+	12, 16	14	—	Bryn and Rietschel (1978)
Shigella flexneri	+	+	12, 14, 16	14	—	Bryn and Rietschel (1978)
Shigella sonnei	+	+	12, 14, 16	14	14	Lugowski and Romanowska, (1974)
Haemophilus	+	+	?	14	—	Flesher and Insel (1978)
Actinobacillus	+	+	14, 16	14	—	Kiley and Holt (1980)
Bordetella	+	+	14	10, 12, 14	—	Le Dur et al. (1980)
Fusobacterium	+	+	14, 16	14, 16	—	Hofstad and Skaug (1980)
Vibrio cholera	+	–	12, 14, 16	12, 14	—	Broady et al. (1981)
Neisseria	+	+	12	12, 14	—	Jennings et al. (1973)
Moraxella	–	+	10, 12	12, 14	—	Bryn (unpublished)
Branhamella	–	+	10, 12	12, 14	—	Bryn (unpublished)
Kingella	–	+	12, 14	12, 14	—	Bryn (unpublished)
Actinetobacter	–	+	12	12, 14	12	Bryn (unpublished)
Brucella abortus	–	+	16, 18	16?	—	Moreno et al. (1978)

[a] 3-hydroxy and 2-hydroxy fatty acids, respectively.
[b] Number of carbon atoms in the chain.

applied for characterization of streptococcal strains (Figure 5). Their technique combines the very high resolving power of a capillary column with the high sensitivity and specificity for monosaccharide trifluoroacetates of an electron capture detector. Thus, carbohydrate profiles as shown in Figure 5, where each peak largely is resolved and identified, may be obtained from material of single colonies grown on solid medium.

In our laboratory, we have employed a similar capillary column technique for analysis of bacterial lipopolysaccharides (Bryn and Jantzen, 1982).

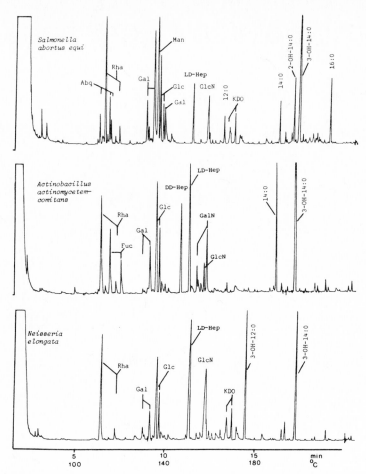

FIGURE 6. GC profiles of 3 LPS preparations after methanolysis and trifluoroacetylation. Column: fused-silica (25 m × 0.2 mm i.d.) coated with methylsilicone. Abbreviations: Abq, abequose; Rha, rhamnose; Fuc, fucose; Gal, galactose; Glc, glucose; Hep, heptose; GlcN, glucosamine; GalN, galactosamine; KDO, 2-keto-3-deoxyoctonate. (Bryn and Jantzen, 1982.)

These structures show a large structural variation among gram-negative bacterial groups (cf. Lüderitz et al., 1971; Rietschel and Lüderitz, 1980) which has been utilized for several years in serological identification. The immunological specificity is directly related to the terminal polysaccharide part (O-antigen) of LPS which show a remarkably structural diversity, especially among strains of Enterobacteriaceae. The so-called core polysaccharide and lipid A part of LPS show less structural variation, but as indicated in Table 4, distinct structural differences in these less variable parts of LPS have also been revealed among more distantly related gram-negative groups. Figure 6 shows three different chromatographic LPS patterns obtained after methanolysis, TFA-derivatization, and GC on a fused-silica capillary column. Such profiles can be obtained from bacterial cells of a single agar plate (Bryn and Jantzen, 1982).

4.3. Influence of Growth Conditions

It is known that the chemical composition of bacteria may be influenced by external factors, such as temperature, substrate composition, age of culture, etc. (Ellwood and Tempest, 1972). The analytical results are also to some degree dependent on the chemical techniques employed.

In a study of nonpathogenic Neisseria, Lambert et al. (1968) found marked differences, both qualitative and quantitative, in the fatty acid composition of the examined bacteria after growth in three different media. Most notably, after growth in one of the media (Difco GC medium base with 1% defined supplement) large amounts of a monoenoic C_{17} acid could be detected in Branhamella catarrhalis and Neisseria caviae. This constituent was not found after growth in the two other media tested. In a study of changes in the fatty acid composition of Staphylococcus aureus under various cultural conditions, Vaczi et al. (1967) also found that the ratio between saturated and unsaturated fatty acids is easily influenced by pH and media composition but less by small variations in temperature. The fatty acid and aliphatic hydrocarbon composition of Sarcina lutea grown in three different media were examined by Tornabene et al. (1967). Only minor differences in the fatty acid content could be observed but GC of the hydrocarbon fractions showed less compounds of odd-numbered carbon chain after growth in one of the media studied.

In a study of fatty acid and monosaccharide composition of species within the family Neisseriaceae, the reproducibility of a standardized technique, using several batches of a single routine medium, was examined (Jantzen et al., 1978). One strain of Branhamella catarrhalis was cultivated, harvested, and analyzed 9 times over a period of 3 years. Most compounds showed a variation in the range of 5%–15%, while those compounds present in very small amounts (0.1%–3%) varied considerably, e.g., up to 50%. This latter variation is probably due to a high content of "background" impurities. The effect of a moderate elevation of growth temperature, from 33°C to 37°C, was also tested and a slight increase in the ratio of saturated to unsaturated fatty acids could be demonstrated. Hence even when the growth conditions and procedures are standardized, a certain day-to-day variation in cellular composition (10%–20%) has to be expected. Better reproducibility requires chemically defined media, which probably is impractical in clinical work.

These studies show clearly that both the growth conditions and the chemical procedures have to be standardized before a meaningful comparison of GC patterns can be obtained. Even then the quantitative differences should be of a certain magnitude to be considered significant.

4.4. Taxonomic Resolution

While several groups of bacteria have been analyzed for fatty acid composition, considerably less GC data are available on other cellular constituents such as sugars and amino acids. Hence, the taxonomic implications of fatty acid profiles may currently be indicated, whereas the utility of GC profiles of sugars and amino acids is more difficult to specify, although recent reports on sugar profiles point to a useful tool for distinguishing closely related bacteria (Pritchard *et al.*, 1981; Bryn and Jantzen, 1982).

Within the bacterial family *Neisseriaceae* the interrelationships between most species are well known since they have been determined semiquantitatively by the genetic method of transformation (Bøvre, 1980). As indicated in Figure 7, the classification obtained by fatty acid analysis, overlapped almost completely the classification based on genetic and phenotypic data. However, as shown in Figure 8, not all species could be differentiated by fatty acid composition.

Fatty acid data allowed considerably less differentiation within the family *Enterobacteriaceae* than in the family *Neisseriaceae* (Jantzen and Lassen, 1980 and unpublished results). Similarly, only minor quantitative differences could be detected among the small gram-negative rod-shaped bacteria belonging to genera *Haemophilus*, *Pasteurella*, and *Actinobacillus* (the HPA-group) (Jantzen *et al.*, 1981). *Pseudomonas*, on the other hand, can be divided into at least eight groups on the basis of their fatty acid composition (cf. Moss, 1978). However, both *Pseudomonas* and *Neisseriaceae* are considerably more phylogenetically heterogeneous than *Enterobacteriaceae* and the HPA-group.

Thus in most instances only identification to a certain group level is possible by the use of fatty acid profiles. On the other hand, the allocation

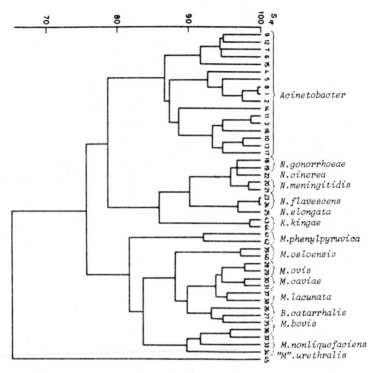

FIGURE 7. Phenogram of *Neisseriaceae* strains based on fatty acid composition (Jantzen *et al.*, 1978).

of "problematic" isolates to group/species level by fatty acid composition can in most cases rapidly be obtained. In practice, however, the usefulness of the technique is largely dependent on a preconceived record of fatty acid data. Certain precautions should also be taken because phylogenetically different bacterial species may occasionally share fatty acid patterns. Thus this GC technique must, like most other techniques, be used in combination with other methods.

5. DIAGNOSTIC IMPLICATIONS

5.1. General Discussion

In practical diagnostic work GC profiles of cellular fatty acids have gained considerable interest since Abel *et al.* (1963) demonstrated the applicability of this chemical approach for the characterization of bacteria.

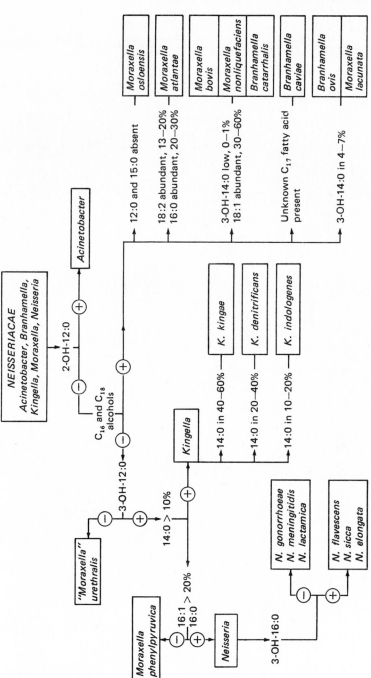

FIGURE 8. Flow chart for differentiation of *Neisseriaceae* species by GC of cellular fatty acids and alcohols.

TABLE 5. Cellular Fatty Acids of Some Genera/Groups of Glucose Nonfermenters[a]

	Bacterial groups															
	Pseudomonas[b]										Achromo-[c] bacter		Alcali-[d] genes			
Fatty acid	1	2	3	4	5	6	7	8	IIk-1	IIk-2	1	2	1	Ve-1[e]	Ve-2[e]	IVe[e]
12:0[f]	5[g]	—	12	—	—	—	3	5	—	—	—	—	1	5	6	—
13:0	—	—	—	—	—	—	3	—	—	—	—	—	—	—	—	10
14:0	1	5	2	4	6	5	4	5	t	3	—	t	6	t	t	—
15:0	1	1	4	t	5	t	7	2	t	t	—	t	—	t	t	—
16:1	18	7	15	3	6	10	16	22	13	27	3	9	15	22	19	3
16:0	20	21	24	29	26	18	8	25	26	10	12	27	25	23	26	26
17:1	—	—	—	—	—	—	16	—	9	—	3	—	—	—	—	—
17:0	t	t	2	t	4	t	2	t	t	t	4	t	—	—	t	—
18:1	25	17	25	22	34	t	4	25	52	t	32	6	16	29	33	45
18:0	t	1	1	t	10	t	t	1	t	t	14	4	5	t	4	t
19:0	t	2	2	—	—	—	—	—	t	—	t	t	—	—	—	—
20:0	—	—	—	—	—	—	—	—	—	—	t	t	—	—	—	—
i11:0	—	—	—	—	—	5	11	—	—	t	—	—	—	—	—	—
i13:0	—	—	—	—	—	2	4	—	—	—	—	—	—	—	—	—
i14:0	—	—	—	—	—	2	—	—	—	—	—	—	—	—	—	—
i15:1	—	—	—	—	—	—	—	—	—	t	—	—	—	—	—	—
i15:0	—	—	—	—	—	30	22	—	—	37	—	—	—	—	—	—
i16:0	—	—	—	—	—	t	—	—	—	—	—	—	—	—	—	—

Fatty acid[f]																	
i17:1	—	—	—	—	—	—	t	—	—	—	—	—	—	—	—	—	—
i17:0	—	—	—	—	—	—	—	—	—	—	—	—	—	—	—	—	—
3-OH-10	5	—	5	—	10	—	—	—	4	—	—	—	—	4	—	—	9
2-OH-12	9	5	—	2	—	5	—	—	—	—	4	—	5	4	5	—	—
3-OH-12	2	2	3	3	2	—	—	—	—	2	—	3	5	4	4	5	3
2-OH-14	—	—	—	—	—	—	—	—	—	—	—	—	2	—	—	2	—
3-OH-14	t	—	—	t	—	t	6	—	—	—	—	—	—	1	—	—	—
3-OH-14	6	—	—	—	—	—	—	—	—	—	—	—	—	9	—	—	—
2-OH-16	5	—	—	—	—	—	—	—	—	—	—	—	—	—	—	—	—
3-OH-16	4	—	—	—	—	—	—	—	—	—	—	—	—	—	—	—	—
3-OH-17	—	—	—	—	—	—	—	t	—	—	—	—	—	—	—	—	3
2-OH-i11	—	—	—	4	—	—	—	—	—	—	—	—	—	—	—	—	—
3-OH-i11	—	—	—	4	—	—	—	—	—	—	—	—	—	—	—	—	—
3-OH-i13	—	—	—	4	—	—	—	—	—	—	—	—	—	—	—	—	—
2-OH-i15	—	—	—	—	—	—	18	—	—	—	—	—	—	—	—	—	—
3-OH-i15	—	—	—	—	—	—	3	—	—	—	—	—	—	—	—	—	—
3-OH-i17	—	—	—	—	—	—	2	—	9	—	—	—	—	18	2	—	—
17:cyclo	5	17	—	t	t	t	t	—	—	t	—	5	30	—	6	—	—
19:cyclo	11	13	29	—	—	—	—	34	—	—	—	—	3	—	—	—	5

---

[a] Extracted from Dees and Moss (1975; 1978); Dees et al. (1979; 1980); Moss and Dees (1976).

[b] 1, *P. aeruginosa*, *P. putida*, and *P. fluorescens*; 2, *P. cepacia* and *P. pseudomallei*; 3, *P. stutzeri*, *P. mendocina*, and *P. pseudoalcaligenes*; 4, *P. diminuta*; 5, *P. vesicularis*; 6, *P. maltophila*; 7, *P. putrefaciens*; 8, *P. acidovorans* and *P. testosteroni*; IIk-1, *P. paucimobilis*; IIk-2, (King group).

[c] 1, *Achromobacter* sp. (King group Vd); 2, *A. xylosoxidans*.

[d] 1, *Alcaligenes faecalis*, *A. denitrificans*, and *A. odorans*.

[e] King groups.

[f] Figure before colon indicates number of carbon atoms in the chain; after, number of double bonds. The prefix "i" indicates an iso-branched chain. 2-OH and 3-OH denote a hydroxyl group and its position.

[g] Mean value (relative amount) t; trace amount (less than 0.5%).

Since that time fatty acid patterns have mainly been used as a general aid in identification of isolates of a few distinguishing characters such as glucose nonfermenters and certain slow-growing organisms, e.g., certain species of genus *Mycobacterium*. The other frequently suggested application is as a means for detection of microbes in clinical material. The latter approach is still in the developmental stage.

5.2. Glucose Nonfermenters

Moss and co-workers have demonstrated the utility of fatty acid profiles in identification of bacterial isolates which are difficult to identify (cf. Moss, 1978, 1981). These investigators have elucidated the fatty acid composition of several groups of nonfermenters, including *Pseudomonas*, *Achromobac-ter*, and *Alcaligenes*. Species of these genera are imprecisely defined by conventional characters but can be distinguished on the basis of their fatty acid composition. Some of these results are summarized in Table 5 (cf. Moss, 1978). The eight GC groups of *Pseudomonas* are differentiated on the basis of the distinct patterns of straight-chain-, branched-chain-, hydroxy-, as well as cyclopropane-substituted fatty acids. The two examined species of *Achromobacter*, i.e., *A. xylosoxidans* and *A.* sp. (King's group Vd), were easily distinguished on the basis of the presence and absence of hydroxy fatty acids and a 17-carbon cyclopropanoic acid.

The fatty acid patterns of the three species of *Alicaligenes*, i.e., *A. faecalis*, *A. denitrificans*, and *A. odorans*, were almost indistinguishable (Dees and Moss, 1975), but as a group they can be differentiated from the *Pseudomonas* and *Achromobacter* species examined (Table 5).

Some unnamed groups of nonfermentative organisms (e.g., IIk-1, IIk-2, Ve-1, Ve-2, and IVe) have also been studied by Moss and co-workers (references in Table 5). Organisms belonging to these groups are frequently isolated in the clinical laboratory and show some resemblance to species of *Pseudomonas*, *Achromobacter*, or *Alcaligenes*. However, their identifications are generally rather difficult and fatty acid analysis seems to be a useful supplementary identification technique for isolates of these groups, also.

Similarly Moss *et al.* (1977) have demonstrated that the classification and identification of the species of *Legionella* can be guided by their very characteristic fatty acid profiles (Figure 9). This unique pattern is easily recognizable by its high level of iso-methyl-branched fatty acids (e.g., i16 : 0). These constituents are uncommon, and as indicated by Fisher-Hoch *et al.* (1979), they represent an interesting potential marker for a rapid and specific detection of legionary disease in analysis of clinical samples.

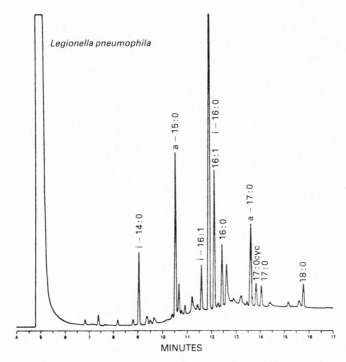

FIGURE 9. GC profile of fatty acids from whole cells of *Legionnella pneumonia*. Column: fused-silica (50 m × 0.2 mm i.d.) coated with methylsilicone (Moss, 1981).

5.3. Species of Neisseriaceae

The family *Neisseriaceae* (presently encompassing the genera *Neisseria, Branhamella, Moraxella, Kingella,* and *Acinetobacter*) includes *Neisseria gonorrhoeae* and *N. meningitidis.* A diagnosis to species level among other species of the family may often be difficult by conventional techniques (Bøvre *et al.*, 1976). Both fatty acid and monosaccharide composition of the species of this taxon have been examined (Jantzen *et al.*, 1974b; 1975; 1976; and 1978). Figure 8 and Table 6 indicate the feasibility of the fatty acid patterns for the distinction among species of this family. Several chemical markers may be used diagnostically. For instance, the presence of 2-hydroxy dodecanoate distinguishes the oxidase-negative members of the family (*Acinetobacter*); the presence of wax esters, represented by fatty alcohols in the GC profiles (Bryn *et al.*, 1977), excludes *Neisseria* and *Moraxella phenylpyruvica*; abundant amounts of n-tetradecanoate are strongly indicative of *Kingella*. These chemotaxonomic patterns have

TABLE 6. Cellular Fatty Acids of Genera/Species of *Neisseriaceae*[a]

	Bacterial groups											
	Neisseria[b]		Moraxella/Branhamella[c]						Kingella[d]			Acinetobacter[e]
Fatty acids	1	2	1	2	3	4	5	6	1	2	3	1
12:0	4	3	1	1	—	1	1	—	7	2	4	3
2-OH-12:0	—	—	—	—	—	—	—	—	—	—	—	3
3-OH-12:0	7	7	7	8	3	14	10	4	3	3	4	10
14:0	5	4	t	t	t	1	2	7	51	11	31	t
3-OH-14:0	6	6	t	4	1	t	2	1	2	3	4	3
15:0	t	2	—	—	—	t	t	2	t	t	t	t
16:1	28	24	15	17	14	9	2	2	18	19	15	16
16:0	35	31	11	9	6	19	31	30	7	22	33	22
3-OH-16:0	—	3	—	—	—	—	—	8	—	—	—	—
17:1	t	t	4	3	2	t	1	—	—	—	—	t
17:0	t	t	t	—	t	t	1	1	—	—	—	2
18:0alc	—	—	5	4	1	—	1	—	t	—	—	t
18:2	t	t	3	5	2	18	16	t	t	1	—	1
18:1	12	16	39	45	66	24	20	46	3	35	—	31
18:0	2	2	4	2	5	7	13	2	1	1	—	1

[a] Extracted from Jantzen et al. (1974b; 1975). [b] 1, *N. gonorrhoeae, N. meningitidis, N. lactamica, N. cinerea*; 2, *N. elongata, N. sicca, N. animalis, N. flavescens, N. mucosa*. [c] 1, *B. catarrhalis, M. bovis, M. nonliquefaciens, M. caviae*; 2, *M. ovis, M. lacunata*; 3, *M. osloensis*; 4, *M. phenylpyruvica*; 5, *M. atlantae*; 6, "*M. urethralis*"; [d] 1, *K. kingae*; 2, *K. indologenes*; 3, *K. denitrificans*. [e] 1, *A. calcoaceticus*.

proved to be stable and useful both in taxonomic research and as a diagnostic tool (Bøvre et al., 1977; unpublished results).

5.4. Slow-Growing Mycobacteria

Identification of mycobacteria with standard methods of biochemical testing is hampered by their slow growth rate. Therefore there is a need for sensitive and more rapid methods to identify clinical isolates. Fatty acid profiles have been suggested by several investigators since Anderson (1943) demonstrated the uniqueness of mycobacterial cell-wall lipids. Thoen et al. (1971 and 1973) demonstrated the utility of whole-cell fatty acid profiles for differentiating mycobacterial species. These results were recently followed up by Tisdall et al. (1979, 1982). Their study included 18 mycobacterial species represented by 128 reference strains, and 79 clinical isolates. Of the isolates tested, 64% were identified to species level by fatty acid profiles. An additional 35% were differentiated to the level of groups consisting of two or three species. An identification scheme could be constructed where only 2 out of the 288 strains tested gave atypical chromatographic patterns.

In recent studies in our laboratory (Andersen et al., 1982; Jantzen and Saxegaard, unpublished results) a number of mycobacterial isolates from humans, as well as from wild and domestic animals, were analyzed for fatty acids by a technique similar to that of Tisdall et al. (1979). Included in these studies were also 18 test strains of well-characterized mycobacterial species. The results generally confirmed those of the American group. The fatty acid patterns could in most cases distinguish the strains tested to species level, whereas members of the so-called MAIS complex (M. avium, M. intracellulare, and M. scrofulaceum) provided identical patterns.

5.5. Detection of Bacterial Constituents in Clinical Material

Ultrasensitive technique such as obtained by GC/MS in selected ion monitoring (SIM) mode and by electron-capture (EC) detection of halogen-enriched derivatives is capable of detecting bacterial constituents in amounts down to about 20 pg (Larsson et al., 1980; Sud and Feingold, 1979). This high sensitivity makes it possible to detect the presence of unique bacterial marker molecules of bacteria in body fluids and specimens. Recently published works show that the sensitivity is sufficient and that the analytical approach is feasible (Odham et al., 1979; Larsson et al., 1980 and 1981; Maitra et al., 1978; Sud and Feingold, 1979).

5.5.1. Lipopolysaccharide Constituents

Two of these recently described techniques are based on the detection of gram-negative cell-wall lipopolysaccharide (LPS) constituents, and are intended for a rapid detection of sepsis or arthritis caused by gram-negative bacteria. As shown in Table 7, the 3-hydroxy fatty acids of the lipid A part of LPS have hitherto been chosen as marker molecule. These lipid constituents do not occur in mammalian lipids, but they are relatively abundant and easily recognized. Several sugar constituents may also be candidates for characterization of infectious agents. Dead organisms as well as viable organisms can be identified to a certain level by chemical analysis of the kind discussed.

Sud and Feingold (1979) suggested the use of electron capture detection of 3-hydroxy-dodecanoic acid (heptafluorobutyrated methylester) as a screening method for detection of gonococcal infection. The lower detection limits of the technique are about 10^5 gonococci, but interfering substances make it problematic to work at maximum sensitivity. However, the method is sensitive enough for a direct detection of gonorrhea in vaginal specimens. In certain diagnostic situations this method may represent a useful alternative to the conventional cultural technique.

Maitra and co-workers (1978) introduced a quantitative technique for determining the Lipid A content of LPS added in experimental serum samples which were analyzed by GC/MS in SIM mode. Their technique was based on the detection of the trimethylsilyl (TMS) derivative of 3-hydroxy tetradecanoic acid methyl ester and the mass spectrometer focused at m/z 315.4 (M-15). Although problems of high lipid background were experienced and a silica gel chromatography step was required, the sensitivity was determined to about 60 pg (Table 7), which is claimed sufficient for detection of gram-negative bacteria in serum.

5.5.2. Mycobacterial Lipids

The potential of capillary GC/MS in the early detection of tuberculosis, studying clinical material, has been demonstrated. Odham *et al.* (1979) and Larsson *et al.* (1979; 1980; 1981) have applied this technique in selected ion monitoring (SIM) mode focused on ions of tuberculostearic acid. As indicated in Table 7, based on electron impact (EI), this technique could detect the marker molecule at a level of about 20 pg. Substitution of EI by chemical ionization (CI) enhanced the sensitivity to about 1 pg. Thus, tuberculostearic acid (a marker molecule of acid-fast bacteria, including *Mycobacterium tuberculosis*), can be detected in sputum samples from tuberculous patients. Tuberculous and nontuberculous cases of

TABLE 7. Characteristics of Some Marker Molecules for Ultrasensitive Detection of Infection by GC/MS

Bacterial substance	Derivative	Method of detection	Diagnostic ion	Sensitivity (pg)	References
3-hydroxy tetradecanoic acid	Methylester, TMS ether	EI, SIM	315.4 (M-15)	60	Maitra (1978)
3-hydroxy dodecanoic acid	Butylester, heptafluorobutyrate	GC, EC-detection	—	25	Sud and Feingold (1979)
10-methyl stearic acid	Methylester	EI, SIM	312 (M)	20	Larsson et al. (1979)
	Methylester	CI (isobutane)	312 (M)	1	Larsson et al. (1980)
C32 mycocerosic acid	Methylester	CI (ammonia)	512 (M + 18)	1	Larsson et al. (1981)
Muramic acid	Alditol acetate	EI, SIM	168, 210, 228		Fox et al. (1980)

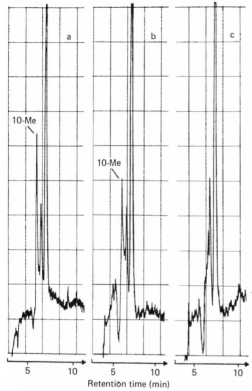

10-Me

10-Me

5   10   5   10   5   10

Retention time (min)

FIGURE 10. EI mass fragmentograms of sputum specimens from a patient with tuberculosis (a) and nontuberculous pneumonia supplemented (b) and not supplemented (c) with 50 pg of authentic tuberculostearic acid, using splitless injections. Column: 25 m glass capillary coated with OV-17. Detection: single ion monitoring at m/z 312. (Larsson *et al.*, 1980.)

pneumonia can also be differentiated on the basis of the chemical method employed (Figure 10).

Brennan and co-workers have recently demonstrated the presence of species-specific glycolipids or peptidoglycolipids as surface structures of several mycobacterial species, including *M. leprae* (Brennan *et al.*, 1981; Hunter and Brennan, 1981). These compounds are present in considerable amounts and can be extracted by nonpolar solvents. They differ mainly in the composition of their *O*-methyl deoxysugars (Figure 11). As shown in Figure 12 these unique deoxysugars provide characteristic MS fragmentation and hence are well suited as diagnostic markers in a GC/MS SIM technique.

Muramic acid is a component of the peptidoglycan moiety of cell walls of all bacteria and is not found elsewhere in nature. A GC/MS assay for muramic acid in tissues of rats with streptococcal cell wall-induced polyarthritis has recently been reported (Fox *et al.*, 1980). Figure 13 shows the GC profile of tissue from arthritic (A) and normal (B) rats. The authors

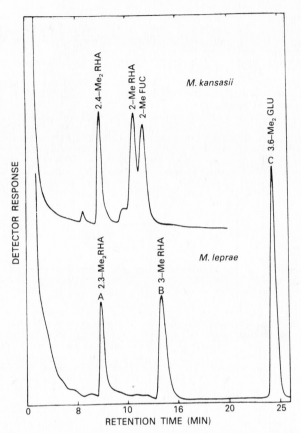

FIGURE 11. GC profile of characteristic O-methyl sugar constituents (alditol acetate derivatives) of phenolic glycolipids of *M. kansasii* and *M. leprae*. (Hunter and Brennan, 1981).

pointed out that the method has the potential to measure total bacterial biomass in tissue and accordingly may become an important assay in elucidating the etiological role of bacterial components in chronic inflammatory diseases in humans.

6. CONCLUSIONS

Chemical analyses of cellular constituents by GC, GC/MS, or other analytical techniques represent a new and highly useful technology in bacterial systematics, which presently make important contributions to classification and identification at a very high rate. Currently, the main

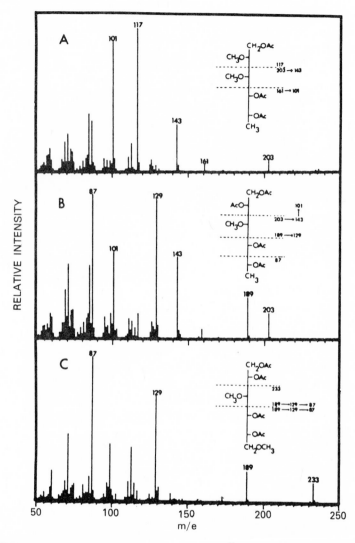

FIGURE 12. Mass spectra (EI) of characteristic *O* methyl sugars (as alditol acetates) derived from the phenolic glycolipid of *M. leprae* (peaks A, B, and C, in Figure 11) (Hunter and Brennan, 1981).

emphasis is put on two main fields of applications. Firstly, as an aid in classification and identification of unusual bacterial isolates. Secondly, as a means for ultrasensitive detection of compounds from infectious agents directly in clinical material.

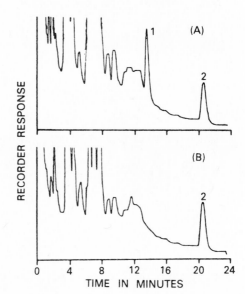

FIGURE 13. GC profile of alditol acetate derivatives of (A) extracts of spleen from animal injected i.p. 6 days previously with cell wall fragments and (B) extract of normal spleen. Peak 1 corresponds to muramic acid; peak 2 corresponds to the internal standard, glucosamine. (Fox *et al.*, 1980.)

Lipids and carbohydrates are generally occurring bacterial cell-wall/membrane constituents. Remarkable structural variations among these classes of substances exist between the various bacterial species. GC and GC/MS analysis of these two groups of substances can usually be achieved without tedious and time-consuming procedures, and they are therefore of particular interest in connection with rapid diagnosis.

Cellular fatty acid analysis seems to be a particularly useful tool in bacterial classification and identification because:

1. Fatty acids are readily and rapidly analyzed by GC and GC/MS. Only a few colonies of bacterial cells are usually required for analysis.
2. Fatty acids are present in almost all bacteria. A large variation among bacterial groups does exist.
3. Classification based on fatty acid composition appears to match reasonably well with the traditional classification.

Fatty acid analysis has a comparatively low resolution in identification. As pointed out (Section 4.4), only identification to a certain "group-level" can usually be obtained. Closely related bacteria are normally not possible to differentiate and the main advantage of fatty acid profiles is as a reliable means for a rapid detection of the correct genus or subgenus group.

The well-resolved profiles of sugars (or sugars and fatty acids) on capillary columns provide a wealth of information and will probably soon be taken into practical use by several microbiological laboratories. The stability of this kind of data has not been fully elucidated, but the limited number of publications indicate that sugar-profiles in many cases are species-, or even subspecies-specific and thus of considerable potential as a diagnostic tool.

Since several bacterial compounds do not occur in human cells and GC and GC/MS are very sensitive techniques, it is feasible to extend this analytical approach to the detection of bacteria in cerebrospinal and synovial fluids. For instance, a positive test of rhamnose, muramic acids, dideoxysugars, heptoses, ketodeoxyoctonate (KDO) as well as hydroxylated fatty acids, would indicate the presence of bacteria. In some cases an allocation to genus or even to a certain species may be indicated. Thus this kind of GC and GC/MS analysis opens yet another dimension to the diagnosis of infection.

REFERENCES

Abel, K., DeSmertzing, H., and Peterson, J. I., 1963, Classification of microorganisms by analysis of chemical composition. I. Feasibility of utilizing gas chromatography, *J. Bact.* **85**:1039–1044.

Amdur, B. H., Szabo, E. I., and Socransky, S. S., 1978, Fatty acids of gram-positive bacterial rods from human dental plaque, *Archs. Oral Biol.* **23**:23–29.

Andersen, O., Jantzen, E., Closs, O., Harboe, M., Saxegaard, F., and Fodstad, F., 1982, Fatty acid and polar lipid analysis as tools in the identification of *M. leprae* and some related slow-growing mycobacterial species, *Ann. Microbiol. (Inst. Pasteur)* **133B**:29–37.

Anderson, R. J., 1943, The chemistry of the lipids of the tubercle bacillus, *Yale J. Biol. Med.* **15**:311–345.

Asselineau, J., 1966, *The Bacterial Lipids*, Herman, Paris.

Blaser, M. J., Moss, C. W., and Weaver, R. E., 1980, Cellular fatty acid composition of *Campylobacter fetus*, *J. Clin. Microbiol.* **11**:448–451.

Bowie, I. S., Grigor, M. R., Dunckley, G. C., Loutit, M. V., and Loutit, J. S., 1972, The DNA base composition and fatty acid constitution of some gram-positive pleomorphic soil bacteria. *Soil Biol., Biochem.* **4**:397–412.

Brennan, P. J., Mayer, H., Aspinall, G. O., and Shin, E. N., 1981, Structures of the glycopeptidolipid antigens from serovars in the *Mycobacterium avium/Mycobacterium intracellulare/Mycobacterium scrofulacelum* complex, *Eur. J. Biochem.* **115**:7–15.

Broady, K. W., Rietschel, E. T., and Lüderitz, O., 1981, The chemical structure of the lipid A component of lipopolysaccharides from *Vibrio cholerae*, *Eur. J. Biochem.* **115**:463–468.

Brooks, B. W., Murray, R. G. E., Johnson, J. L., Stackebrandt, E., Woese, C. R., and Fox, G. E., 1980, Red-pigmented micrococci: a basis for taxonomy, *Int. J. System. Bact.* **30**:627–646.

Bryn, K., and Jantzen, E., 1982, Analysis of lipopolysaccharides by methanolysis, trifluoroacetylation, and gas chromatography on a fused-silica capillary column, *J. Chromatog.* **240**:405–413.

Bryn, K., Jantzen, E., and Bøvre, K., 1977, Occurrence and patterns of waxes in *Neisseriaceae*, *J. Gen. Microbiol.* **102**:33–43.

Bryn, K., and Rietschel, E. T., 1978, L-2-Hydroxytetradecanoic acid as a constituent of *Salmonella* lipopolysaccharides (Lipid A), *Eur. J. Biochem.* **86**:311–315.

Buchanan, R. E., and Gibbons, N. E. (eds.), 1974, *Bergey's Manual of Determinative Bacteriology*, 8th ed., The Williams and Wilkins Co., Baltimore.

Bøvre, K., 1980, "*Neisseriaceae—Genetic Affinities*," in *Microbiological Classification and Identification* (M. Goodfellow, and R. G. Board, eds.), pp. 189–256, Academic Press, New York.

Bøvre, K., Fuglesang, J. E., Hagen, N., Jantzen, E., and Frøholm, L. O., 1976, *Moraxella atlantae* sp. nov. and its distinction from *Moraxella phenylpyrouvica*, *Int. J. System. Bact.* **26**:511–521.

Bøvre, K., Hagen, N., Berdal, B. P., and Jantzen, E., 1977, Oxidase positive rods from cases of suspected gonorrhoea. A comparison of conventional, gas chromatographic and genetic methods of identification, *Acta Path. Microbiol. Scand., Sect. B* **85**:27–37.

Chomarat, M., Flandrois, J. P., and Viallier, J., 1981, Analyse des acides gras des mycobactéries par chromatographie gaz–liquide. Étude de *M. avium* et de 20 autre espéces, *Zbl. Bakt. Hyg., I. Abt. Orig. C*, **2**:21–32.

Clamp, J. R., Bhatti, T., and Chambers, R. E., 1971, "Determination of Carbohydrates in Biological Materials by Gas–Liquid Chromatography," in *Methods of Biochemical Analysis*, Vol. 19 (D. Glick, ed.), pp. 229–344, Wiley-Interscience, New York.

Cohen, P. G., Moss, C. W., and Farshthi, D., 1970, Cellular fatty acids of treponemes, *Br. J. Vener. Dis.* **46**:10–12.

Cummins, C. S., and Harris, H., 1956, The chemical composition of the cell wall in some gram-positive bacteria and its possible value as a taxonomic character, *J. Gen. Microbiol.* **14**:583–600.

Dees, S. B., Hollis, D. G., Weaver, R. E., and Moss, C. W., 1981, Cellular fatty acids of *Brucella canis* and *Brucella suis*, *J. Clin. Microbiol.* **14**:111–112.

Dees, S. B., and Moss, C. W., 1975, Cellular fatty acids of *Alcaligenes* and *Pseudomonas* species from clinical specimens, *J. Clin. Microbiol.* **1**:414–419.

Dees, S. B., and Moss, C. W., 1978, Identification of *Achromobacter* species by cellular fatty acid and production of keto acids, *J. Clin. Microbiol.* **8**:61–66.

Dees, S. B., and Moss, C. W., 1978, Identification of the total cellular fatty acids of *Staphylococcus* species of human origin, *Int. J. System. Bact.* **28**:223–228.

Dees, S. B., Moss, C. W., Weaver, R. E., and Hollis, D., 1979, Cellular fatty acid composition of *Pseudomonas paucimobilis* and groups IIk-2, Ve-1, and Ve-2, *J. Clin. Microbiol.* **10**:206–209.

Dees, S. B., Thanabalasundrum, S., Moss, C. W., Hollis, D. G., and Weaver, R. E., 1980, Cellular fatty acid composition of group IVe, a nonsaccharolytic organism from clinical sources, *J. Clin. Microbiol.* **11**:664–668.

Drücker, D. P., 1981, *Microbial Application of Gas Chromatography*, Cambridge University Press, Cambridge, England.

Drücker, D. P., 1976, "Gas–Liquid Chromatographic Chemotaxonomy," in *Methods of Microbiology*, Vol. 9 (J. R. Norris, ed.), pp. 51–125, Academic Press, New York.

Ellender, R. D., Jr., Hidalgo, R. J., and Grumbles, L. C., 1970, Characterization of five clostridial pathogens by gas–liquid chromatography, *Am. J. Vet. Res.* **31**:1863–1866.

Elsden, S. R., Hilton, M. G., Parsey, K. R., and Self, R., 1980, The lipid fatty acids of proteolytic clostridia, *J. Gen. Microbiol.* **118**:115–123.

Ellwood, D. C., and Tempest, D. W., 1972, Effects of environment on bacterial wall content and composition, *Adv. Microbial. Physiol.* **7**:83–117.

Farshy, D. C., and Moss, C. W., 1970, Characterization of clostridia by gas chromatography: differentiation of species by trimethylsilyl derivatives of whole-cell hydrolysates, *Appl. Microbiol.* **20**:78–84.

Fautz, E., Rosenfelder, G., and Grotjahn, L., 1979, Iso-branched 2- and 3-hydroxy fatty acids as characteristic lipid constituents of some gliding bacteria, *J. Bact.* **140**:852–858.

Fischer, W., 1977, The polar lipids of group B streptococci. II. Composition and positional distribution of fatty acids, *Biochim. Biophys. Acta* **487**:89–104.

Fisher-Hoch, S., Hudson, M. J., and Thompson, M. H., 1979, Identification of a clinical isolate as *Legionella pneumophila* by gas chromatography and mass spectrometry of cellular fatty acids, *Lancet*, **viii**:323–325.

Flesher, A. R., and Insel, R. A., 1978, Characterization of lipopolysaccharides of *Haemophilus influenzae, J. Inf. Dis.* **138**:719–730.

Fox, A., Schwab, J. H., and Cochran, T., 1980, Muramic acid detection in mammalian tissues by gas–liquid chromatography/mass spectrometry, *Infect. Immun.* **29**:526–531.

Goldfine, H., 1972, Comparative aspects of bacterial lipids, *Adv. Microbial Physiol.* **8**:1–57.

Guerrant, G. O., Lambert, M. S., and Moss, C. W., 1979, Identification of diaminopimelic acid in the Legionnaires disease bacterium, *J. Clin. Microbiol.* **10**:815–818.

Hofstad, T., and Jantzen, E., 1982, Fatty acids of *Leptotrichia buccalis*: taxonomic implications, *J. Gen. Microbiol.* **128**:151–153.

Hofstad, T., and Skaug, N., 1980, Fatty acids and neutral sugars present in lipopolysaccharides isolated from *Fusobacterium* species, *Acta Path. Microbiol. Scand., Sect. B* **88**:115–120.

Hunter, S. W., and Brennan, P. J., 1981, A novel phenolic glycolipid from *Mycobacterium leprae* possibly involved in immunogenicity and pathogenicity, *J. Bact.* **147**:728–735.

Jantzen, E., Berdal, B. P., and Omland, T., 1979, Cellular fatty acid composition of *Francisella tularensis, J. Clin. Microbiol.* **10**:928–930.

Jantzen, E., Berdal, B. P., and Omland, T., 1980 Cellular fatty acid composition of *Haemophilus* species, *Pasteurella multocida, Actinobacillus actinomycetemcomitans* and *Haemophilus vaginalis* (*Corynebacterium vaginale*), *Acta Path. Microbiol. Scand. Sect. B* **88**:89–93.

Jantzen, E., Berdal, B. P., and Omland, T., 1981, "Fatty Acid Taxonomy of *Haemophilus, Pasteurella*, and *Actinobacillus*," in *Haemophilus, Pasteurella, and Actinobacillus* (M. Kilian, and W. Fredriksen, eds.), pp. 197–203, Academic Press, New York.

Jantzen, E., Bergan, T., and Bøvre, K., 1974a, Gas chromatography of bacterial whole cell methanolysates. VI. Fatty acid composition of strains within *Micrococcaceae, Acta Path. Microbiol. Scand., Sect. B* **82**:785–798.

Jantzen, E., Bryn, K., Bergan, T., and Bøvre, K., 1974b, Gas chromatography of bacterial whole cell methanolystates. V. Fatty acid composition of neisseriae and moraxellae, *Acta Path. Microbiol. Scand., Sect. B* **82**:767–779.

Jantzen, E., Bryn, K., Bergan, T., and Bøvre, K., 1975, Gas chromatography of bacterial whole cell methanolysates. VII. Fatty acid composition of *Acinetobacter* in relation to the taxonomy of *Neisseriaceae, Acta Path. Microbiol. Scand., Sect. B* **83**:569–580.

Jantzen, E., Bryn, K., and Bøvre, K., 1974c, Gas chromatography of bacterial whole cell methanolysates. IV. A procedure for fractionation and identification of fatty acids and monosaccharides of cellular structures, *Acta Path. Microbiol. Scand., Sect. B,* **82**:753–766.

Jantzen, E., Bryn, K., and Bøvre, K., 1976, Monosaccharide patterns of *Neisseriaceae, Acta Path. Microbiol. Scand., Sect. B* **84**:177–188.

Jantzen, E., Bryn, K., Hagen, N., Bergan, T., Bøvre, K., 1978, Fatty acids and monosaccharides of *Neisseriaceae* in relation to established taxonomy, *NIPH Annals (Oslo)* **1**:59–71.

Jantzen, E., Frøholm, L. O., Hytta, R., and Bøvre, K., 1972, Gas chromatography of bacterial whole cell methanolysates. I. The usefulness of trimethylsilyl- and trifluoroacetyl derivatives for strain and species characterization, *Acta Path. Microbiol. Scand., Sect. B* **80**:660–671.

Jantzen, E., and Hofstad, T., 1981, Fatty acids of *Fusobacterium* species: taxonomic implications, *J. Gen. Microbiol.* **123**:163–171.

Jantzen, E., Knudsen, E., and Winsnes, R., 1982, Fatty acid analysis for differentiation of *Bordetella* and *Brucella* species, *Acta Pathol. Microbiol. Scand., Sect. B,* **90**:353–359.

Jantzen, E., and Lassen, J., 1980, Characterization of *Yersinia* species by analysis of whole-cell fatty acids, *Int. J. System. Bact.* **30**:421–428.

Jennings, H. J., Hawes, G. B., Adams, G. A., and Kenny, C. P., 1973, The chemical composition and serological reactions of lipopolysaccharides from serogroups A, B, X, and Y *Neisseria meningitidis, Can. J. Biochem.* **51**:1347–1354.

Julak, J.,Turecek, F., and Mikova, Z., 1980, Identification of characteristic branched-chain fatty acids of *Mycobacterium kansasii* and *gordonae* by gas chromatography–mass spectrometry, *J. Chromatog.* **190**:183–187.

Kamio, Y., Kanegasaki, S. and Takahashi, H., 1969, Occurrence of plasmalogens in anaerobic bacteria, *J. Gen. Microbiol.* **15**:439–451.

Kaneda, T., 1977, Fatty acids of the genus *Bacillus*: An example of branched-chain preference, *Bact. Rev.* **41**:391–418.

Keddie, R. M., and Bousfield, I. J., 1980, "Cell-Wall Composition in the Classification and Identification of Coryneform Bacteria," in *Microbial Classification and Identification,* Society for Applied Bacteriology (M. Goodfellow, and R. G. Board, eds.), pp. 167–187, Academic Press, New York.

Keddie, R. M., and Cure, G. L., 1978, "Cell-Wall Composition of Coryneform Bacteria," in *Coryneform Bacteria,* Society for General Microbiology (I. J. Bousfield, and A. G. Calley, eds.), pp. 47–83, Academic Press, New York.

Kiley, P., and Holt, S. C., 1980, Characterization of the lipopolysaccharide from *Actinobacillus actinomycetemcomitans* Y4 and N27, *Infect. Immun.* **30**:862–873.

Knudsen, E., Jantzen, E., Bryn, K., Ormerod, J., and Sirevåg, R., 1982, Quantitative and structural characteristics of lipids in *Chlorobium* and *Chloroflexus, Arch. Microbiol.* **132**:149–154.

Lambert, M. A., and Armfield, A. Y., 1979, Differentiation of *Peptococcus* and *Peptostreptococcus* by gas–liquid chromatography of cellular fatty acids and metabolic products, *J. Clin. Microbiol.* **10**:464–476.

Lambert, M. A., Hollis, D. G., Moss, C. W., Weaver, R. E., and Thomas, M. L., 1971, Fatty acid composition of nonpathogenic *Neisseria, Can. J. Microbiol.* **17**:1491–1502.

Lambert, M. A., and Moss, C. W., 1976, Cellular fatty acid composition of *Streptococcus mutans* and related streptococci, *J. Dent. Res. Spec. Iss.A.* **55**:A96–A102.

Larsson, L., Bergman, R., and Mårdh, P.-A., 1979, Gas chromatographic characterization of porcine and human strains belonging to the *Mycobacterium avium-intracellulare* complex, *Acta Path. Microbiol. Scand., Sect. B* **87**:205–209.

Larsson, L., and Mårdh, P.-A., 1976, Gas chromatographic characterization of mycobacteria: Analysis of fatty acids and trifluoroacetylated whole-cell methanolysates, *J. Clin. Microbiol.* **3**:81–85.

Larsson, L., and Mårdh, P.-A., 1977, Application of gas chromatography to diagnosis of microorganisms and infectious diseases, *Acta Path. Microbiol, Scand., Sect. B, Suppl.* **259**:5–15.

Larsson, L., Mårdh, P.-A., and Odham, G., 1979, Detection of tuberculostearic acid in mycobacteria and nocardia by gas chromatography and mass spectrometry using selected monitoring, *J. Chromatog.* **163**:221–224.

Larsson, L., Mårdh, P.-A., Odham, G., and Westerdahl, G., 1980, Detection of tuberculostearic acid in biological specimens by means of glass capillary gas chromatography–electron and chemical ionization mass spectrometry, utilizing selected ion monitoring, *J. Chromatog.* **182**:402–408.

Larsson, L., Mårdh, P.-A., Odham, G., and Westerdahl, G., 1981, Use of selected ion monitoring for detection of tuberculostearic and C_{32} mycocerosic acid in mycobacteria and in five-day-old cultures of sputum specimens from patients with pulmonary tuberculosis, *Acta Path. Microbiol. Scand. Sect. B* **89**:245–251.

Lechevalier, M. P., 1977, Lipids in bacterial taxonomy—A taxonomist's viewpoint, *Crit. Rev. Microbiol.* **5**:109–210.

Le Dur, A., Caroff, M., Chaby, R., and Szabo, L., 1980, A novel type of endotoxin structure present in *Bordetella pertussis*. Isolation of two different polysaccharides bound to Lipid A, *Eur. J. Biochem.* **84**:579–589.

Livermore, B. P., and Johnson, R. C., 1974, Lipids of the *Spirochaetales*, *Spirochaeta*, *Treponema*, and *Leptospira*, *J. Bact.* **120**:1268–1273.

Lüderitz, O., Westphal, O., Staub, A. M., and Nikaido, H., 1971, "Isolation and Chemical and Immunological Characterization of Bacterial Lipopolysaccharides," in *Microbial Toxins*, Vol. IV (G. Weinbaum, S. Kadis, and S. Ajl, eds.), pp. 145–224, Academic Press, New York.

Lugowski, C., and Romanowska, E., 1974, Chemical studies on *Shigella sonnei* lipid A, *Eur. J. Biochem.* **48**:319–323.

Maitra, S. K., Scholtz, M. C., Yoshikawa, T. T., and Guze, L. B., 1978, Determination of lipid A and endotoxin in serum by mass spectrometry, *Proc. Natl. Acad. Sci. USA* **75**:3993–3997.

Mandel, M., 1969, New approaches to bacterial taxonomy: Perspective and prospects, *Ann. Rev. Microbiol.* **23**:239–274.

Matthews, H., Yang, T.-K., and Jenkin, H., 1979, Unique lipid composition of *Treponema pallidum* (Nichols virulent strain), *Inf. Immun.* **24**:713–719.

Mayberry, W. R., 1980, Cellular distribution and linkage of D-(-)-3-hydroxy fatty acids in *Bacteroides* species, *J. Bact.* **144**:200–204.

Mayberry, W. R., 1981, Dihydroxy and monohydroxy fatty acids in *Legionella pneumophila*, *J. Bact.* **147**:373–381.

Minnikin, D. E., and Goodfellow, M., 1980, "Lipid Composition in the Classification and Identification of Acid-Fast Bacteria," in *Microbiological Classification and Identification*, Society for Applied Bacteriology (M. Goodfellow, and R. G. Board, eds.), pp. 189–256, Academic Press, New York.

Minnikin, D. E., Goodfellow, M., and Collins, M. D., 1978. "Lipid Composition in the Classification and Identification of Coryneform and Related Taxa," in *Coryneform Bacteria*, Society for General Microbiology (I. J. Bousfield, and A. G. Calley, eds.), pp. 84–160, Academic Press, New York.

Miyagawa, E., and Suto, T., 1980, Cellular fatty acid composition in *Bacteroides oralis* and *Bacteroides ruminicola*, *J. Gen. Appl. Microbiol.* **26**:331–343.

Mitruka, B. M., 1974, *Gas Chromatographic Applications in Microbiology and Medicine*, John Wiley and Sons, New York.

Moreno, E., Pitt, M. W., Jones, L. M., Schurig, G. G., and Berman, D. T., 1979, Purification and characterization of smooth and rough lipopolysaccharides from *Brucella abortus*, *J. Bact.* **138**:361–369.

Morrison, S. J., Tornabene, T. G., and Kloos, W. E., 1971, Neutral lipids in the study of relationships of members of the family *Micrococcaceae*, *J. Bact.* **108**:353–358.

Moss, C. W., 1978, "New Methodology for Identification of Nonfermenters: Gas–Liquid Chromatographic Chemotaxonomy," in *Glucose Nonfermenting Gram-Negative Bacteria in Clinical Microbiology* (G. L. Gilardi, ed.), pp. 171–201, CRC Press, Boca Raton, Florida.

Moss, C. W., 1981, Gas-liquid chromatography as an analytical tool in microbiology, *J. Chromatog.* **203**:337–347.

Moss, C. W., and Dees, S. B., 1976, Cellular fatty acids and metabolic products of *Pseudomonas* species obtained from clinical specimens, *J. Clin. Microbiol.* **4**:492–502.

Moss, C. W., Dowell, V. R., Farshtchi, D., Raines, L. J., and Cherry, W. B., 1969, Cultural characteristics and fatty acid composition of propionibacteria, *J. Bact.* **97**:561–570.

Moss, C. W., Dowell, V. R., Lewis, J. T., and Schekter, M. A., 1967, Cultural characteristics and fatty acid composition of *Corynebacterium acnes*, *J. Bact.* **94**:1300–1305.

Moss, C. W., and Dunkelberg, W. E., 1969, Volatile and cellular fatty acids of *Haemophilus vaginalis*, *J. Bact.* **10**:544–546.

Moss, C. W., Lambert, M. A., and Merwin, W., 1974, Comparison of rapid methods for analysis of bacterial fatty acids, *Appl. Microbiol.* **28**:80–85.

Moss, C. W., Thomas, M. L., and Lambert, M. A., 1971, Amino acid composition of treponemes, *Br. J. Vener. Dis.* **47**:165–168.

Moss, C. W., Weaver, R. E., Dees, S. B., and Cherry, W. B., 1977, Cellular fatty acid composition of isolates from Legionnaires disease, *J. Clin. Microbiol.* **6**:140–143.

Niskanen, A., Kiutamo, T., Raisanen, S., and Raevuori, M., 1980, Determination of fatty acid composition of *Bacillus cereus* and related bacteria: a rapid gas chromatographic method using glass capillary column, *Appl. Environ. Microbiol.* **35**:453–455.

Odham, G., Larsson, L., and Mårdh, P.-A., 1979, Determination of tuberculostearic acid in sputum from patients with pulmonary tuberculosis by selected ion monitoring, *J. Clin. Invest.* **63**:813–819.

Oyaizu, H., and Komagata, K., 1981, Chemotaxonomic and phenotypic characterization of the strains of species in the *Flavobacterium–Cytophaga* complex, *J. Gen. Microbiol.* **27**:57–107.

Raines, L. J., Moss, C. W., Farshtchi, D., and Pittman, B., 1968, Fatty acids of *Listeria monocytogenes*, *J. Bact.* **96**:2175–2177.

Pritchard, D. G., Coligan, J. E., Speed, S. E., and Gray, B. M., 1981, Carbohydrate fingerprints of streptococcal cells, *J. Clin. Microbiol.* **13**:89–92.

Rietschel, E. T., Gottert, H., Lüderitz, O., and Westphal, O., 1972, Nature and linkages of the fatty acids present in the lipid-A component of *Salmonella* lipopolysaccharides, *Eur. J. Biochem.* **28**:166–173.

Rietschel, E. T., and Lüderitz, O., 1980, Struktur von Lipopolysaccharid und Taxonomie gram-negativer Bakterien, *Forum Microbiol.* **1/80**:12–20.

Schleifer, K. H., and Kandler, O., 1972, Peptidoglycan types of bacterial cell walls and their taxonomic implications, *Bact. Rev.* **36**:407–477.

Shaw, N., 1974, Lipid composition as a guide to the classification of bacteria, *Adv. Appl. Microbiol.* **17**:63–108.

Stocker, B. A. D., and Mäkelä, P. H., 1971, "Genetic Aspects of Biosynthesis and Structure of *Salmonella* Lipopolysaccharides," in *Microbial Toxins*, Vol. IV (G. Weinbaum, S. Kadis, and S. J. Ajl, eds.), pp. 369–438, Academic Press, New York.

Sud, I. J., and Feingold, D. S., 1979, Detection of 3-hydroxy fatty acids at picogram levels in biologic specimens. A chemical method for the detection of *Neisseria gonorrhoeae*? *J. Investig. Dermat.* **73**:521–525.

Tabor, M. W., MacGee, J. and Holland, J. W., 1976, Rapid determination of dipicolinic acid in the spores of *Clostridium* species by gas–liquid chromatography, *Appl. Environ. Microbiol.* **31**:25–28.

Tadayon, R. A., and Carrol, K. K., 1971, Effect of growth conditions on the fatty acid composition of *Listeria monocytogenes* and comparison with the fatty acids of *Erysipelothrix* and *Corynebacterium*, *Lipids* **6**:820–825.

Thoen, C. O., Karlson, A. G., and Ellefson, R. D., 1971, Comparison by gas–liquid chromatography of the fatty acids of *Mycobacterium avium* and some other photochromogenic mycobacteria, *Appl. Microbiol.* **22**:560–563.

Thoen, C. O., Karlson, A. G., and Ellefson, R. D., 1973, Differentiation between *Mycobacterium kansasii* and *Mycobacterium marinum* by gas–liquid chromatographic analysis of cellular fatty acids, *Appl. Microbiol.* **24**:1009–1010.

Tisdall, P. A., De Young, D. R., Roberts, G. D., and Anhalt, J. P., 1982, Identification of clinical isolates of mycobacteria by gas–liquid chromatography alone: A 10-month follow-up study. *J. Clin. Microbiol.* **16**:400–402.

Tisdall, P. A., Roberts, G. D., and Anhalt, J. P., 1979, Identification of clinical isolates of mycobacteria with gas–liquid chromatography alone, *J. Clin. Microbiol.* **10**:506–514.

Tornabene, T. G., Gelpi, E., and Oro, J., 1967, Identification of fatty acids and aliphatic hydrocarbons in *Sarcina lutea* by gas chromatography and combined gas chromatography–mass spectrometry, *J. Bact.* **94**:333–343.

Uchida, K., and Mogi, K., 1972, Cellular fatty acid spectra of *Pediococcus* species in relation to their taxonomy, *J. Gen. Appl. Microbiol.* **18**:109–129.

Vaczi, L., Redai, I., and Rethy, A., 1967, Changes in the fatty acid composition of *Staphylococcus aureus* under various cultural conditions, *Acta Microbiol. Acad. Sci. Hung.* **14**:293–298.

Veerkamp, J. H., 1971, Fatty acid composition of *Bifidobacterium* and *Lactobacillus* strains, *J. Bact.* **108**:861–867.

Wexler, H., and Oppenheim, J. D., 1980, "Listerial LPS: An Endotoxin from a Gram-Positive Bacterium," in *Bacterial Endotoxins and Host Response* (M. K. Agarwal, ed.), pp. 27–50, Elsevier/North-Holland Biomedical Press, Amsterdam.

Wilkinson, S. G., 1977, "Composition and Structure of Bacterial Lipopolysaccharides," in *Surface Carbohydrates of the Procaryotic Cell* (I. W. Sutherland, ed.) pp. 97–175, Academic Press, New York.

Zanetta, J. P., Breckenridge, W. C., and Vincendon, G., 1972, Analysis of monosaccharides by gas–liquid chromatography of the *O*-methyl glycosides as trifluoroacetate derivatives. Application to glycoproteins and glycolipids, *J. Chromatogr.* **69**:291–304.

Yamakawa, T., and Ueta, N., 1964, Gas chromatographic studies of microbial components. I. Carbohydrate and fatty acid constitution of *Neisseria*, *Jpn. J. Exp. Med.* **34**:361–374.

Yano, I., Ohno, Y., Masui, M., Kato, K., Yabuuchi, E., and Ohyama, A., 1976, Occurrence of 2- and 3-hydroxy fatty acids in high concentrations in the extractable and bound lipids of *Flavobacterium meningosepticum* and *Flavobacterium* IIb, *Lipids* **11**:685–688.

9

QUANTITATIVE MASS SPECTROMETRY AND ITS APPLICATION IN MICROBIOLOGY

Göran Odham, Laboratory of Ecological Chemistry, University of Lund, Ecology Building, Helgonavägen 5, S-223 62 Lund, Sweden

Lennart Larsson and Per-Anders Mårdh, Department of Medical Microbiology, University of Lund, Sölvegatan 23, S-223 62 Lund, Sweden

1. INTRODUCTION

Advanced chemoanalytical techniques are now becoming introduced in microbiological laboratories, thereby adding new dimensions to traditional methods of studying microbial metabolic processes.

Gas chromatography (GC), in the early 1970s a fairly exclusive technique for quantitative analysis in microbiology, is now routinely employed in many research and diagnostic laboratories. Among its uses are the study of respiratory function and nitrogen fixation (reduction of acetylene to ethylene) and the identification of bacteria (see Chapters 6, 7, 8, 11, and 12).

GC in combination with mass spectrometry (MS), though constituting a versatile and powerful tool in chemoanalysis, is not yet widely used in microbiology. Mass spectrometers, with or without gas chromatographic inlet systems, have been used over the past two decades to establish the chemical structure of constituents of microbial cells and extracellular products, including antibiotics.

Quantitative MS, the basis of which was first described by Sweeley *et al.* (1966) and Hammer *et al.* (1978), is a more recent adaptation of MS, made possible by instrumental progress and computerized data handling. The rapid development of MS as a quantitative technique has met requirements from workers in applied biochemistry and pharmacology. Most applications of quantitative GC/MS have so far concerned drugs and drug metabolites. (For reviews, see Falkner *et al.*, 1975; Millard, 1979; de Leenher and Cruyl, 1980.)

The purpose of this chapter is to present some microbiological applications of quantitative GC/MS, where the unrivalled sensitivity and selectivity of the technique are a prerequisite.

2. MASS SPECTROMETRY

2.1. Methods of Ionization

Since the basic concept of MS was dealt with in Chapter 2, the following presentation concerns only aspects of particular significance for quantitative MS.

Both electron (EI) and chemical (CI) ionization have been employed in quantitative MS analyses of biological substances. EI, probably for reasons of availability, is hitherto the more commonly used of the two. Factors to be considered in the choice of ionization method include the number and intensity of the fragments formed from a given compound and their specificity with respect to chemical structure. With most substances of biological origin, electron bombardment—usually at 70 eV—results in extensive cleavages and thus in high total ion currents. The intensity of the characteristic ions, which usually are of high mass, may therefore represent only a very small fraction of the total ion intensity.

The total ion current produced in CI is significantly less than that in EI. In CI, adducts representing the nonfragmented molecule and/or some simple cleavage thereof usually account for most of the total ion signals. Specificity is generally higher with CI than with EI, but whether or not there is a parallel increase in sensitivity remains to be experimentally determined. The reactant gases most commonly employed in quantitative CI–MS are methane, isobutane, and ammonia.

2.2. Mass Filters

Commercially available mass spectrometers used in quantitative MS are either of magnetic or of quadrupolar type, each type having its own advantages. From the well-known basic equation $m/z = H^2 r^2 / 2V$ of mass selection in magnetic instruments, where r is the radius of the path described by the ion (instrument-dependent), H the magnetic field strength, and V the acceleration potential, several conclusions can be made. Scanning a complete mass spectrum implies a continuous sweep, either of the magnetic field at a constant accelerating potential, or of the accelerating voltage at a constant field strength. To increase sensitivity, the continuous scanning may be replaced by measurements of ions exclusively at a limited number of preselected masses (selected ion monitoring, SIM), or measurements at constant field and accelerating voltage of a single mass number (single ion detection). In practice this can be achieved by adjusting the accelerating voltage alternator to operate either stepwise or at a constant value. A disadvantage of magnetic instruments is the rather limited number of ions that can be studied "simultaneously" because of the relatively low speed of the alternator.

High-resolution quantitative MS offers unique selectivity. By incorporating an electrostatic analyzer in the mass filter (double-focusing magnetic instruments), the distribution of the initial kinetic energy of the ions of a certain m/z is controlled, permitting accuracy to a few parts per million of the mass measurements. This means, in practice, that ions of given mass numbers can be defined to their elemental compositions.

The nonmagnetic mass spectrometer employs as the mass filter a radiofrequency (rf) quadrupole electric field. Although they operate only at low resolution, the quadrupole instruments have achieved wide use in quantitative MS, largely because their use requires comparatively little experience. Other positive features of the rf instruments are their ability to scan a mass spectrum in less than a second, thus permitting "real-time" visualization of mass spectra or chromatograms on an oscilloscope. The fast scanning speed is advantageous also when capillary GC is used as an inlet system to the spectrometer. Continuous or stepwise scanning is made by changing the voltage at a constant rf/dc ratio, which results in a linear mass scale in the output. The latter feature facilitates computer routines.

2.3. Extracted Ion Current Profile (EICP)

In computerized MS, complete or partial mass spectra may be obtained by repetitive scanning of the material entering the ion source. The computer can be programmed to extract from the data mass only ions of specific m/z values (EICP). This may be done during the runs. Quantification is performed on the fragment ion current as generated by the computer and visualized on the oscilloscope. Although its sensitivity is comparatively low (time is spent on measuring ions for the analysis of no significance), the technique is valuable for screening unknown mixtures. It permits identification and quantification of virtually all components of such mixtures (provided that suitable calibration is performed). Figure 1 illustrates application of the EICP technique in an analysis of the relative amounts of total fatty acids in cells of *Legionella pneumophila* (Fisher-Hoch *et al.*, 1979). The upper two chromatograms show the total ion current tracings of the

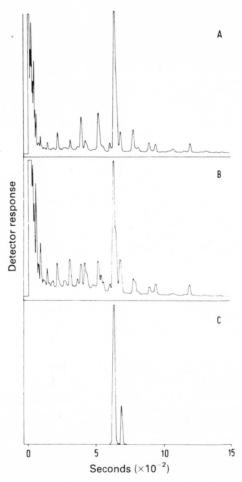

FIGURE 1. Gas chromatogram of fatty-acid methyl esters from Bloomington-2 isolate (A) and Oxford isolate (B) of *Legionella pneumophila*, and EICP (C) of the GLC profile of the fatty-acid esters from the Oxford isolate in terms of the molecular ion (m/z 270) characteristic of $C_{16:0}$ fatty acids (Fischer-Hoch *et al.*, 1979).

serotypes Bloomingdale 2 and Oxford of *L. pneumophila*. The bottom chromatogram shows a tracing at m/z 270 characteristic of $C_{16:0}$ fatty acid methyl esters.

2.4. Selected Ion Monitoring (SIM)

Although EICP is valuable for screening mixtures of unknown composition, its sensitivity for the individual components is not greater than that achieved by the mass spectrometer in recording full spectra (1–10 ng). However, by allowing the spectrometer to focus exclusively at preselected ions (SIM), thus maximizing the instrument's duty cycle at these ions, the sensitivity may be increased by several orders of magnitude. The degree of increase will be dependent on the relative intensity of the ions studied. The ultimate response is achieved when only one ion is studied (single ion detection). Like EICP, SIM offers excellent specificity. The main disadvantage of SIM is that the selection of ions requires a decision prior to data collection on which components are to be analyzed.

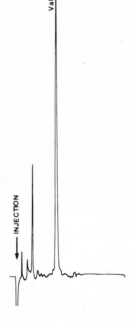

FIGURE 2. Mass fragmentogram (EI) of HFB *i*-butyl ester derivatives of free amino acids in running water of a Swedish creek monitored at m/z 268, characteristic of valine. The peak represents about 50 pg of valine. (Bengtsson and Odham, 1979).

As already mentioned, switching between masses when using magnetic instruments is made by altering the acceleration voltage. Computerized control of the alternator has enabled magnetically scanned instruments to monitor up to eight ions at 100 ms for each.

Switching of the mass filter of quadrupole instruments is easier, because electrostatic voltages—from a technical viewpoint—are simpler to change than magnetic fields or accelerating voltages at high speed. Several quadrupole instrumentations allow selection of any m/z within the mass range.

Figure 2 illustrates the determination by single ion detection of a single amino acid, that is valine, produced by fungi (*Hyphomycetes*) on leaves in a running water system. The detection limit when monitoring at m/z 268 (EI) was below 1 pg (Bengtsson and Odham, 1979).

3. SAMPLE HANDLING

3.1. Pretreatment of Sample

Analysts working with material of microbial origin frequently face problems of two major categories when designing a method for mass spectrometry. Firstly, the chosen extraction procedure may result in an unresolved complex mixture, and secondly, the components of interest in this mixture often are not sufficiently volatile for ionization in the ion source by EI or CI.

There is therefore a great damand for simple and efficient methods of pretreatment (extraction, purification, derivatization) of samples to be introduced into the ion source. These processes are more fully dealt with in Chapters 3–5. Several comprehensive textbooks on MS also deal with problems of pretreatment.

The extraction medium is one important factor in designing the analytical procedure. Various hydrophobic solvents, such as hexane and a chloroform:methanol mixture (4:1 v/v) can be used for selective removal of lipids from microbial cellular material. Hexane dissolves only lipophilic substances such as free fatty acids, glycerides, waxes, and steroids. Amino acids and peptides are commonly extracted with acidified aqueous solutions, whereas many carbohydrates dissolve more or less selectively in an ethanol:water mixture, for example, in the proportions 90 : 10 (v/v). Rapid and simple chromatographic techniques, e.g., preparative thin-layer chromatography or ion exchange chromatography may occasionally be used as a complement to the extraction procedure to give further purification prior to the MS analysis. To achieve volatility adequate for MS analysis, particularly when GC is used as an inlet technique, the sample is commonly derivatized.

3.1.1. Fatty Acids

The methyl esters represent the most commonly used derivatives of fatty acids in MS analysis (cf. Chapters 3 and 8). However, other transformations of the free carboxylic group can sometimes be advantageous. Hintze *et al.* (1973) showed that the benzyl esters may by used in investigations of sensitive biological materials because quantitative derivatization can be effected under very mild conditions, apparently without isomerization or hydrolysis of other, coexistent ester bonds. The esterification may be performed with phenyldiazomethane (Klem *et al.*, 1973).

Trimethylsilyl (TMS) esters of fatty acids have recently been used in GC/MS (Markey, 1974; Kuksis *et al.*, 1976). A mass spectrum of an unsaturated TMS ester (of an eicosenoic acid) is shown in Figure 3. Of particular interest for quantitative EICP and SIM is the comparatively large $M^{\ddagger}$ accompanied by an ion representing loss of a methyl group (M-15). Trimethylsilylation allows complete derivatization of free acids also in the presence of other types of esters. In addition, the TMS esters yield proportionally large amounts of high-mass fragments, including the parent ions, as compared with the methyl esters. This renders TMS derivatization particularly useful in high-sensitivity, quantitative, and qualitative MS. The *tert*-butyldimethylsilyl (*t*-BDMS) esters have also been used, for the same reasons (Phillipou *et al.*, 1975; Larsson *et al.*, 1980).

3.1.2. Carbohydrates

The most widely used carbohydrate derivatives in MS analyses are the methyl and TMS ethers and the acetates. TMS ethers are highly volatile and separate well on many types of GC columns (Dutton *et al.*, 1973;

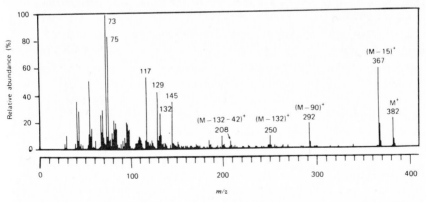

FIGURE 3. Mass spectrum (EI) of the trimethylsilyl ester of arachidonic acid, 80 eV (Kuksis *et al.*, 1976).

Dutton *et al.*, 1974). Their EI spectra are dominated by ions derived from the TMS groups, but molecular peaks of low abundance accompanied by more prominent peaks from $(M - CH_3)^+$ ions are usually also found.

In EI spectra of the methyl ethers and acetates the molecular ion peaks and other peaks of high molecular weight are often small or absent, which makes these derivatives rather unsuitable for high-sensitivity quantitative MS. However, the use of CI (NH_3 or CH_4 as a reactant gases) gives improvement in this respect (Radford and deJongh, 1980). The CI (CH_4) spectrum of 1,2,3,4,6-penta-*O*-acetyl-β-D-glucopyranose shows prominent peaks at m/z 391 (=M + 1), m/z 331 (=M + 1 − AcOH), m/z 271 (=M + 1 − 2 × AcOH), etc. Although only a few studies on permethylated carbohydrates using CI have so far been published, available data suggest that CI ionization of the methylated carbohydrates produces a series of ions analogous to those obtained for the acetylated derivatives (Solov'yov *et al.*, 1977).

Systematic investigations with respect to the usefulness of these derivatives in quantitative MS (EI and CI) remain to be performed.

Recently several other derivatives, notably the pertrifluoroacetate (TFA) and cyclic boronates, have been used for the facile determination of molecular weights of carbohydrates in connection with GC–MS analyses (for reviews, see Radford and deJongh, 1980). The per-TFA derivatives are significantly more volatile than the corresponding per-TMS ethers (Vilkas *et al.*, 1966). The TFA derivatives generally yield much simpler spectra than the peracetates. Moreover, molecular ions and other high-mass ions are more intense for the per-TFA derivatives than for the peracetates, a factor of importance in quantitative MS.

As regards fatty acids, silylated derivatives of carbohydrates other than TMS may offer advantages for their quantitative determination by MS. For example, *tert*-butyldimethylsilyl ethers yield simple EI spectra which often contain base peaks of high mass, originating from loss of the butyl group from the molecular ions (Kraska *et al.*, 1974).

3.1.3. Amino Acids

In qualitative GC–MS on amino acids, the best derivatives hitherto developed are probably those of the *N*-perfluoroacyl alkyl esters (Kaiser *et al.*, 1974; March, 1975; Bertilsson, 1976). *N*-heptafluorobutyryl *n*- or *i*-butyl and *n*-propyl esters are the derivatives most commonly used. These derivatives are equally well suited for quantitative MS. EI spectra of the *N*-perfluoroacyl alkyl ester of an amino acid show prominent peaks (sometimes base peaks) specific to the amino acid. Figure 4 shows the EI spectrum of the *N*-trifluoroacetyl *n*-butyl ester of valine (Leimer *et al.*, 1977). The

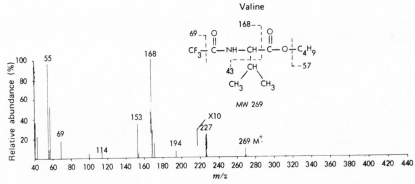

FIGURE 4. Mass spectrum (EI) of the *N*-trifluoroacetyl-*n*-butyl ester of valine, 70 eV (Leimer *et al.*, 1977).

base peak, characteristic of valine, is at m/z 168. SIM (EI) monitoring at this ion results in excellent specificity and sensitivity. By using single ion detection focusing on m/z 268 of the *N*-heptafluorobutyryl *i*-butylester of valine, it is possible to detect as little as 0.7 picogram of this amino acid (Bengtsson and Odham, 1979), even in a complex mixture.

CI mass spectra have recently been published which represent acylated amino acid esters with methane, isobutane, and NH_3 employed as reactant gases (Petty *et al.*, 1976; Vetter, 1980). The fragmentation of the, valine derivative (*N*-trifluoroacetyl *n*-butyl) after protonation by methane appears to be remarkably extensive, whereas protonation by isobutane leads to spectra with significantly high $(M + 1)^+$ peaks (base peak). Use of ammonia results in abundant formation of $(M + NH_4)^+$ ions and only very little fragmentation. CI (NH_3) SIM as an alternative to EI SIM thus appears to represent highly sensitive means for the detection of *N*-perfluoracylated

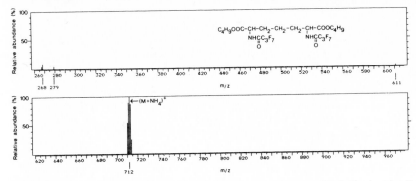

FIGURE 5. Mass spectrum (CI, NH_3) of the *N*-HFB-*i*-butyl ester of 2,6-diaminopimelic acid, 70 eV, 1 Torr (Tunlid and Odham, 1982).

amino acid esters. For instance, the nonprotein amino acid 2,6-diaminopimelic acid, a unique compound of most bacterial cell walls, can be detected in amounts of approximately 1 pg. Its CI mass spectrum is reproduced in Figure 5 (Tunlid and Odham, 1982).

The fragmentation of the TMS derivatives strongly resembles that of alkyl esters. These derivatives, however, are very easily hydrolyzed and, for example, require special GC columns to avoid loss of the protecting groups. For this reason, the N-peracylated alkyl esters of the amino acids are preferable.

3.2. Introduction of Sample Using Gas Chromatography (GC)

The combination of a gas chromatograph with the mass spectrometer has had a tremendous impact on the development of quantitative MS. Consequently, GC/MS instruments are nowadays used in most applications of quantitative MS analysis. This apparatus combines the separating power of the chromatographic column with the unique ability of the mass spectrometer to separate and analyze the ions formed. According to the vast experience hitherto gained, there is little risk that two different compounds should give both identical retention times and undifferentiable mass spectra.

To optimize the specificity of the analytical procedure, the conventional packed columns, coupled via interfaces consisting of various types of molecular separators, may be replaced by glass or fused silica capillary columns coupled directly to the ion source. The outstanding separation quality of these columns, combined with the recent development of solvent-free, unsplit injections, make capillary GC particularly attractive in quantitative GC/MS (cf. Chapters 1 and 6).

Although in quantitative GC/MS there is a tendency towards increasing gas chromatographic resolution, the mass spectrometer is commonly used at a resolution of less than 1000 (10% valley definition). That utmost performance, in terms of specificity, is attainable with a double-focusing magnetic instrument working in the high-resolution mode is a noteworthy point.

3.3. Introduction of Sample Using Liquid Chromatography (LC)

Liquid chromatography (LC) has become widely used as an analytical procedure in the past decade. In contrast to GC, LC has the advantage of the ability to process directly material having little or no volatility, thus reducing the need for derivatization. Although LC technology continues to develop instrumentally, it is still hampered by lack of detector systems matching those of GC with respect to sensitivity and specificity. Attempts have therefore been made to evolve on-line LC/MS systems capable of identifying and quantifying compounds, corresponding to current possibilities with GC/MS.

The principal difficulty in designing LC systems to meet these demands is to make the large solvent effluent volumes compatible with the low pressure in the ion source, i.e., selective removal of the solvent. Two such systems are commercially available. One of them uses a moving-belt device from which the effluent is heat-evaporated during transport to the ion source via a differential pumping system. It was designed by Scot *et al.* (1974), Erdahl and Privett (1977), and McFadden *et al.* (1977). Both EI and CI can be used to ionize the analyte from the heated belt in the ion source.

The second system was proposed and introduced by Arpino *et al.* (1974), and McLafferty and Baldwin (1976). It utilizes a fraction of the effluent (e.g., a 0.01–0.1 ml/min portion), which is injected via a diaphragm acting as a neubilizer directly into the ion source. The ion source meanwhile operates at a pressure of 1 Torr. The instrument, thus functioning in the CI mode, utilizes the vaporized solvent as reactant gas. Common LC solvents such as water, methanol, chloroform, and hexane, have all been successfully employed.

With the latter LC/MS technique a mass spectrum can be obtained with only a few nanograms entering the source. Because of the necessity of splitting the column output, however, the quantities injected for a full spectrum must be in the low microgram range. When using SIM the detection limit can be lowered by two or three orders of magnitude, depending on the chemical character of the compound and the purity of the solvent used. If microcolumns are used, it is possible to operate without a splitter. Figure 6 shows, as an example, an LC–MS analysis of saccharose in methanol at 1 ml/min (Arpino *et al.*, 1981).

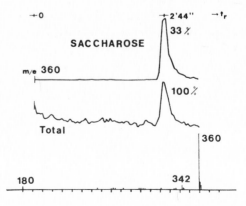

FIGURE 6. LC/MS analysis of saccharose in methanol at 1 ml/min. Reacting ion species are from methanol (solvent) and ammonia (added gas). Top trace: EICP for ion at m/z 360 $(M + NH_4^+)$. Middle trace: total ion current. Bottom: mass spectrum (CI, NH_3) of saccharose (Arpino *et al.*, 1976).

3.4. Direct Sample Introduction

The direct inlet probe system in most modern mass spectrometers may be utilized for quantification. A small glass or gold crucible containing the analyte is introduced into the ion source. Upon heating, the analyte is vaporized. The resulting total or selected ion current is recorded. Quantitative calibration may be performed by using either a reference compound of known partial pressure, introduced at constant rate from a separate reservoir inlet system, or an internal standard of known amount, which is thoroughly mixed with the sample in the crucible.

In the former case the signal of the reference compound, normally perfluorokerosene, remains constant during the generation of ions from the sample in the evaporation process. In the latter case two integrated ion currents are recorded at the respective m/z values of the sample molecules and the internal standard, both being evaporated simultaneously. After calibration, calculation of the ratio of the peak heights directly gives quantification of the sample (cf. Section 4.2).

Quantification of free fatty acids using CI and directly comparing the peak height of the protonated molecular ion with that of the corresponding deuterium-labeled fatty acid, was described by Mee *et al.* (1976). By using fixed amounts of standards deuterated in position 2, calibration curves were constructed for the saturated C_{12}, C_{14}, C_{16}, and C_{18} acids. Linear relationship was found between $(M + 1)$ peak height ratio (D_0/D_2) of the unlabeled and labeled acid and the amounts of fatty acid present. In this manner the concentration of a given fatty acid in a mixture could be calculated from its CI spectrum. The authors also used the technique to determine free fatty acids in microsamples of dried blood spots and to

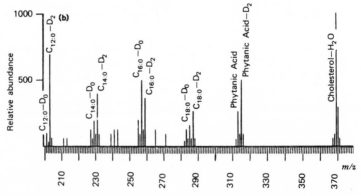

FIGURE 7. Mass spectral profile (CI, isobutane) of a blood spot from a patient suffering from Refsum's disease; 70 eV, 1 Torr (Mee *et al.*, 1976).

detect phytanic acid, a metabolite characteristically associated with Refsum's disease (Figure 7).

4. QUANTIFICATION

4.1. Calibration

The principle of calibration of the mass spectrometer for quantitative determinations is that known amounts of the compounds of interest are introduced and the resulting response is measured. The data are used to plot a calibration curve. Since the response is recorded during a finite period of time, it is preferable to use the integrated signal, i.e., the area under the peak, rather than the height of the peak. In computerized MS systems the integration procedure presents no problems. For manual treatment, planimetry, weighing of cut-out peaks, triangulation, height × width at half-height are among the methods from which a choice must be made.

After the linearity of the system has been established, the calibration curve is commonly plotted by a computer using the linear regression technique. A calibration plot for authentic methyl tuberculostearate (the ester of 10-methyloctadecanoic acid) a characteristic compound of microorganisms of the genus *Actinomycetales*, is shown in Figure 8 (Odham *et al.*,

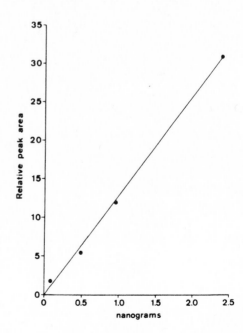

FIGURE 8. Integrated ion intensity vs. sample size of 10-methyl-tuberculostearate (EI, 70 eV) (Odham *et al.*, 1979).

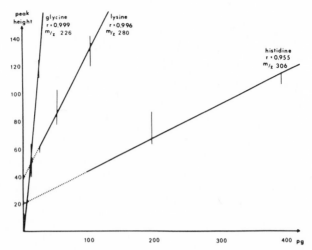

FIGURE 9. Peak heights (mean and s.d.) vs. sample size of the *N*-HFB-*i*-butyl esters of glycine, lysine, and histidine (EI, 70 eV). Selected ion monitoring at m/z 226, 280, and 306, respectively (Bengtsson *et al.*, 1981).

1979). The peak areas, from monitoring at m/z 312 (EI, molecular peak), were obtained by triangulation. The linearity of the plot indicates that direct comparison of peak areas can be used for quantification. This plot, however, was based on a synthesized substance. With samples of biological origin, corrections must be made for losses during extraction, work-up, and derivatization. This calibration plot is thus valid only for the GC/MS determinations from which it was constructed.

Special attention should be paid to calibration plots of amounts near the instrument's limits of detection, i.e., in the low picogram range. Signals in this range may be unexpectedly small or totally absent. This phenomenon can be explained by adsorption or partial thermal decomposition of the sample, either during the chromatographic procedure or in the interface between the GC and the MS.

Figure 9 (Bengtsson *et al.*, 1981) shows calibration plots for the pure *N*-heptafluorobutyryl-*iso*-butyl ester derivatives of glycine, lysine, and histidine. The plots for lysine and histidine do not intercept the *y*-axes at zero, but instead indicate a finite positive value. The finding reflects the lack of precautions taken during the derivatization procedure to remove the normal background amounts of these common amino acids.

4.2. Method of Standardization

To compensate for variabilities in the mass spectrometric system, such as fluctuations of the pressure in the ion source and minor changes of the electron multiplier gain, the measurements must be normalized against a suitable standard. One means consists of interfoliating the analyses with several measurements on known amounts of authentic material (*external* standard, or absolute calibration) to compensate changes in response of the MS system. The technique presumes, however, that exact amounts are introduced into the system.

To circumvent this critical step, it is an advantage to add to the sample a known amount of an *internal* standard, whose response signal will not interfere with that of the compound under study. Prior to the measurements a calibration curve is constructed by using the standard/reference weight ratio and response ratio. The response ratio for the material under study is measured, while the weight ratio can be calculated from the calibration curve.

Since the mass spectrometric response depends both on the chemical structure and the amount of the compound to be measured, the choice of internal standard is critical. One way of compensating for losses during purification and derivatization of the sample, and differences in injected volumes, is to add a *structural analog* to the analyte. Such a compound should possess the same structural elements that cause fragmentation after ionization, but yet differ by its gas chromatographic retention time. Position-isomeric compounds are examples of structural analogs that can be used as internal standards.

The optimal internal standard is the *stable-isotope-labeled analog* which, from both chemical and mass spectrometric viewpoints, comes very close to the behavior of the analyte. Ions from the analyte and standard can then be readily distinguished by MS, even when the gas chromatographic separation is poor or nonexistent.

The availability of labeled reference material is often a limiting factor in use of this isotopic technique. Isotopic purity and location of label are important. Significant amounts of unlabeled reference compounds are frequently present as impurities in the labeled reference and define, in practice, the blank level and hereby the limits of detection. The location of the label should be such that it is retained in the ions of interest.

Some stable isotopes of biological interest, of which deuterium is the most commonly employed in quantitative MS, are ^{2}H, ^{13}C, and ^{15}N. Deuterium is the most inexpensive of the isotopes and is available in high isomeric purity. Furthermore, a large variety of labeled small molecules, suitable for buildup to more complex molecules, may be purchased. The recent reduction in cost of ^{13}C-enriched material has increased interest in this isotope for internal standardization. Few data are as yet available on the use of ^{15}N in quantitative MS.

The increase of the masses in the standard due to labeling should be considered. In nature the elements C, H, N, and O contain small amounts of the heavy isotopes ^{13}C, ^{2}H, ^{15}N, ^{17}O, and ^{18}O. Therefore, for each combination $C_v H_x N_y O_z$ of nominal mass (m), ions will occur also at the masses ($m + 1$), ($m + 2$), etc. In practice, only the ($m + 1$) and ($m + 2$) ions are intense enough to be considered in the mass spectrum. This circumstance implies that a mass increase of 3 or more in the standard due to labeling is desirable.

Quantification using GC/MS of trace amounts of glutamic acid in water, involving the use of a 2,3,3,4,4-d_5 glutamic acid derivative, is illustrated in Figure 10 (Coutts *et al.*, 1979). The ion current trace of ions m/z 212 and m/z 217 is recorded, the ions being formed from loss of a methoxycarbonyl group from the dimethyl ester *N*-trifluoroacetylated amino acids. The difference in gas chromatographic retention time between the unlabeled and the labeled derivative is clearly negligible.

The isotope may also be incorporated in the molecule during the derivatization step. For example, acids may be esterified with trideuteromethanol to form labeled methyl esters (Sjöqvist and Änggård, 1972), compounds with active hydrogen atoms may be silylated with deuterated silylating reagents, etc. With this technique, however, possible losses during manipulations prior to the derivatization step cannot be controlled.

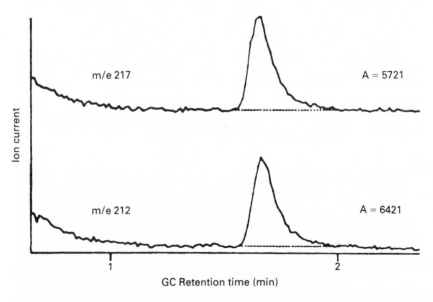

FIGURE 10. Mass fragmentography of the *N*-trifluoroacetyl dimethyl ester derivative of glutamic acid with the 2,3,3,4,4,-d_5 analog as internal standard. Monitoring at m/z 212 and 217, respectively (EI, 70 eV) (Coutts *et al.*, 1979).

4.3. Accuracy and Precision

Error, by definition, is the difference between an observed value and the true value. The error cannot be established if the true value is not known. The error of a measurement defines the *accuracy* of that measurement. GC/MS analyses of material of microbiological origin usually involve a number of manipulations such as extraction, derivatization, and chromatographic separation. In quantitative MS, therefore, it is important to take the error into account and to make correction for it.

The accuracy of the analytical procedure can be tested by adding a fixed amount of a structural analog of the analyte to the origin matrix and determining recovery. The yield in this highly recommendable test serves as an indication of the losses during the complete procedure.

Precision expresses the reproducibility of a measurement. In the analytical procedure, precision is analyzed by performing a series of determinations under specified conditions. The arithmetical mean (M) of the results and the standard deviation (s.d.) are then calculated. Precision is often expressed in percentage as (s.d.)/$M \times 100$. The literature on quantitative MS often is deficient in data on the precision of the analytical procedure that has been used. When data are given, they usually relate to one MS apparatus.

TABLE 1. ^{15}N Incorporation into Dog Plasma Amino Acids during Continuous Infusion of (^{15}N) Leucine

Amino acid	m/z monitored[a]	Isotope ratio at 0 h, $\times 100$[b]	Ratio Difference 9 h–0 h, $\times 100$[c]
ala	(201, 202)	11.236 ± 0.047	11.872 ± 0.062
val	201, 202	11.354 ± 0.064	12.054 ± 0.107
gly	(188, 189)	10.242 ± 0.058	10.344 ± 0.066
ile	216, 217	12.462 ± 0.063	13.464 ± 0.106
leu	216, 217	12.450 ± 0.091	18.110 c
pro	200, 201	11.402 ± 0.095	11.506 ± 0.050
thr	246, 247	12.460 ± 0.133	12.612 ± 0.053
ser	232, 233	11.370 ± 0.055	11.636 ± 0.062
asp	260, 261	13.658 ± 0.093	13.748 ± 0.074
met	234, 235	12.338 ± 0.072	12.350 ± 0.037
phe	250, 251	15.672 ± 0.044	15.760 ± 0.060
glu	(302, 303)	16.684 ± 0.123	17.030 ± 0.068
try	308, 309	18.072 ± 0.270	18.072 ± 0.115
orn	259, 260	13.914 ± 0.096	14.234 ± 0.066
lys	(301, 302)	17.224 ± 0.098	17.144 ± 0.134

[a] The m/z values in parentheses are the $(M + C_2H_5)^+$ ion pair rather than the $(M + H)^+$ ion pair.
[b] Mean ± standard deviation; $n = 5$.
[c] For $n = 2$, rather than $n = 5$.

The precision of the analytical method can be exemplified by data given in Table 1 (Matthews *et al.*, 1979) taken from a study of ^{15}N-incorporation into dog plasma amino acids during continuous infusion of ^{15}N-leucine. The isotopic enrichment in 15 common amino acids (^{14}N- and ^{15}N-labeled) was determined, using CI-SIM-GC/MS (methane). The precision $[(s.d.)/M \times 100]$ is well below 2% for each isotope ratio measurement. This figure is representative for current MS methodology. As the detection limit of the instrument (picogram range) is approached, however, the percentage usually increases significantly.

4.4. Sensitivity and Detection Limits

Among the various factors that can influence the sensitivity in quantitative MS are (a) the chemical nature of the compound under study, (b) the inlet system, and (c) the mass spectrometric conditions. At any given ionizing energy and type of ionization, the number and intensity of the generated ions can differ considerably between different compounds or their derivatives. Specifications of sensitivity, therefore, are meaningful only for a defined analyte and instrumental setup.

A gas chromatograph is the most common inlet system to mass spectrometers. Sample injection, chromatographic conditions, and the design and temperature of the interface between the GC and the ion source of the MS all affect the sensitivity of the analysis. With direct or splitless injections (cf. Chapter 1) at temperatures at which no thermal breakdown occurs, and optimized interfaces (including direct capillary connections), the GC column as such strongly influences detectability.

As mentioned, the mass spectrometric response is defined by the number of ions (current per time unit). In order to optimize the sensitivity, therefore, it is important to aim at sharp gas chromatographic peaks (strong signals per time unit). This may be achieved by designing the chromatographic procedure to favor short retention times and by careful selection of column types. Capillary columns are outstanding in this respect and consequently should be chosen when optimal sensitivity is expedient (cf. Section 4.5). For high sensitivity in SIM work, low column bleed (temperature and nature of stationary phase) is also of importance.

With respect to the conditions for mass spectrometry, the type of ionization and the ionization energy are of decisive importance. The sensitivity that can be attained for a given analyte, using SIM, is dependent on the total ion current represented by the monitored ion. Some biological compounds or their derivatives produce specific ions which are proportionately abundant in the high-mass range when EI is used. Amino acids are examples of such compounds (cf. Chapter 5). Other molecules, such as common derivatives of fatty acids, yield spectra which are dominated by uncharacteristic hydrocarbon ions in the low-mass range, whereas characteristic ions, including the molecular ion, have much lower relative intensities (e.g., ~10%). Monitoring at such less abundant ions obviously results in a significantly raised limit of detection.

Apart from altering the derivative, there are at least two instrumental possibilities of increasing the relative intensities of the selected ions and hence the sensitivity of the technique. A convenient method, still using EI, is to alter the ionization energy. Conventionally, c. 70 eV is the energy used in MS to ensure good fragmentation for purposes of structure determination. Lowering the energy to 20–40 eV, however, may result in lessened fragmentation to small, often insignificant ions and thus favor large, characteristic ions, including the molecular ion. Because of overall decrease in ionization (total ion current) when the electron energy is lowered, an optimum value must be sought.

Use of chemical ionization (CI) is the second means of influencing the fragmentation pattern. CI results in very little fragmentation and simpler spectra. The adduct ions formed, usually related to the entire molecule or to a simple fragment thereof, often represent the majority of the total ions (~75%). Figure 11 (Larsson *et al.*, 1980) illustrates the fragmentation of

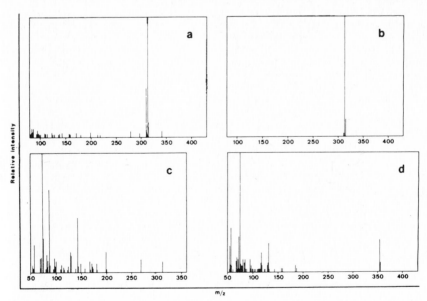

FIGURE 11. Mass spectra of derivatized tuberculostearic acid. (a) CI methane spectrum of methyl ester (1 Torr), (b) CI isobutane spectrum of methyl ester (1 Torr), (c) EI spectrum of methyl ester, (d) EI spectrum of *t*-BDMS ester (70 eV) (Larsson *et al.*, 1980).

the methyl ester of tuberculostearic acid using (a) CI methane, (b) CI isobutane, and (c) EI (70 eV) ionization. The effect of a change of derivative using EI is shown in (d) (*t*-BDMS ester).

Chemical ionization is doubtless a much "softer" method than ionization with 70-eV electrons. The total ion current, however, is considerably less with CI than with EI, and whether or not sensitivity is, in fact, increased must be experimentally determined for each individual compound.

4.5. Specificity

Particularly when combined with data on GC retention time, the complete mass spectrum will in most cases ensure the identity of the analyte, even when it is present in complex mixtures. Most modern mass spectrometers permit recording of full spectra from as little as a few nanograms of test material. Such amounts are frequently handled in quantitative MS using the EICP or SIM techniques.

SIM is often used for selected measurements in the picogram range. Since such quantities are too small to allow registration of complete spectra, to ensure the identity of the compound under study, special consideration must be given to the GC retention time.

An example of the determination of the methyl ester of tuberculostearic acid (Larsson *et al.*, 1980) is given in Figure 12. Of the isomeric C_{19} acids present in bacterial cells, the 2-, 16-, and 17-methyl-substituted

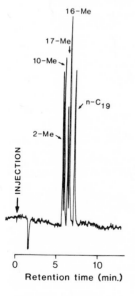

FIGURE 12. EI mass fragmentogram of a mixture of equal amounts of the methyl esters of 2-, 10-, 16-, and 17-methyloctadecanoic acid and *n*-nonadecanoic acid monitored at m/z 312 (70 eV). A 25-m glass capillary column with OV-17 as stationary phase was used at an isothermal column temperature of 210°C. Each peak represents about 300 pg of methyl ester. (Larsson *et al.*, 1980.)

C_{18} acids and the straight-chain C_{19} acids can be expected to be by far the most common. With respect to the detection of tuberculostearic acid for the diagnosis of tuberculosis, it was essential in this analysis to establish the identity of the finding. For analysis in the low picogram range a wall-coated glass capillary column was used. Clearly, there was no difficulty in establishing the identity of methyl tuberculostearate from the GC retention time, using SIM monitoring at m/z 312 (EI), although the MS response was about the same for all four structural isomers. Use of capillary columns in SIM work therefore can be highly recommended, particularly for analyses of compounds present in the picogram range.

The specificity offered by EI as compared with CI merits some comment. In general, specificity decreases with decreasing mass number of the monitored fragment. High mass of the ions being recorded, moreover, reduces possible interference from the instrumental background and from column bleed. As mentioned, ionization using CI generally gives much less fragmentation than EI-ionization, and also higher specificity. The CI mode of procedure, therefore, is important when biological samples of complex nature are to be analyzed by quantitative MS.

5. APPLICATIONS

5.1. Clinical Microbiology

Quantitative MS has been employed as a tool in diagnostic microbiology for identification of microbial metabolites in cultures and in body fluids of infected persons.

Odham et al. (1979), Larsson et al. (1981), and Mårdh et al. (1983) used SIM to detect tuberculostearic acid and C_{32}-mycocerosic acid in patients with tuberculosis. The diagnosis of the disease by culture studies requires several weeks. There is therefore a need for more rapid methods for diagnosing tuberculosis as well as other types of mycobacterial infections.

(R)-10-methyloctadecanoic acid (tuberculostearic acid) is considered to be a cellular constituent unique for organisms of the order *Actinomycetales*, including mycobacteria. C_{32}-mycocerosic acid (2R, 4R, 6R, 8R-tetramethyloctacosanoic acid) has been found exclusively in lipids of *Mycobacterium bovis*, *M. tuberculosis*, *M. kansasii*, and *M. africanum*. The relative amounts of tuberculostearic and C_{32}-mycocerosic acids in each of these four species vary (Larsson et al., 1981). Demonstration of tuberculostearic acid in body fluids and in samples from cultures of such fluids incubated for some days only would constitute a rapid means of establishing the presence of mycobacteria. Furthermore, the finding of mycocerosic acids would pinpoint a limited number of mycobacterial species, including the most important of those pathogenic for humans. Further, determination of the relative amount of these two acids would add to the diagnostic specificity.

Figure 13 shows mass fragmentograms of samples obtained by scraping from the surface of Löwenstein–Jensen slants inoculated with sputum from a patient with active pulmonary tuberculosis. The scrapings were made after 5 days of incubation of the cultures at 37°C. The sample was studied, after methanolysis (Odham et al., 1979), using single ion monitoring at (a) m/z 312 (EI, molecular ion of tuberculostearate), (b) m/z 101 (EI, fragment ion, characteristic of the methyl group in position 4 of C_{32}-mycocerosic acid), and (d) m/z 495 (CI, isobutane, protonated molecular ion of the methyl ester of C_{32} mycocerosic acid).

The results of such analyses of sputum cultures from 14 patients, six with culture-verified tuberculosis and eight with nontuberculous pneumonia, are summarized in Table 2. Tuberculostearic acid was demonstrated in sputum cultures from all of the six patients with tuberculosis but in none from the other eight patients. C_{32}-mycocerosic acid was detected in sputum

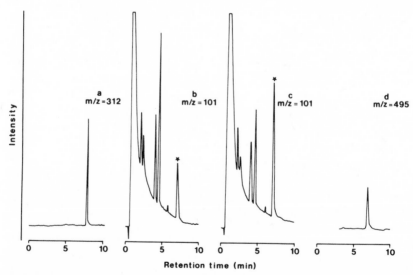

FIGURE 13. Mass fragmentogram obtained by analysis of 5-day-old cultures of a sputum specimen from a patient with pulmonary tuberculosis, demonstrating the presence of the methyl esters of tuberculostearic acid, EI m/z 312 (a), C_{32} mycocerosic acid (*), EI m/z 101 (b), C_{32} mycocerosic acid in a sample to which 50 pg of the authentic ester had been added (*), EI m/z 101 (c), and C_{32} mycocerosic acid, CI, isobutane m/z 495 (d). (Larsson et al., 1981).

TABLE 2. Contents of Tuberculostearic and C_{32} Mycocerosic Acids in 5-Day-Old Cultures of Sputum Specimens

Patient	Patient infected with	Tuberculostearic acid (ng)	C_{32} mycocerosic acid (ng)	Ratio C_{32} mycocerosic/ tuberculostearic acid
A	M. tuberculosis	230	10	0.04
B	M. tuberculosis	69	—	0
C	M. tuberculosis	822	10	0.01
D	M. tuberculosis	5015	211	0.04
E	M. tuberculosis	551	13	0.02
F	M. avium	1026	—	—
G–N	No mycobacteria	—	—	—

cultures from four of the six tuberculotic patients. In five of these patients the culture studies had shown *M. tuberculosis* (Cases A–E). In Case F the infecting organism was *M. avium*, a species in which C_{32}-mycocerosic acid is absent. In Case B microscopy of Ziehl–Neelsen stained sputum samples suggested paucity of organisms in the samples, which may explain why only tuberculostearic acid was detected in the EI-SIM analyses (Larsson *et al.*, 1981).

Demonstration of C_{32}-mycocerosic and tuberculostearic acids in clinical specimens thus seems to offer a means for rapid diagnosis of mycobacterial infections. Determination of the relative amounts of these acids in such specimens (Table 2) can be used for presumptive diagnosis of pulmonary tuberculosis, even of the species identity of the mycobacterium involved.

Disseminated fungal infections in compromised hosts often are not diagnosed early enough to permit curative treatment, due to lack of reliable diagnostic methods. Quantitative MS may be useful for this purpose. Thus, Roboz *et al.* (1980) presented a GC–MS method for the quantification of D-arabinitol in serum from patients infected with *Candida*. D-arabinitol has been found to be a metabolite of the two *Candida* species studied, viz., *C. albicans* and *C. tropicalis*. Elevated levels of this alcohol in serum might therefore indicate fungemia by these organisms. D-arabinitol, erythritol, and 2-deoxygalactitol (the two latter alcohols used as internal standards) were determined as their TMS or pentaacetate derivatives, using a combined GC–MS computer system with CI (isobutane). Figure 14 shows a chromatogram representing the silylated components in serum infected by *C. albicans*. (A) indicates the total ion current tracing and (B) EICP at m/z 513 (protonated molecular ion of the TMS derivative of arabinitol, base peak).

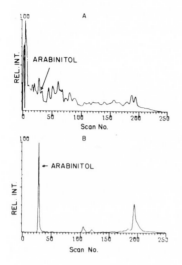

FIGURE 14. Total ion current tracing of silylated components in serum infected by *Candida albicans* (A) and EICP at m/z 513 (protonated molecular ion of the trimethylsilyl derivative of arabinitol (Roboz *et al.*, 1980).

Monitoring of arabinitol levels in two samples of serum from patients with acute myelocytic leukaemia is illustrated in Figure 15. These tests were made with single ion monitoring, focusing at m/z 513. (A) shows a normal endogenous arabinitol level (0.7 μg/ml) and (B) a raised level (2.5 μg/ml) of the alcohol coinciding with the diagnosis of disseminated candidiasis in this patient, a diagnosis confirmed by culture studies.

The study by Roboz *et al.* included 39 controls, whose mean arabinitol level was 0.52 μg/ml (standard deviation ± 0.34). Of 11 patients with confirmed disseminated infections with *Candida*, nine had arabinitol levels >1.2 μg/ml, while two showed values within the normal range. In one patient with acute myelocytic leukaemia the serum arabinitol was repeatedly measured during a period of one year. Maximum levels coincided with the diagnosis of fungemia.

The possibility that other microorganisms which frequently infect immuno-suppressed patients also produce appreciable amounts of

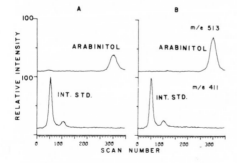

FIGURE 15. Single ion monitoring at m/z 513 (arabinitol) and m/z 411 (erythritol, internal standard) in a leukemic patient without infection (normal range), (A). Data for the same patient as in A at the time of diagnosed disseminated candidiasis, (B). (Roboz *et al.*, 1980.)

arabinitol was investigated by comparing control sera with serum samples which had been inoculated with *Aspergillus fumigatus*, *A. niger*, *Escherichia coli*, *Klebsiella pneumoniae*, and *Pseudomonas aeruginosa*. In no case, however, were enhanced levels of arabinitol detected.

The authors concluded that SIM readily permit monitoring of arabinitol content in blood, and that the technique may provide a means of diagnosing disseminated infections with *Candida*.

Muramic acid, a peptidoglycan moiety of bacterial cell walls, has been subjected to GC/MS assays. Fox *et al.* (1980) demonstrated use of the method to detect muramic acid in tissues of rats with polyarthritis induced by streptococcal cell wall samples. The acid was determined as the alditol (pentaacetyl muramicitol) derivative obtained after reduction with sodium borohydride followed by peracetylation. SIM (EI) monitoring at the characteristic ions m/z 168, 210, and 228 was performed.

Maitra *et al.* (1978) developed a potential method for the diagnosis of endotoxinemia, using 3-hydroxy tetradecanoic acid [β(OH) myristic acid] as marker in GC/MS analyses. Endotoxin, which constitutes part of the cell wall of gram-negative bacteria, is composed of a polysaccharide core and a lipid moiety called lipid A, which on hydrolysis yields 3-hydroxy-fatty acids as the major portion of the total fatty acids. The most common 3-hydroxy acid is β(OH) myristic acid. Detection of this acid would provide one means of diagnosing endotoxinemia.

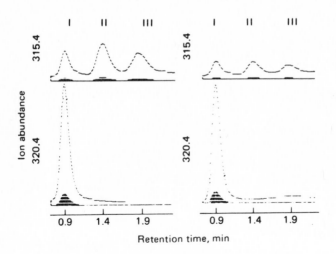

FIGURE 16. Selected ion monitoring (EI) at m/z 315 of the trimethylsilyl ether methyl ester derivative of 3-hydroxymyristic acid in hydrolyzed normal human (left) and rabbit (right) sera. 1000 pmole of the pentadeuterated analog was added as internal standard (m/z 320). (Maitra *et al.*, 1978.)

Using SIM, Maitra *et al.* determined the β(OH) myristic acid contents of lipopolysaccharides of the gram-negative species of *Salmonella minnesota*. Monitoring was done at m/z 315.4, which is the base peak in the EI spectrum of the TMS ether methyl ester derivative of this acid. The detected amounts of β(OH) myristic acid were as low as 200 fmol. Sera normally contain small (c. 1.5 pmol) amounts of β(OH) myristic acid, as indicated in Figure 16. This shows SIM scans of normal human and rabbit sera, with 1000 pmol of added pentadeuterated acid as internal standard. Data from studies involving liquid chromatography on silica acid indicated that the background levels of β(OH) myristic acid in such sera originated, not from endotoxin or lipid A, but from other serum lipids, probably phospholipids.

Under optimized conditions about 1 ng of endotoxin per milliliter of serum can be detected.

5.2. Ecological Microbiology

As yet there are few reports on the use of quantitative MS for study of problems pertaining to ecological microbiology. Some data, however, indicate that GC/MS can provide possibilities for investigating the chemical nature of specific interactions in biological microenvironments, or in environments containing minute amounts of active components of, for example, nutritional or physiological interest.

The determination of valine in running water at the femtomole level has already been mentioned (Bengtsson and Odham, 1979).

Bengtsson and Mårdh (1978) and Bengtsson and Odham (1981) monitored the formation of ornithine by mycoplasmas. The purpose of the study was to develop a rapid method by which to confirm contamination of tissue cell cultures by such organisms. Valid information on this point may be highly relevant under a number of conditions such as in the use of tissue cell cultures in diagnostic virology, but also in biochemical, cytogenetic, and virological research. Conventional techniques for detecting mycoplasmas in cell cultures can be tedious and unreliable. Certain species of *Mycoplasma* are known to catalyze, by ornithine desmidase, enzymatic hydrolysis of arginine to citrulline and ornithine—a phenomenon which has not yet been demonstrated in mammalian tissues. The ornithine exuded by *Mycoplasma hominis* was selectively examined by GC/MS. After 2 h of incubation in mycoplasma medium containing added arginine, samples were derivatized to the *N*-heptafluorobutyryl-*i*-butyl esters and analyzed by EI-SIM at m/z 266 [M-($C_3F_7COH_2$ + $OCOC_4H_9$)] (Figure 17). The presence of ornithine was scarcely recognizable by conventional GC using flame ionization, or even by electron capture detection.

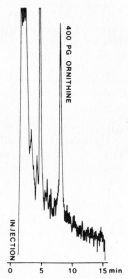

FIGURE 17. Mass fragmentogram of the *N*-HFB-*i*-butyl ester derivative of ornithine produced by *Mycoplasma hominis* during incubation in a phosphate buffer. The peak was measured at m/z 266 and represents 400 pg of ornithine. (Bengtsson and Odham, 1981.)

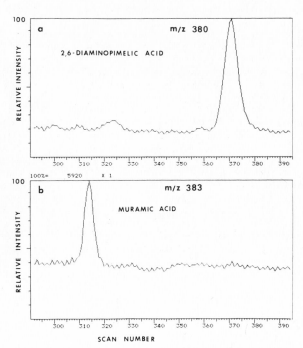

FIGURE 18. Mass fragmentogram of the *N*-HFB-*i*-butyl ester derivatives of 2,6-diaminopimelic (m/z 380) (a), and muramic (m/z 383) (b) acids in a total cell hydrolysate of *Escherichia coli*. EI, 70 eV. (Tunlid and Odham, 1982.)

Tunlid and Odham (1982) developed GC/MS methods which permit detection of bacteria and/or fragments thereof, and possibly quantification of bacterial biomass, by simultaneous determination of muramic acid and 2,6-diaminopimelic acid. Using the N-heptafluorobutyryl-i-butyl ester derivatives and EI SIM monitoring at m/z 383 (muramic acid) and m/z 380 (2,6-diaminopimelic acid), they concluded that about 5 ng of a gram-positive (*Bacillus subtilis*) and 8 ng of a gram-negative (*E. coli*) bacterium (dry weight) could be detected. Figure 18 shows a mass fragmentogram of derivatives of the two cell wall characteristic amino acids in a hydrolysate of a *E. coli* culture. The peaks corresponding to muramic and diaminopimelic acids represent about 580 and 740 pg, respectively.

6. CONCLUSIONS

Quantitative MS is characterized by analytical results of high reliability. It combines sensitivity with extraordinary specificity. In microbiology, quantitative MS has its own, though complementary, place among other analytical techniques, particularly for problems of varying intractability. Its main drawbacks are the comparatively high cost of the equipment and the continuous need for highly trained personnel.

The use of quantitative MS in microbiology can be expected to continue to expand. An area of particular interest in clinical, diagnostic microbiology should be the evolution of MS techniques by which metabolites or constituents specific for a given microbial genus or species other than the examples referred to can be detected in body fluids or tissues. Such evolution should extend current diagnostic methods and provide alternatives for the diagnosis of a number of infectious diseases. In ecological microbiology, quantitative MS may provide a highly sensitive means of identifying a wide variety of microbial populations of minute biomass. It can also be used to establish the presence and quantities of noncultivable microorganisms, or fragments thereof, in biological environments.

Interactions between microorganisms and various hosts are often governed by chemical parameters in the interfaces. Provided that suitable sampling techniques are developed, the unique sensitivity and selectivity of quantitative MS can open up possibilities for chemical studies of such small environments.

REFERENCES

Arpino, P. J., Baldwin, M. A., and McLafferty, F. W., 1974, Liquid chromatography–mass spectrometry. II. Continuous monitoring, *Biomed. Mass Spectrom.* **1**:80.

Arpino, P. J., Krien, P., Vajta, S., and Devant, G., 1981, Optimization of the instrumental parameters of a combined liquid chromatograph–mass spectrometer, coupled by an interface for direct introduction. II. Nebulization of liquids by diaphragms. *J. Chromatogr.* **203**:117.

Bengtsson, G., and Mårdh, P. A., 1978, The application of gas chromatography–mass spectrometry for detection of free amino acids in cultures of Mycoplasmatales and in tissue cell cultures contaminated by mycoplasmas, p. 127, *Abstracts of the XIIth Int. Congress of Microbiology*, Munich, West Germany.

Bengtsson, G., and Odham, G., 1979, A micromethod for the analysis of free amino acids by gas chromatography and its application to biological systems, *Anal. Biochem.* **92**:426.

Bengtsson, G., Odham, G., and Westerdahl, G., 1981, Glass capillary gas chromatographic analysis of free amino acids in biological microenvironments using electron capture or selected ion-monitoring detection, *Anal. Biochem.* **111**:163.

Bertilsson, L., and Costa, E., 1976, Mass fragmentographic quantitation of glutamic acid and γ-aminobutyric acid in cerebellar nuclei and sympathetic ganglia of rats, *J. Chromatogr.* **118**:395.

Coutts, R. T., Jones, G. R., and Liu, S. F., 1979, Quantitative gas chromatography–mass spectrometry of trace amounts of glutamic acid in water samples, *J. Chromatogr. Sci.* **17**:551.

deLeenheer, A. D., and Cruyl, A. A., 1980, "Quantitative Mass Spectrometry," in *Biochemical Applications of Mass Spectrometry*," first supplementary volume (G. R. Waller, and O. C. Dermer, eds.), p. 1169, John Wiley and Sons, New York.

Dutton, G. G. S., 1973, Application of gas–liquid chromatography to carbohydrates, *Adv. Carbohyd. Chem. Biochem.* **28**:11.

Dutton, G. G. S., 1974, Application of gas–liquid chromatography to carbohydrates. II, *Adv. Carbohyd. Chem. Biochem.* **30**:10.

Erdahl, W. L., and Privett, O. S., 1977, A new system for lipid analysis by liquid chromatography–mass spectrometry, *Lipids* **22**:797.

Falkner, F. C., Sweetman, B. J., and Watson, J. T., 1975, Biomedical applications of selected ion monitoring, *Appl. Spectrosc. Rev.* **10**:51.

Fischer-Hoch, S., Hudson, M. J., and Thompson, M. H., 1979, Identification of a clinical isolate as *Legionella pneumophila* by gas chromatography and mass spectrometry of cellular fatty acids, *Lancet* **8138**:323.

Fox, A., Schwab, J. H., and Cochran, T., 1980, Muramic acid detection in mammalian tissues by gas liquid chromatography–mass spectrometry, *Infection and Immunity* **29**:526.

Hammar, C. G., Holmstedt, B., and Ryhage, R., 1968, Mass fragmentography. Identification of chloropromazine and its metabolites in human blood by a new method, *Anal. Biochem.* **25**:532.

Hintze, U., Röper, H., and Gercken, G., 1963, Gas chromatography–mass spectrometry of C_1–C_{20} fatty acid benzyl esters. *J. Chromatog.* **87**:481.

Kaiser, F. E., Gehrke, C. W., Zumwalt, R. W., and Kuo, K. C., 1974, Amino acid analysis. Hydrolysis, ion-exchange clean up, derivatization, and quantitation by gas–liquid chromatography, *J. Chromatog.* **94**:113.

Klem, H. P., Hintze, U., and Gercken, G., 1973, Quantitative preparation and gas chromatography of short and medium chain fatty acid benzyl esters (C_1–C_{22}), *J. Chromatog.* **75**:19.

Kraska, B., Klemer, A., and Hagedorn, H., 1974, Umsetzung von Kohlenhydraten mit *tert*-Butylchlordimethylsilan, *Carbohyd. Res.* **36**:398.

Kuksis, A., Myher, J. J., Marai, L., and Geher, K., 1976, Estimation of plasma free fatty acids as the trimethylsilyl (TMS) esters, *Anal. Biochem.* **70**:302.

Larsson, L., Mårdh, P. A., Odham, G., and Westerdahl, G., 1980, Detection of tuberculostearic acid in biological specimens by means of glass capillary gas chromatography—electron and chemical ionization mass spectrometry, utilizing selected ion monitoring, *J. Chromatog. Biomed. Appl.* **182**:402.

Larsson, L., Mårdh, P. A., Odham, G., and Westerdahl, G., 1981, Use of selected ion monitoring for detection of tuberculostearic and C_{32}-mycocerosic acid in mycobacteria and in five-day-old cultures of sputum specimens from patients with pulmonary tuberculosis, *Acta Pathol. Microbiol. Scand., Sect. B.* **89**:245.

Leimer, K. R., Rise, R. H., and Gehrke, C. W., 1977, Complete mass spectra of *N*-trifluoroacetyl-*n*-butyl esters of amino acids, *J. Chromatog.* **141**:121.

McFadden, W. H., Bradford, D. C., Games, D. E., and Gower, J. L., 1977, Applications of combined liquid chromatography–mass spectrometry, *Am. Lab.* October 55.

McLafferty, F. W., and Baldwin, M. A., 1976, Liquid chromatography–mass spectrometry system and method. US Pat. 3,997,298.

Maitra, S. K., Scholz, M. C., Yoshikawa, T. T., and Guze, L. B., 1978, Determination of lipid A and endotoxin in serum by mass spectroscopy, *Proc. Natl. Acad. Sci. USA* **75**: 3993.

March, J. F., 1975, A modified technique for the quantitative analysis of amino acids by gas chromatography using heptafluorobutyric *n*-propyl derivatives, *Anal. Biochem.* **69**:420.

Mårdh, P.-A., Larsson, L., Odham, G., and Engbaek, H. C., 1983, Tuberculostearic acid as a diagnostic marker in tuberculous meningitis, *Lancet* **1**:367.

Markey, S. P., 1974, in *Applications of Gas Chromatography–Mass Spectrometry to the Investigation of Human Disease* (O. A. Mamer, W. J. Mitchell, and C. R. Scriver, eds.), p. 239, McGill University, Montreal Children's Hospital Research Institute, Montreal.

Matthews, D. E., Ben-Galim, E., and Bier, D. M., 1979, Determination of stable isotopic enrichment in individual plasma amino acids by chemical ionization mass spectrometry, *Anal. Chem.* **51**:80.

Mee, J. M. L., Korth, J., and Halpern, B., 1976, Rapid and quantitative blood analysis for free fatty acids by chemical ionization mass spectrometry, *Anal. Lett.* **9**:1075.

Millard, B. J., 1979, *Quantitative Mass Spectrometry*, Heyden, London.

Odham, G., Larsson, L., and Mårdh, P. A., 1979, Demonstration of tuberculostearic acid in sputum from patients with pulmonary tuberculosis by selected ion monitoring, *J. Clin. Invest.* **63**:813.

Petty, F., Tucker, H. N., Molinary, S. V., Flynn, M. W., and Wanter, J. D., 1976, Quantitation of glycine in plasma and urine by chemical ionization mass fragmentography, *Clin. Chim. Acta* **66**:111.

Phillipou, G., Bigham, D. A., and Seamark, R. F., 1975, Subnanogram detection of *t*-butyldimethylsilyl fatty acid esters by mass fragmentography, *Lipids* **10**:714.

Radford, T., and deJongh, D. C., 1980, "Carbohydrates," in *Biochemical Applications of Mass Spectrometry* (G. R. Waller, and O. C. Dermer, eds.), p. 255, John Wiley and Sons, New York.

Roboz, J., Suzuki, R., and Holland, J. F., 1980, Quantification of arabinitol in serum by selected ion monitoring as a diagnostic technique in invasive Candidiasis, *J. Clin. Microbiol.* **12**:594.

Scott, R. P. W., Scott, G. G., Munroe, M., and Hess, J., Jr., 1974, Interface for on-line liquid chromatography–mass spectrometry analysis, *J. Chromatog.* **99**:395.

Sjöqvist, B., and Änggård, E., 1972, Gas chromatographic determination of homovanillic acid in human cerebrospinal fluid by electron capture detection and by mass fragmentography with a deuterated internal standard, *Anal. Chem.* **44**:2297.

Solov'yov, A. A., Kadentsev, V. I., and Chizhov, O. S., 1976, *Izv. Akad. Nauk. Arm. SSR, Khim. Nauki,* 2256; *Chem Abstr.* **86**, 106974 m (1977).

Sweeley, C. C., Elliott, W. H., Fries, I., and Ryhage, R., 1966, Mass spectrometric determination of unresolved components in gas chromatographic effluents, *Anal. Chem.* **38**:1549.

Tunlid, A., and Odham, G., 1983, Capillary gas chromatography using electron capture or selected ion monitoring detection for the determination of muramic acid, diaminopimelic acid and the ratio of D/L alanine in bacteria. *J. Microbiol. Methods.*, in press.

Vetter, W., 1980, "Amino Acids", in *Biochemical Applications of Mass Spectrometry* (G. R. Waller, and O. C. Dermer, eds.), p. 439, John Wiley and Sons, New York.

Vilkas, M., Jan, H. I., Boussac, G., and Bonnard, M. C., 1966, Chromatographie en phase vapeur de sucres a L'etat de trifluoroacetates, *Tetrahedron Lett.* 1441.

10

ANALYTICAL PYROLYSIS IN CLINICAL AND PHARMACEUTICAL MICROBIOLOGY

Gerhan Wieten, Department of Biomolecular Physics, FOM-Institute for Atomic and Molecular Physics, Kruislaan 407, 1098 SJ Amsterdam, The Netherlands

Henk L. C. Meuzelaar, Biomaterials Profiling Center, University of Utah, 391 South Chipeta Way, Research Park, Salt Lake City, Utah 84108

Johan Haverkamp, Department of Biomolecular Physics, FOM-Institute for Atomic and Molecular Physics, Kruislaan 407, 1098 SJ Amsterdam, The Netherlands

1. INTRODUCTION

The applicability of gas chromatography (GC) and mass spectrometry (MS) to the analysis of organic samples is restricted to volatile substances. Chemical derivatization can be used for some classes of less volatile compounds as a powerful, yet complicated and laborious way to bring them to a suitable form for GC or MS analysis. However, the analysis of highly complex biological compounds such as macromolecules and in particular complete bacterial cells remains out of reach of these separation and analysis techniques, unless preceded by an additional fragmentation method such as pyrolysis.

Pyrolysis has been defined as a chemical transformation through the agency of heat alone (Levy, 1966) generally resulting in a series of more or less volatile fragment molecules (pyrolyzate). Molecules tend to break at specific points and therefore the final mixture of low-molecular-weight products can be expected to contain characteristic information. The constitution of the pyrolyzate is analyzed by methods such as gas chromatography

(Py-GC) or mass spectrometry (Py-MS). Py-GC and Py-MS provide analytical methods with nearly universal applicability for biological materials (Meuzelaar *et al.*, 1982).

In the following sections of this chapter various aspects are described with regard to analytical pyrolysis of microbiological samples, viz., instrumental systems, methods for data handling, and applications to clinically relevant bacterial samples.

2. METHODS

2.1. Pyrolysis Techniques

Certain aspects should be taken into consideration that tend to decrease the characteristicity of the pyrolyzate. For instance, pyrolysis should not be performed in an oxidative atmosphere since this results in the formation of small, noncharacteristic fragments (Levy, 1966). Moreover it is to be noted that primary bond-scission is not the only process during pyrolysis. Some pyrolysis products may become involved in further reactions, e.g., secondary pyrolysis processes or recombination reactions. These secondary reactions tend to decrease both characteristicity and reproducibility of the pyrolyzate. Secondary reactions deserve careful attention in the selection or design of pyrolyzers. Important pyrolyzer features with respect to the reduction of secondary reactions are low dead volume, short residence time of the pyrolyzate in the hot pyrolysis region, well defined temperature/time profile (TTP) and—less strictly related—the amount of sample that has to be analyzed (Simon and Giacobo, 1965).

Pyrolyzers can be divided into continuous mode (e.g., furnaces) and pulse mode pyrolyzers. In modern analytical pyrolysis, pulse mode pyrolyzers are mainly used (Irwin, 1979a). Three different types should be mentioned: (1) resistively heated pyrolyzers, (2) Curie point pyrolyzers, and (3) laser pyrolyzers.

2.1.1. Resistive Pyrolyzers

Resistive pyrolyzers have been used frequently to analyze microorganisms as well as a broad range of other organic materials (Irwin, 1979a; 1979b; Gutteridge and Norris 1979, Meuzelaar *et al.*, 1982). The filament usually consists of a Pt-ribbon or coil. The sample is positioned either within a quartz boat inside the coil or—preferably—deposited directly on the filament. For microbiological applications sample quantities usually range from 10 up to several hundreds of micrograms. Resistive pyrolyzers can be connected directly to the GC injection port or the MS ion source, or they can even be positioned within the MS ion source. In the latter configuration pyrolyzate transfer problems are minimized at the cost of increased ion source contamination.

Modern electronics have made the TTP of resistive pyrolyzers easily programmable. Temperature rise times (TRT), i.e., the time needed to reach equilibrium temperature range from 15 ms up to several minutes. The latter is being used in linear programmed thermal degradation mass spectrometry (LPTD/MS) (Risby and Yergey, 1978). Fast heating rates may enhance reproducibility, while slow rates facilitate time-resolved recording of the degradation of specific compounds.

2.1.2. Curie-Point Pyrolyzers

Curie point pyrolysis employs high-frequency inductive heating of a ferromagnetic sample carrier, generally a simple wire (Meuzelaar *et al.*, 1982). The sample is deposited directly on the wire surface. Typical sample amounts range from 5 to 20 µg. Preferably the pyrolysis wires are made from slowly oxidating metals (e.g., Ni, $T_c = 358°C$) or alloys (Fe–Ni; Fe–Co–Ni) covering a broad range of equilibrium temperatures (T_{eq}) (see Figure 1). A typical figure for the TRT is 100 ms, but values up to 6 s were also used (Huff *et al.*, 1982) to facilitate time-resolved recording of pyrolysis phenomena. Total heating time (THT) values are typically 1 to 10 s. The range of TTP variations obtainable with Curie-point pyrolyzers is restricted when compared to resistive-type pyrolyzers. However, Curie-point pyrolysis techniques excel in simplicity, reliability, and TTP-reproducibility. These and other advantages of the Curie-point method, e.g., the good prospect for interlaboratory standardization, led Meuzelaar to construct fully automated systems for Curie-point py–GC (Meuzelaar *et al.*, 1975a) and Curie point py–MS (Meuzelaar *et al.*, 1976). Disposable pyrolysis wires are mounted in glass reaction tubes (see Figure 2) which are removed with each sample. Reaction tubes can be cleaned and reused. The low temperature (e.g., ambient) of the reaction tubes helps to minimize secondary reactions in the pyrolysate, whereas condensation of less volatile pyrolysis products on the cold glass walls prevents the analytical system from heavy contamination.

2.1.3. Laser Pyrolyzers

The application of laser pyrolyzers for microbiological samples represents the latest developments in pyrolysis techniques. Laser pyrolysis is

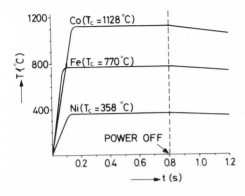

FIGURE 1. Temperature/time profiles of different ferromagnetic sample wires, as used for automated Curie-point pyrolysis (see also Figure 4).

still in its early stages of development, yet this approach has some attractive features. Very short TRT's can be achieved which may enhance the characteristicity of the pyrolyzate by minimizing char formation (Levy, 1966). Moreover, extremely low sample quantities representing very small surface areas can be pyrolyzed, which offers the possibility of characterizing single cells and specific cellular constituents. Wechsung *et al.* (1978) developed a specially designed instrument for laser pyrolysis mass spectrometry. This commercially available instrument (LAMMA) is equipped with a Nd-YAG laser focused on the sample, which can be inspected and centered by means of a normal optical microscope. Spatial resolution is better than 0.5 μm and typical sample quantities are 10^{-12} g. The system has been developed for the analysis of intracellular electrolytes and trace metal concentrations. The potential of LAMMA for typing of organic polymers and cells is, however, not yet fully established (Seydel and Heinen, 1979; Heinen *et al.*, 1980; Dunoyer *et al.*, 1982; Hercules *et al.*, 1982).

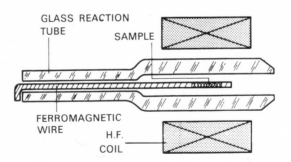

FIGURE 2. Scheme of the Curie-point wire/reaction tube system located within the high-frequency induction coil.

2.2. Analytical Systems

2.2.1. Introduction

The aim of the analytical system is to provide the operator with interpretable, characteristic data with regard to the composition of the pyrolysate. Intact bacterial cells are among the most complex biochemical samples, but at the same time the basic building blocks of different bacterial species are often quite similar. Inevitably, the pyrolysis process reduces the biochemical specificity of the sample, e.g., through the possible formation of identical fragment molecules from originally different compounds. Simmonds (1970) carried out a Py-GC/MS study on *Bacillus subtilis* and *Micrococcus luteus* and demonstrated the overall qualitative similarity of bacterial pyrolysates. For *B. subtilis* only six compounds out of the 60 that were eluted proved to be specific and for *M. luteus* only 3. All other differences were of quantitative nature. Often, quantitative data are the only source of information available so that a high degree of reproducibility of the analytical system is required. Therefore, quantitative reproducibility has to be considered at least equally important to specificity and sensitivity. Two other features which deserve to be mentioned with respect to analytical system requirements are the compatibility of the analytical data with computerized processing techniques and the speed of analysis. In the next paragraphs some frequently used techniques for analytical pyrolysis will be briefly discussed.

2.2.2. Py–GC Systems

For a long time gas–liquid chromatography has been the method of choice in analytical pyrolysis for separation and fingerprinting of bacterial pyrolysates (Irwin, 1979a, 1979b; Gutteridge and Norris, 1979). In view of the enormous numbers of pyrolysis fragments [at least several hundreds (Schulten *et al.*, 1973a)] and the wide variety of their polarities, the choice of stationary phase is crucial. Polar liquids such as Carbowax 20M still represent the most popular stationary phases. Needleman and Stuchbery (1977) reported the use of packed columns with nonpolar Apiezon L as stationary phase and claimed slightly better resolution than with Carbowax 20M. According to Simmonds' findings (1970), similar numbers of pyrolysis components are eluted from both phases. More characteristic pyrolysis components of higher molecular weight can be eluted by using stationary phases of higher thermostability, e.g., terephtalate-terminated Carbowax (Reiner and Kubica, 1969). Packed columns resolve only approximately 60 peaks, which indicates clearly the need for better resolving systems,

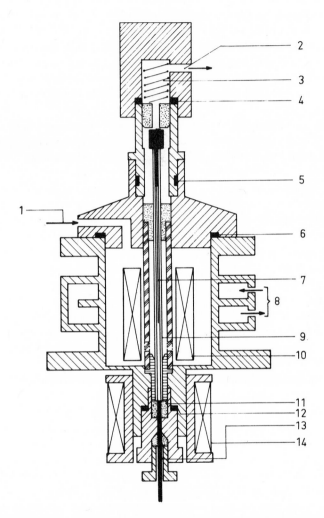

FIGURE 3. Schematic representation of a Curie-point pyrolysis reactor for Py–GC systems. (1) carrier gas inlet, (2) purge gas outlet to needle valve, (3) pressure spring, (4, 5, 6) O-ring (Viton), (7) pyrolysis wire, (8) cooling water, (9) glass reaction tube, (10) rf coil, (11) reaction tube–column seal (glass-filled PTFE), (12) washer (gold), (13) WCOT column (glass), (14) heating element (from Van de Meent, 1980).

e.g., capillary columns (Quinn, 1974; Needleman and Stuchbery, 1977). Meuzelaar and In't Veld (1972) developed a Curie-point pyrolysis reactor to be used with capillary columns (see Figure 3). A modification of this reactor was applied in a fully automated Py-GC system (Meuzelaar et al., 1975a). New developments in sample injection techniques and capillary column technology, e.g., (chemically bonded) fused silica columns, will increase the possibilities of highly reproducible separation of several hundreds of pyrolysate components of strongly different polarities.

2.2.3. Py-MS Systems

Renewed interest in Py-MS after Zemany's early report (1952) was stimulated by a number of considerations which tend to restrict the applicability of GC. Among these were difficulties in encoding data for computerized processing and relatively low sample analysis rates. Moreover many low-volatile and/or unstable compounds may more readily be detected by MS.

Meuzelaar and Kistemaker (1973) coupled a specially designed Curie point pyrolyzer to a small quadrupole mass spectrometer. Pyrolyzer and electron-impact (EI) ion source were interfaced through a gold-coated, heated expansion volume. This setup allowed repetitive scanning of the pyrolysis products over periods up to 30 s (time-averaged accumulation of data). Spectra were found to be highly reproducible (within 5%). Based on this design a fully automated Py-MS system was developed (Meuzelaar et al., 1976) as is schematically represented in Figure 4. This instrument has automated sample changing with an analysis capacity of 30 samples/h. Sample analysis, data acquisition, and mass assignment were controlled by a dedicated minicomputer. From January 1977 to January 1982 this system produced about 25,000 mass pyrograms for over 100 different users (Meuzelaar et al., 1982). Structural complexity of the spectra is considerable, due to the formation of fragments with the same nominal mass from originally different components and the absence of reliable separation procedures. Also, part of this complexity may be due to secondary fragmentation during ionization, and various techniques were used that minimize these effects efficiently, e.g., low-energy EI-ionization with 10–15-eV electrons instead of 70 eV (Meuzelaar et al., 1973), chemical ionization (CI) (Risby and Yergey, 1978), and field ionization (FI) (Schulten et al., 1973). The presence of high-molecular-weight fragments in mass pyrograms depends largely on the conditions of transfer between the pyrolysis zone and the ionization region where adsorption, condensation, and decomposition of large polar fragments may occur. Pyrolysis close to or inside the ion source (direct-probe-Py) is favorable in this respect (Anhalt and

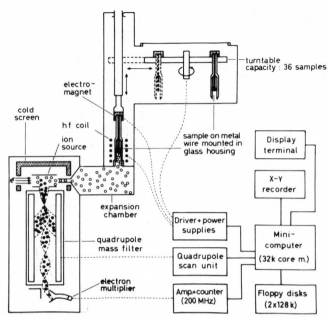

FIGURE 4. Scheme of an automated Curie-point Py–MS system with dedicated minicomputer. Samples are selected by a pickup arm from the turntable and positioned inside the high-frequency coil. The pyrolysate diffuses via a buffer volume into the mass spectrometer where the molecules are ionized and mass-analyzed.

Fenselau, 1975; Risby and Yergey, 1976). Despite the complexity of low-resolution mass pyrograms, structural information is usually directly available through the presence of characteristic ions or ion series. Notably this information is sometimes ambiguous or subjective, but a wide variety of specialized techniques can be used to provide clarity, e.g., LPTD/MS, high-resolution mass spectrometry (HRMS), and tandem (collisionally activated dissociation) mass spectrometry (CADMS). LPTD/MS of cells, using the CI mode (Risby and Yergey, 1976) provided highly characteristic time-resolved spectra of molecular fragments. FI was coupled to HRMS (Schulten *et al.*, 1973) and also proved to give highly informative data. In an off-line bacterial pyrolysate close to 200 components were detected including nearly all products that were previously reported by Simmonds (1970) using Py-GC/MS. In addition, many products too polar or unstable for Py-GC/MS detection were monitored. Optimum conditions for the detection of large and highly informative fragments from nanogram samples can be achieved by the use of field-desorption, Py-FDHRMS (Schulten, 1979). With this method residence time in the pyrolysis zone is extremely short (10^{-11} s), thus suppressing secondary reactions. Py-FDHRMS was not yet used for microorganisms but the impressive results in biochemistry and biomedicine (Schulten, 1979) hold a promise for future applications in microbiology.

CADMS provides a method for structural analysis of ion species that includes differentiation between chemical isomers. Basically two mass spectrometers are interfaced by a collision cell. The first mass spectrometer can be tuned to select an ion species with specific mass value, e.g., representing an interesting component of the pyrolysate. Subsequent collisions with inert gas molecules cause ion-specific dissociations and the resulting fragment ions are mass-analyzed by the second mass spectrometer (McLafferty, 1981). Louter et al., (1980a) developed a Py-CADMS system which operates under essentially the same pyrolysis conditions as used in the automatic Py-MS system described earlier (see Figure 4). Extremely sensitive detection of fragment ions was achieved by using a nonscanning magnet and simultaneous electro-optical ion detection. The application of this instrument for structural elucidation of components of bacterial pyrolysates was reported (Louter et al., 1980b).

3. REPRODUCIBILITY

Three levels of reproducibility can be distinguished: short-term, long-term, and interlaboratory reproducibility. Short-term reproducibility is not a basic problem; in both Py-GC and Py-MS reproducibility of peak intensities within 5% can be easily achieved (Meuzelaar et al., 1975a; Meuzelaar and Kistemakker, 1973). However, long-term and interlaboratory reproducibility of analytical pyrolysis systems still need to be properly evaluated and proved. In the next paragraphs factors that influence reproducibility in analytical pyrolysis of microbiological samples will be discussed, viz., culturing and sampling, pyrolysis, pyrolysate transfer, and analysis.

3.1. Culturing and Sampling

There is little doubt that changing the growth medium alters the pyrograms of microorganisms (e.g., Reiner and Ewing, 1968; Oxborrow et al., 1977b). In fact, such changes may exceed differences between pyrograms of different species that were cultured on the same medium (Gutteridge and Norris, 1980). Variations due to differences between batches of one medium have little or no effect (Oxborrow et al., 1977b). However, as an example of changes caused by differences in culturing conditions, Wieten (unpublished results) observed that leaky stoppers on culture tubes may greatly affect the mass-pyrograms of mycobacterial species. Culturing time was also reported to have a species-dependent influence on pyrograms (Oxborrow *et al.*, 1977a); Gutteridge and Norris, 1980), whereas a systematic rather than a species-dependent shift in the mass pyrograms was observed as the result of autoclaving (see Section 5.12). Notably this shift did not alter discrimination between species. Gutteridge and Norris (1980) reported that the characteristicity of features used to descriminate three bacterial species remained sufficiently stable upon variation in culturing conditions, to allow differentiation of the species under all conditions tested. Yet it is doubtful whether for large numbers of strains or species this will still be true. Especially in view of interlaboratory reproducibility, the selection of generally applicable media and standardized culturing conditions should be considered.

Also the way of sample preparation prior to pyrolysis may influence both long-term and interlaboratory reproducibility. Commonly used are methods that avoid contamination of the samples by medium components. These procedures, e.g., as described by Wickman (1977), involve washing of harvested bacteria, often followed by lyophylization and resuspension in sterile water. However, it is likely that washing and lyophylization can cause uncontrolled changes in the samples. On the other hand it has to be noted that the microsurrounding of bacteria (e.g., the presence of excreted biopolymers and/or metabolites) may contribute positively to the samples (Kistemaker *et al.*, 1975). Based on this assumption, sampling directly from the solid medium, i.e., without any further physicochemical manipulation, is propagated. Both reproducibility and characteristicity obtained by this sampling method proved to be superior to methods involving washing procedures (Meuzelaar *et al.*, 1976). Oxborrow *et al.*, (1976) reported direct sampling of bacteria from membrane filters that were placed on top of the culture medium. However, this procedure may not be applicable to species growing into the agar, rather than forming colonies on top of the nutrient medium.

3.2. Pyrolysis

Reliable pyrolyzers of the Curie-point and resistive type are nowadays available and especially for Curie-point systems the influence on reproducibility of changes in pyrolysis conditions is well documented (Meuzelaar, 1978; Windig *et al.*, 1979). Sample sizes should be in the lower μg range (1–20 μg), since for larger samples controlled heat transfer and uniform sample heating is difficult to achieve and intensified secondary reactions can be expected (Simon and Giacobo, 1965). High reproducibility of temperature/time profiles has been obtained with both pyrolyzer types.

However, for resistive pyrolyzers (e.g., C.D.S. "Pyroprobe") it may be difficult to reach stable equilibrium temperatures when short THTs (less than 1 s) are combined with pyrolysis temperatures in excess of several hundreds of degrees centigrade (Wells *et al.*, 1980). The half-life time of many pyrolytic decomposition reactions are in the submillisecond range (Farré-Rius and Guiochon, 1968). Thus, for Py-GC, i.e., pyrolysis at or above atmospheric pressure, TRT is supposed to be a critical parameter (Levy *et al.*, 1972). TRT-variation seems to be less critical under high vacuum conditions, as used in Py-MS. Meuzelaar (1978) and Windig *et al.*, (1980) observed little or no effect when varying TRT in a range from 0.1 to 1.5 s. When extremely long TRT's are used (up to several minutes), increasing char formation can be expected (Shafizadeh, 1982; Lincoln, 1965) with resulting loss of information and reproducibility. In accordance with the above-mentioned speed of decomposition reactions, Meuzelaar (1978) and Windig *et al.* (1980) observed little variation in pyrograms upon variation of THT from 0.3 to 1.2 s. This suggests that most of the sample (5 μg) was completely decomposed before the equilibrium temperature of the pyrolysis wire was reached. However, it should be pointed out that in Py-GC, changes in THT may cause significant quantitative changes (Levy, 1966). Variation of T_{eq} over a wide range, e.g., 358 and 770°C (Meuzelaar, 1978; Windig *et al.*, 1980) can be expected to change pyrograms significantly. Heating characteristics and, consequently, pyrograms can also be influenced by bad positioning of Curie point wires or deterioration of repetitively used resistive filaments. Finally, proper cleaning of filaments can contribute positively to (long-term) reproducibility (Meuzelaar, 1978; Windig *et al.*, 1980).

3.3. Pyrolyzate Transfer

Pyrolyzate conditions can seriously influence long-term reproducibility and because there is little uniformity in inlet design, these effects may hamper interlaboratory reproducibility as well. Systems with short or no transfer lines, e.g., direct on-column pyrolysis systems or in-source pyrolysis systems, may suffer from uncontrolled build-up of column or ion source contamination. Lowering the sample size can reduce the effect but especially for Py-GC possibilities for using smaller samples are restricted due to relatively low sensitivity. Long transfer lines are preferably heated (Meuzelaar, 1978; Windig *et al.*, 1980) to prevent loss of information due to condensation of less volatile products. Transfer line temperatures equal to or less than 175°C is recommended for most microbiological applications.

3.4. Analytical System

Modern GC and MS systems offer good control of analytical parameters. The main contribution to irreproducibility will be from column or ion source contamination, respectively. Rapid contamination of GC columns after relatively few analyses, e.g., 200, was reported by several authors (Quinn, 1974; Needleman and Stuchbery, 1977). In addition the relatively high operating temperatures needed may cause a gradual column degradation. Thus, a rather frequent exchange of columns will be needed. In this respect the use of chemically bonded fused silica columns may be advantageous.

In mass spectrometric systems slight variations in ion source parameters, due to either contamination, electronic drift, or bad tuning, can cause changes in ionization and ion-optical conditions which can result in reduced reproducibility (Meuzelaar *et al.*, 1982). For Py-MS low-energy EI ionization is normally used. Meuzelaar (1978) reported that minor variations in electron energy (± 1 eV) resulted in an averaged change of 4% in peak intensities, which has to be considered a marked effect when compared to the effect of relatively large variations in pyrolysis conditions. Minimum influence on long-term reproducibility was observed as the result of mechanical overhaul (Meuzelaar *et al.*, 1976; Wieten *et al.*, 1981b). Nevertheless it can be expected that reproducibility of the analytical systems—GC as well as MS—would benefit from the use of microbiologically relevant standard samples.

To achieve some degree of interlaboratory reproducibility, culturing, and sampling, pyrolyzate transfer and analytical conditions need to be standardized. International correlation trials have been organised for Py-GC of synthetic polymers (Gough and Jones, 1975; Walker, 1977). However, up till now only a few, rather crude, interlaboratory comparisons of pyrograms from microorganisms were reported (Quinn, 1974; Stack *et al.*, 1977). For Py-MS analysis, data on interlaboratory comparisons are even more scarce (Meuzelaar *et al.*, 1982). As a first step to standardization Meuzelaar (1978) and Windig *et al.* (1980) presented a list of recommended

TABLE 1. Recommended Standard Pyrolysis Conditions

Wire cleaning method	Reductive
Suspending liquid	Methanol
Sample size	5–20 μg
Equilibrium temperature	510°C
Temperature rise time	0.1–1.5 s
Total heating time	0.3–1.2 s
Inlet temperature	150°C

pyrolysis conditions (see Table 1). Under these conditions long-term reproducibility better than 20% can be expected. Given the complexity of microbiological samples it will certainly take considerable effort and time to establish any significant degree of interlaboratory reproducibility. Alternatively for specific applications, e.g., identification of unknowns, selected references can be used effectively to bypass factors that influence long-term reproducibility (Wieten *et al.*, 1981b). To achieve interlaboratory reproducibility a similar approach may prove profitable.

4. DATA PROCESSING

4.1. Introduction

In general terms the objectives of data processing are twofold: quantitative comparison and qualitative (bio)chemical interpretation (Meuzelaar *et al.*, 1982). Data analysis can be either visual or computer-assisted. Recent developments in computer-assisted evaluation have contributed greatly to the utility of analytical pyrolysis of complex organic samples, e.g., microorganisms. For Py-MS data, computer-assisted evaluation is more commonly used than for Py-GC data. A reason for this is that mass pyrograms are relatively easy to convert into computer-readable format. In addition, the high complexity of mass pyrograms makes computer-assisted evaluation almost indispensible. For Py-GC data, computer assisted handling is hampered by difficulties encountered in conversion of pyrograms into computer-readable format despite the availability of electronic peak integration methods. Peak-integration has to be used judiciously in relation to resolution and base-line drift. Notwithstanding the high resolution of modern, capillary, GC-columns, pyrograms from bacterial samples are complicated and the existence of qualitative differences between pyrograms is at least disputed (Quinn, 1974). Therefore, there can be little doubt that the evaluation of Py-GC patterns will also greatly benefit from the intensified use of computer-assisted data processing.

4.2. Data Handling Packages

At present a wide variety of data handling packages suitable for the evaluation of pyrolysis data are available, e.g., the FOMPYR package developed specifically for Py-MS data by Eshuis (1977) and Windig *et al.* (1981b), the ARTHUR package (Kowalski, 1975), SPSS (Nie *et al.*, 1975), the BMDP package (Frane, 1976), and the CLUSTAN package (Wishart, 1978). Most packages offer a comprehensive choice of routines such as data preprocessing, univariate and multivariate statistical analysis, feature selection, factor and discriminant analysis, cluster analysis, and visualisation procedures. A special program (NORMA) for interactive preprocessing of Py-MS data was recently developed by Eshuis (cf. Huff *et al.*, 1982). This program can be used to transform data into a format compatible with the SPSS, ARTHUR, and FOMPYR packages. An advanced procedure for cluster analysis, called SIMCA, was developed by Wold (cf. Blomquist *et al.*, 1979) and is included in the ARTHUR package. Apart from these packages a number of relatively simple procedures with restricted applicability are used, e.g., one-way analysis of variance (Wieten *et al.*, 1981a), conformity (Emswiler and Kotula, 1978) and similarity calculations (Meuzelaar *et al.*, 1975b). It is beyond the scope of this chapter to describe in detail all routines presently in use for evaluation of analytical pyrolysis data. The most commonly used procedures, i.e., visual interpretation, the FOMPYR routines, and the factor/discriminant analysis routines from the SPSS and BMDP packages, will be discussed in the following sections. To illustrate these processes of data evaluation a set of mass pyrograms from 18 strains of *E. coli* will be used. This data set was taken from the FOM Institute data base and contains two types of *E. coli* strains: one type possessing the capsular polysaccharide K1 ($K1^+$), the other type lacking this virulence factor ($K1^-$). All strains were analyzed in quadruplicate, in order to obtain an estimate of pyrogram reproducibility.

4.3. Visual Interpretation

All data processing usually begins with a visual inspection of pyrograms. Typical pyrograms of the *E. coli* $K1^+$ and $K1^-$ strains are given in Figure 5. Although the overall similarity of the patterns is evident, quantitative dissimilarities may be observed, e.g., at m/z 56, 59, 67, 80, 97, 109, 135, and 151. However, the significance of these differences can only be judged when peak reproducibilities are taken into account. Moreover, on visual inspection alone, small but characteristic dissimilarities may well be overlooked and objective comparison of large numbers of pyrograms (e.g., 72 in the present case) is almost impossible. Hence, for the evaluation of differences in large data sets, the application of computer-assisted procedures will be indispensable.

4.4 The FOMPYR Package

4.4.1. Data Preprocessing

A pattern scaling procedure precedes the actual comparison of pyrograms. The aim of pattern scaling is to compensate for nonrelevant

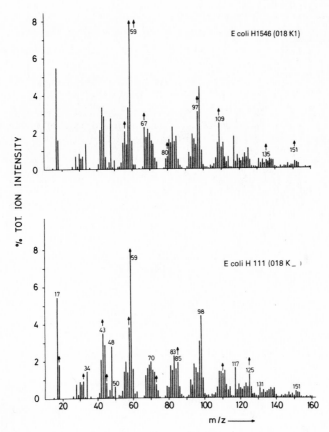

FIGURE 5. Typical mass pyrograms of an *E. coli* K1$^+$ strain and K1$^-$ strain of the same *O*-serotype (averaged pattern from four replicate analyses). Arrows indicate significant peak intensity differences.

variations in peak intensities, such as induced by differences in sample size or changes in instrument sensitivity. A useful procedure is to express individual feature intensities as percentage of total pyrogram intensity. Notably pattern scaling ("normalization") can introduce unwanted, sometimes confusing, effects into the data set (Meuzelaar *et al.*, 1982), e.g., when large and strongly variable peaks are present. High intensities of features will lead to a relative reduction of the intensities of all other features. This effect can be suppressed by temporary exclusion of such features from the data set during the process of pattern scaling, adding them again afterwards. Fortunately the more similar the pyrograms in a data set are, the less problematic pattern scaling will be.

4.4.2. Numerical Comparison of Pyrograms

Numerical comparison of pyrograms is one of the major aspects of data evaluation. In the FOMPYR program emphasis is on displaying dissimilarities rather than similarities between pyrograms. It is convenient to think of mass pyrograms in a slightly different way than the usual bar-graph representation depicted in Figure 5. In principle the n intensities in a pyrogram can be regarded as n coordinates within an n-dimensional orthogonal feature space, wherein each axis is assigned to one of the n features (i.e., mass values). In such a space individual pyrograms are represented by single points. A simple way of calculating the numerical dissimilarity between patterns is to calculate the Euclidean distance, i.e., to measure the distance along a straight line connecting spectral points (Eshuis *et al.*, 1977):

$$D(X, Y) = \left[\frac{1}{n} \sum_{i=1}^{n} (x_i - y_i)^2\right]^{1/2} \tag{1}$$

where X and Y are 2 pyrograms, x_i, y_i are peak intensities, and n is the total number of peaks in the pyrograms.

4.4.3. Feature Scaling and Feature Selection

In general, application of Eq. (1) does not lead to an optimal display of dissimilarities among pyrograms. Use of absolute—although normalized—feature intensities leads to an overestimation of the importance of high-intensity features. To avoid this, feature intensities can be expressed in units of standard deviation, which can be calculated from the replicate analysis of each bacterial strain. (Such a set of replicate pyrograms will be referred to as a "group"). A considerable contribution to adequate dissimilarity calculation is provided by the application of feature selection procedures. In the FOMPYR package mean within-group deviations are calculated for all features in the data set and are used to select reliable, i.e., highly reproducible, features. Obviously reliable features are not necessarily the most specific to reveal differentiation between groups, Calculation and comparison of mean between-group deviations can be used to judge specificity of features. The ratio of both criteria, the "characteristicity" value, may be used as a selective factor that brings out those features in a data set that disciminate best between groups of pyrograms. Notably, whenever between-group deviations are used to select features, the ultimate result of dissimilarity calculation becomes dependent on the particular data set that was analyzed (Wieten *et al.*, 1981a). This may hamper a direct

TABLE 2. Peak Intensity Parameters for the 10 Peaks with Highest Characteristicity in the Pyrograms of *E. coli* K1$^+$ and K1$^-$ Strains

m/z	% Total ion intensity	Within-group deviation (a)	Between-group deviation (b)	Characteristicity (b/a)
67	1.75	0.11	1.18	10.65
109	1.84	0.15	1.18	7.96
73	0.72	0.08	0.38	4.68
126	0.49	0.04	0.19	4.17
102	0.13	0.01	0.06	4.17
45	0.77	0.05	0.22	4.15
140	0.19	0.02	0.08	3.92
84	1.55	0.07	0.29	3.90
75	0.04	0.00	0.02	3.59
82	1.49	0.04	0.13	3.59

comparison of results of different data sets. The result of applying the feature selection procedures to the *E. coli* data-set is given in Table 2. From this table m/z 67 and m/z 109 appear as highly characteristic, which provides a sound basis for the earlier results from visual interpretation (Section 4.3). For subsequent univariate or multivariate dissimilarity calculation, features with high characteristicity values can be selected from this table.

4.4.4. Univariate and Multivariate Analysis

Commonly one-, two- or three-dimensional scatter-plots are used to represent differences within a set of pyrograms. This type of analysis is simple and quick; however, it should be used judiciously when complex patterns are to be compared. Proper feature selection procedures have to be used to reduce the danger of misjudging the importance of single features.

Figure 6 shows an example of bivariate analysis taken from the *E. coli* data set. The features were selected according to their characteristicity (see Table 2) and reveal a clear dissimilarity between K1$^+$ and K1$^-$ strains. Note the strong correlation between the two features, indicating a common molecular origin.

In view of the complexity of microbiological pyrograms, multivariate analysis can be considered superior to univariate analysis (Meuzelaar *et al.*, 1982). Thus, dissimilarities between pyrograms may be calculated using formula (1) and a set of features selected according to the ratios of between- and within-group deviations (see Table 2). However, Eshuis *et al.* (1977)

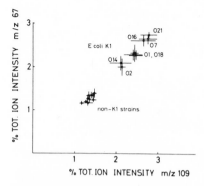

FIGURE 6. Scatter diagram of ion intensities at m/z 67 and 109 in the mass pyrograms of a series of 18 *E. coli* strains. Bars indicate SD values.

demonstrated that separation between groups of *Listeria* pyrograms improved dramatically once these ratios were also used as weight factors in dissimilarity calculation. Hence, the Euclidean distance formula (1) was modified accordingly, leading to formula (2):

$$D(X, Y) = \left[\frac{1}{\sum w_i} \sum_{i=1}^{n} w_i \frac{(x_i - y_i)^2}{\sigma_i} \right]^{1/2} \tag{2}$$

where w_i is a weight factor for peak i, n is the number of peaks used, and σ_i is the mean within-group standard deviation for peak i. This formula is routinely used in the FOMPYR package for multivariate dissimilarity calculation.

Dissimilarity values may be listed in matrix format, thus providing an integral view on the numerical relationships between mass pyrograms. To give an impression, part of the dissimilarity matrix calculated for the *E. coli* data set is represented in Table 3. The dissimilarity values were calculated by using the masses listed in Table 2. Reproducibility of replicate analysis can be judged from dissimilarity values in the diagonal blocks. Off-diagonal values represent dissimilarities between pyrograms of different groups, i.e., different bacterial strains. From the dissimilarity values in the complete matrix a clustering into two major groups, i.e., $K1^+$ and $K1^-$ is evident.

The information content of a dissimilarity matrix is highly comprehensive, but with increasingly large data sets it becomes difficult to form a clear mental picture of the relationships among pyrograms. Additional visualization techniques are therefore used, e.g., group averaged dissimilarity matrices, reordered matrices, nearest-neighbor tables, dendrograms, and nonlinear mapping (Meuzelaar *et al.*, 1982). Of these procedures, nonlinear mapping as developed by Kruskal (1964) and adapted for Py-MS data by Eshuis *et al.* (1977) is the most commonly used visualization

TABLE 3. Part of Dissimilarity Matrix of *E. coli* $K1^+$ and $K1^-$ Strains

Group/ spec. No.	Sero group	1/1	1/2	1/3	1/4	2/1	2/2	2/3	2/4	3/1	3/2	3/3	3/4	4/1	4/2	4/3	4/4	5/1	5/2	5/3	5/4	6/1	6/2	6/3
1/1	$021K1^+$	—																						
1/2		0.9	—																					
1/3		1.0	1.2	—																				
1/4		1.7	1.4	1.0	—																			
2/1	$021K1^-$	7.8	7.6	7.3	6.6	—																		
2/2		8.1	7.8	7.5	6.8	1.0	—																	
2/3		7.6	7.3	7.0	6.3	1.0	0.9	—																
2/4		8.1	7.9	7.6	6.8	1.0	1.0	1.0	—															
3/1	$021K1^-$	7.4	7.3	6.7	6.1	2.5	2.8	2.6	2.7	—														
3/2		7.5	7.3	6.8	6.1	1.8	1.8	1.5	1.9	1.4	—													
3/3		8.0	7.8	7.3	6.7	1.8	2.1	1.7	1.6	1.7	1.0	—												
3/4		7.4	7.2	6.7	6.0	1.7	2.0	1.4	1.7	2.0	0.8	1.0	—											
4/1	$07K1^+$	1.2	1.2	1.3	1.9	7.9	8.1	7.6	8.2	7.3	7.4	8.0	7.4	—										
4/2		1.9	1.6	1.6	1.8	7.4	7.5	7.0	7.6	6.8	6.8	7.4	6.8	1.0	—									
4/3		2.1	1.9	1.4	1.0	6.3	6.4	5.9	6.5	5.8	5.8	6.3	5.7	1.9	1.4	—								
4/4		1.7	1.7	1.3	1.3	6.7	6.9	6.4	7.0	6.1	6.2	6.8	6.2	1.4	1.2	1.2	—							
5/1	$07K1^-$	8.3	7.7	7.1	7.1	2.2	2.2	2.0	2.0	2.0	1.7	2.4	1.7	8.2	7.6	6.6	7.1	—						
5/2		8.0	7.4	6.7	6.3	1.8	1.7	1.6	1.7	1.8	1.5	1.4	1.9	7.9	7.3	6.3	6.8	0.7	—					
5/3		7.7	7.5	6.3	6.3	2.0	1.8	1.6	1.9	1.6	0.8	1.2	1.3	7.6	6.9	5.9	6.4	1.4	0.9	—				
5/4		7.9	7.8	6.6	6.6	1.9	1.9	1.8	1.7	1.9	1.4	1.3	1.7	7.9	7.3	6.2	6.8	1.0	0.7	0.9	—			
6/1	$014K1^+$	3.2	3.1	2.7	2.3	5.1	5.4	4.9	5.4	4.3	4.6	5.1	4.7	3.2	3.0	2.0	2.3	5.3	5.0	4.8	5.0	—		
6/2		2.7	2.5	2.1	1.8	6.1	6.3	5.8	6.4	5.1	5.4	6.1	5.6	2.4	2.2	1.6	1.6	6.3	6.0	5.6	6.0	1.4	—	
6/3		4.5	4.4	3.9	3.2	4.1	4.3	3.9	4.5	3.0	3.3	4.0	3.5	4.4	3.9	2.9	3.2	4.2	3.9	3.5	3.8	1.7	2.3	—
etc.																								

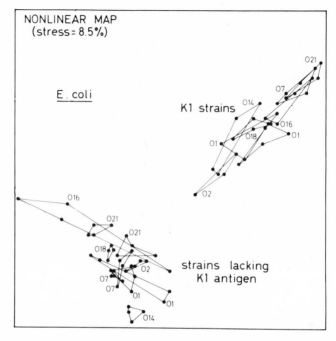

FIGURE 7. Nonlinear map of relations among mass pyrograms of the set of 18 *E. coli* strains as based on Euclidean dissimilarity values (see also Table 3). Each point represents a pyrogram; replicate pyrograms are interconnected.

technique in the FOMPYR package. Nonlinear mapping is an iterative data-fitting procedure which, starting from a multidimensional data set, e.g., a multifeature based dissimilarity matrix, proceeds towards a representation of lower dimensionality (usually $n = 2$). The goodness-of-fit (stress factor) is determined by a set of rules that emphasize relative dissimilarity (nearest neighbors) rather than absolute dissimilarity. Consequently, a nonlinear map can only be a qualitative, visual help; for quantitative interpretation the underlying dissimilarity matrix has to be used. Nevertheless nonlinear mapping has proved to be of extreme value for visual inspection of relationships among large sets of pyrograms. Application of nonlinear mapping is not restricted by any assumptions with respect to the statistical distribution of the data or size of the data set. Figure 7 shows the nonlinear map that corresponds to the *E. coli* dissimilarity matrix. Note the better resolution of the $K1^+$ and $K1^-$ cluster, when compared to the result of bivariate analysis (see Figure 6).

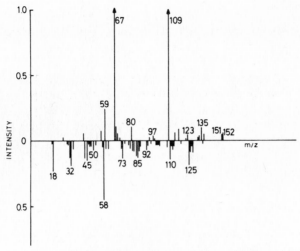

FIGURE 8. Difference spectrum obtained by subtracting the mass pyrograms of the K1$^+$ and K1$^-$ strains represented in Figure 5.

4.4.5. Computer-Assisted Chemical Interpretation

Thus far the emphasis of data processing has been to provide information on dissimilarities among (mass) pyrograms, irrespective of underlying (bio-)chemistry. Obviously chemical interpretation of observed dissimilarities is of high value and sometimes the main reason for analysis.

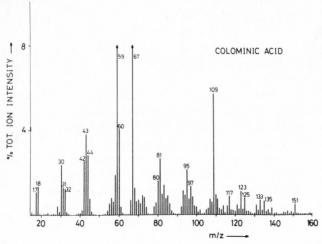

FIGURE 9. Mass pyrogram of the K1 capsular polysaccharide of *E. coli*.

In simple cases, e.g., the *E. coli* data set, where apparently one major component is varying, subtraction of pyrograms can provide important information. Figure 8 shows the difference spectrum obtained by subtracting a pyrogram of a $K1^+$ and $K1^-$ strain of the same O-serotype. A straightforward way to interpret such difference spectra is by comparison with pyrograms of model compounds held in some sort of library system. Figure 9 shows the mass pyrogram of the purified K1 antigen (colominic acid). The set of peaks at m/z 67, 80, 109, 123, 135, and 151 is clearly observed in the positive part of the difference spectrum, indicating that the dissimilarity between $K1^+$ and $K1^-$ strains directly reflects the presence or absence of this polysaccharide in the bacterial cells.

4.5. Factor/Discriminant Analysis

Very often the problem of chemical interpretation of pyrogram patterns is more complicated, i.e., various dissimilarities caused by multiple components are observed. In such cases a procedure is needed that describes the relative contribution of individually changing components. Windig *et al.* (1981b, 1982) proposed the application of factor analysis/discriminant analysis (FA/DA) combined with graphical rotation as a method for chemical interpretation of dissimilarities among pyrograms. This procedure is partly incorporated in the SPSS data handling package and may be used in addition to the FOMPYR package. However, DA can also be used as an independent, highly effective method to display dissimilarities among groups of pyrograms, i.e., dissimilarities among bacterial strains or species (e.g., French *et al.*, 1980; MacFie *et al.*, 1978).

In the multidimensional mass-axes space (see Section 4.4.2) the mass pyrogram of a chemical compound, e.g., the K1-polysaccharide is represented by a single spectral point. Varying concentrations of such a component in the set of pyrograms can be viewed on the "component axis" which passes through this spectral point. The corresponding changes in peak intensities along this axis will be strictly correlated. Thus in principle, a pattern of correlated changing feature intensities may be used to identify a varying component in a multicomponent mixture, i.e., a bacterial pyrogram.

The correlations among feature intensities in a set of pyrograms can be calculated by either FA or DA. In both procedures new variables, i.e., factors or discriminant functions, respectively, are used to describe these correlations; each of the new variables represents a linear combination of the original features (e.g., mass values). Factors are calculated in such a way that the first factor (F1) accounts for the largest proportion of total variance in the data set, which means that the orientation of F1 is in the direction of maximal dissimilarity between individual pyrograms. F2 accounts for the largest proportion of the remaining variance, etc. The calculation of discriminant functions is conditioned by relationships of "between-group" and "within-group" variances in such a way that the first discriminant function (D1) discriminates as well as possible groups of pyrograms, rather than individual pyrograms. D2 represents the linear combination of features that discriminates next best, etc. The basic difference between the calculation of discriminant functions and factors is analogous to the previously mentioned calculation of dissimilarity values with or without the use of weighting factors (see Sections 4.4.2 and 4.4.4), and explains why in general discriminant analysis is the better method to display dissimilarities among groups of pyrograms (Gutteridge et al., 1980; Windig et al., 1982). Discriminant functions can be visually inspected by a bar-graph representation referred to as discriminant spectra. The contribution of a discriminant function to an individual pyrogram is called the "discriminant score" and is easily calculated. Comparison of discriminant scores provides a direct view on dissimilarities among pyrograms.

Notably the feature pattern represented by a discriminant function does not necessarily correspond to the feature pattern of a pure chemical component. However, a search for a specific combination of correlated features, i.e., a component axis, can be performed by rotation of discriminant functions. In practice a stepwise rotation is used and visual inspection of the succesive discriminant spectra is used to locate component axes. Subsequently the component contribution to the individual pyrograms can be deduced from the corresponding discriminant scores. Searching for component axes may be hampered by the fact that in a pyrogram different chemical components may contribute to the intensities of similar sets of features. Furthermore the set of pyrolysis products generated from a particular component can be influenced by the total matrix of organic and inorganic sample components. Consequently, changing feature intensities in a data set will be partly rather than fully correlated and chemical component specificity of the correlated patterns will be somewhat reduced. This may imply that often only classes of related components rather than individual components can be identified.

It was shown by various authors (Windig et al., 1981a, 1982; Van de Meent et al., 1982) that upon application of discriminant analysis and graphical rotation improved qualitative as well as quantitative interpretation of dissimilarities could be obtained. Moreover, discriminant analysis has proved to be a powerful method to discriminate between (bacterial) species (Gutteridge and Norris, 1980; French et al., 1980; Windig et al., 1981a).

The results of applying discriminant analysis to the *E. coli* data set are given in Figure 10. Figure 10a represents a plot of the discriminant scores of D1 and D2. As before (see Figure 7) K1$^+$ and K1$^-$ strains are clearly differentiated, but in addition each cluster is better resolved. In general, display of dissimilarities by DA is found to be better than results from modified Euclidean dissimilarity calculation (see Section 4.4.4). This is explained by the fact that in the latter procedure correlations among features, reflecting a common origin, are fully disregarded.

Rotation of D1 over 45° provides the optimum view on the pattern of changes in intensities that account for the discrimination of *E. coli* K1$^+$ and *E. coli* K1$^-$. The corresponding discriminant spectrum, given in Figure 10b, shows clear similarity to the antigen model compound pyrogram (see Figure 9). Mass 59 is absent in this discriminant spectrum, probably because a peak at this *m/z* value is very common in a great variety of biological molecules (Meuzelaar *et al.*, 1982); a high correlation to some components other than the K1 antigen can therefore be expected. The scattering within the K1$^+$ and K1$^-$ clusters is caused by a complex of other molecular components of cell wall and cell plasma, e.g., proteins, polysaccharides, and lipopolysaccharides, as was evidenced by comparing the rotated discriminant spectra with patterns of reference biopolymers (cf. Meuzelaar *et al.*, 1982).

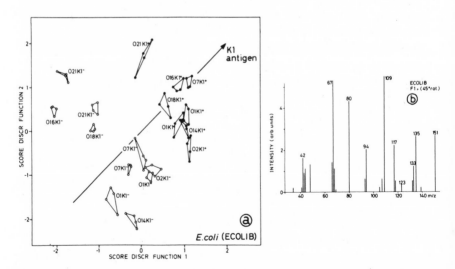

FIGURE 10. Data evaluation of the series of *E. coli* mass pyrograms by discriminant analysis. (a) Plot of the discriminant scores of the first and second discriminant function. (b) Discriminant spectrum after rotation of the first discriminant function over 45° counterclockwise.

5. APPLICATIONS

5.1. Introduction

Since the first reports were published on the applicability of analytical pyrolysis techniques to clinical and pharmaceutical microbiology (Oyama, 1963; Reiner, 1965) this subject has been covered in more than 100 articles. Py-GC and Py-MS are most commonly applied for bacterial fingerprinting. Other, highly specialized techniques, e.g., Py-GC/MS, Py-CADMS, and Py-HRMS have only been used occasionally to obtain detailed structural information about the composition of complex pyrolysates. Bacteria as well as fungi have been subject to investigation. Although various groups of fungi may also be of clinical relevance (e.g., Vincent and Kulik, 1970, 1973; Sekhon and Carmichael, 1972; Oyama and Carle, 1967; Burns *et al.*, 1976; Stretton *et al.*, 1976; Taylor, 1967; Windig and De Hoog, 1982) the applications of analytical pyrolysis mentioned in the following sections will be restricted to bacteriology.

Thusfar, the analysis of about 35 different bacterial genera has been reported. Most intensively investigated were the genera *Mycobacterium*, *Streptococcus*, *Neisseria*, *Escherichia*, *Pseudomonas*, *Bacillus*, and *Clostridium* and, somewhat less frequently, *Salmonella*, *Klebsiella*, *Listeria*, *Yersinia*, *Enterobacter*, and *Vibrio*. From a number of genera, e.g., *Staphylococcus*, *Bacteroides*, *Serratia*, *Leptospira*, *Bordetella*, and *Mycoplasma*, only incidental, although sometimes interesting, reports are available (Gutteridge and Norris, 1979; Meuzelaar *et al.*, 1982).

The objectives of most projects were fairly uniform, i.e., differentiation and classification of bacterial strains at either the genus, species or subspecies level and subsequent identification of unknown strains. Most often, evaluation of results was restricted to fitting the data into already existing taxonomical or diagnostic schemes, thereby providing proof for the feasibility of analytical pyrolysis in clinical bacteriology. Application of Py-GC and/or Py-MS as independent analytical tools represents an alternative approach that may provide important information difficult to obtain by other techniques. The structural information which can be obtained from the pyrograms and the quantitative nature of the data allow them to be successfully used for chemical characterization of whole cells and cellular components, but also for purposes of quality control and screening. Important advantages especially of Py-MS are the small sample sizes required, the high sample throughput, and the ready applicability of the data for computerised evaluation.

For a selection of bacterial groups analyzed, results are summarized in the following sections.

5.2. Streptococcus

In the first report on Py-GC of microorganisms, Reiner (1965) included four types of group A streptococci which were easily differentiated from strains of *E. coli, Shigella*, and mycobacteria. Fully automated Py-GC and Py-MS was used by Meuzelaar *et al.* (1975b) and Kistemaker *et al.* (1975) for the analysis of *S. mutans, S. sanguis*, and *S. feacalis*, including two *Streptococcus* strains differing in just one polysaccharide antigen (viz., Z_3III and Z_3). Visual differentiation was possible with both techniques, not only between species but also between the Z_3III and Z_3 strains. *S. mutans* proved to be a heterogeneous species. A more objective comparison of pyrograms was based on similarity coefficients. This procedure combined with a small library of spectra was used to match unknown strains.

Also Huis In't Veld *et al.* (1973) reported the differentiation by Py-GC of *Streptococcus* strains with regard to the presence or absence of type III antigen. Pyrograms of whole cell samples of Z_3III, FIII, and corresponding mutants lacking the type III polysaccharide showed clear dissimilarities. Pyrograms of purified cell walls (Figure 11) were even more dissimilar. Full account of the observed differences was provided by the analysis of the purified type III antigen. This polysaccharide pyrolysis pattern was shown to be different from those of the types I, II, and IV polysaccharide antigens and also from a (serologically identical) type III antigenic variant which was isolated from culture medium. It was concluded that Py-GC could well be used to screen streptococci for the presence of the type III antigen, but also as quality control procedure for antigen preparations.

Comparison of Py-GC patterns by similarity coefficients was also used by Stack *et al.* (1977, 1978) in order to differentiate between *S. mitior, S. mutans*, and *S. sanguis*. In addition a simple form of discriminant analysis was successfully used to differentiate between these species. Gross heterogeneity among strains of *S. mutans* was observed and also strains that were intermediate to *S. mitior* and *S. sanguis*. Library spectra were successfully used to identify pyrograms of 25 unknown dental plaque cultures as either *S. mitior* or *S. sanguis*. Streptococcal isolates from blood were observed to be highly dissimilar. Finally it was shown that characteristicity of the discriminating features in Py-GC was unaffected by prolonged column use or even column exchange.

5.3. Neisseria

Up till now data on differentiation and classification of *Neisseria* strains are few. Borst *et al.* (1978) reported the differentiation of meningococci and gonococci by Py-MS. Haverkamp *et al.* (1979, 1980b) used Py-MS as

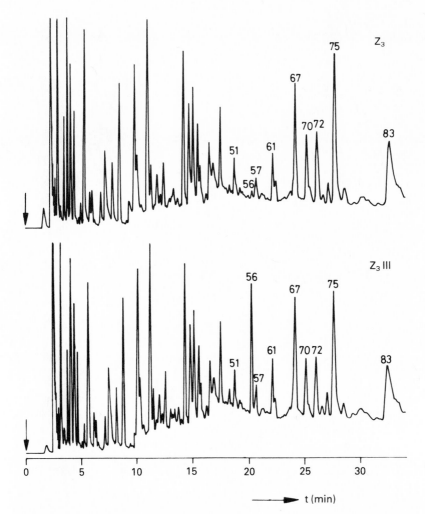

FIGURE 11. Py–GC pyrograms of purified *Streptococcus* Z_3 and Z_3III cell walls. Note intensity differences at peaks 56, 61, and 75 which represent the major peaks of the pure type III polysaccharide. (From Huis in't Veld *et al.*, 1973).

an independent tool for biochemical characterization of *N. meningitidis* capsular polysaccharides, a project closely related to development and production of vaccines. Capsular polysaccharides of serogroups A, B, C, 29E, W-135, X, and Y as well as partially *O*-acetylated variants were analyzed. Differentiation based on multivariate data analysis was observed between polysaccharides built up from hexosyl-sialic acid repeating units (serogroups W-135 and Y) and sialo-homopolymers (serogroups B and C),

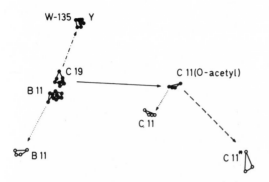

FIGURE 12. Nonlinear map of mass pyrograms of eight purified (●) and three partially purified (○) capsular polysaccharides of *Neisseria meningitidis* groups B, C, W-135, and Y. The partially purified preparations B11 and C11 contain protein; the preparation C11* contains in addition lipopolysaccharide contaminants; shifts on the map correlate to differences in structure and composition: (-·-·-), hexose residue shift; (———), *O*-acetyl substitution shift; (····), protein contamination shift; (---), lipopolysaccharide/protein shift.

whereas the presence of *O*-acetyl substituents can easily be recognized. Even differentiation between $\alpha(2 \to 8)$- and $\alpha(2 \to 9)$-linked sialopolymers (serogroups B and C, respectively) turned out to be possible. Structural considerations with respect to the chemical origin of characteristic features in the spectra were given. Partly purified groups B and C polysaccharides were shown to be contaminated by protein and lipopolysaccharide components in different concentrations. A correlation was made between the Py-MS data and results from immunological and biochemical analyses. The above described differences between a number of capsular polysaccharide preparations are clearly visualized in the nonlinear map presented in Figure 12. In a comparable way Py-MS is presently being used for biochemical characterization of polysaccharide-outer membrane complexes of *N. meningitidis* (Beuvery *et al.*, 1982a) and of covalently linked capsular polysaccharide–protein conjugates (Beuvery *et al.*, 1982b) to be used as vaccines against meningitis. These results demonstrate the feasibility of automated Py-MS for quality control during purification of capsular polysaccharide preparations and screening for structural dissimilarities.

5.4. Escherichia

Reiner (1965) has demonstrated that small, quantitative dissimilarities on just a few pyrogram features were sufficient to differentiate between 18 antigenically different strains of *E.coli* and one strain of *Shigella*. This work was extended by analyzing 49 coded strains of *E. coli* (Reiner and Ewing, 1968). The ratio of two visually selected features was used to differentiate all pyrograms into five groups which were shown to represent nine serotypes. Characteristicity was stable, even when culturing conditions were changed. Gutteridge and Norris (1980) made similar observations and reported that the characteristicity of four features which differentiated *E. coli*, *Ps. putida*, and *S. aureus* was unaffected by variation of culturing temperature, time, or medium. Notably, when a larger number of features were used, drastic differences were observed on medium variation among pyrograms of any of the mentioned species. Application of factor analysis proved to be highly effective in displaying systematic trends in the data (Gutteridge *et al.*, 1979).

A proposal to use computerized data processing was made by Oyama and Carle (1967). Although these authors were able to differentiate between strains of *E. coli*, *Pseudomonas*, *Bacillus*, *Cellumonas*, and *Azotobacter* by visual comparison of pyrograms, they indicated that an objective procedure was needed for selection of quantitatively discriminating features and comparison of pyrograms. The possibility of discriminating between *E. coli* strains differing in single antigenic components, e.g., the flagellar and H antigens, was indicated by Reiner and Ewing (1968). Haverkamp *et al.* (1980a) using automated Py-MS and multivariate data analysis, discriminated between *E. coli* strains possessing the capsular K1-antigen (colominic acid) and strains lacking this polysaccharide (see also Section 4.4.4). In a set of 18 patient strains clear differentiation was observed between the two types. The spectral differences were readily proved to be caused predominantly by the presence or absence of the capsular polysaccharide. From the results of data processing highly discriminating features were deduced to be used for fast routine screening. Up till now close to 150 different strains were typed in this way. Capsular polysaccharides other than K1 appear not to disturb the detection (except for K92 which also is a sialic acid polymer, however, of a different linkage type) (Haverkamp and Guinée, unpublished results).

5.5. Klebsiella

In a study by Abbey *et al.* (1981), who used high-resolution Py-GC–MS, it was shown that serologically nontypable strains of *K. pneumoniae* could be readily differentiated. Coded duplicate strains were all matched correctly by visual inspection. MS was used to identify characteristic features.

In a hospital epidemiological study by Meuzelaar *et al.* (1982) 20 isolates from 6 patients were tested. Conventional typing failed to differentiate the strains; however, Py-MS analysis revealed possible cross-infections for two pairs of patients only (see Figure 13). Furthermore, it has been shown that *Klebsiella* strains of different serotypes can often clearly be differentiated on the basis of the Py-mass spectra (Haverkamp *et al.*, 1979). Capsular polysaccharides play an important part in this differentiation.

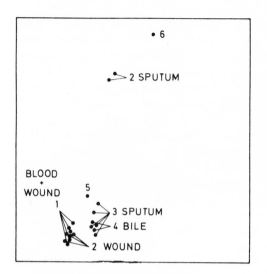

FIGURE 13. Nonlinear map of mass pyrograms of 20 *Klebsiella* isolates from six patients in one hospital (the centroids of quadruplicate analyses are shown). Samples from patients 1 and 2 as well as from patients 3 and 4 indicate possible cross-infections.

An interesting phenomenon with regard to the constitution of bacterial pyrolysates was described by Hudson *et al.* (1982). In the pyrolysates of *Klebsiella* strains the product acetamide was found to be nearly absent. This corresponds with observations made by Abbey *et al.* (1981), who used a similar technique. It should be noted that cell wall pyrolysates of almost all other bacterial genera do contain acetamide, which should be considered as derived mainly from the N-acetyl substituents of the peptido-glycan network (Simmonds, 1970; Louter *et al.*, 1980*b*).

5.6. Yersinia

The high-resolution Py-GC patterns of 12 strains of *Yersinia enterocolitica* and individual strains of *Salmonella typhi, Hafnia alvei,* and *Proteus vulgaris* were analyzed by discriminant analysis (Stern *et al.*, 1979). All genera were clearly differentiated. In a later study, stepwise discriminant analysis was used to test for virulence of 14 *Yersinia* strains (Stern *et al.*, 1980). Hela-cell invasiveness was used as a criterion for virulence, and pyrograms could be differentiated accordingly. When tested on replicate cultures which were treated as unknowns all invasive strains were identified correctly; however, 5 out of 8 noninvasive strains were misidentified.

5.7. Salmonella

The analysis of a large number of strains (55 coded samples) revealed dissimilarities in pyrograms that were shown to correspond to serological and biochemical typing (Reiner *et al.*, 1972). Characteristic pyrogram features were related to the presence of specific carbohydrate components (e.g., 3,6-dideoxyhexoses) and also a relationship was found between Py-GC peaks and somatic and flagellar antigens. Emswiler and Kotula (1978) showed that by stepwise discriminant analysis applied to pyrograms of purified flagella from 10 different *Salmonella* serotypes a 100% correct classification of the serotypes was possible. The analysis of whole cells resulted in a 90% correct classification of the serotypes. The feasibility of Py-GC for identifying unknown *Salmonella* strains was demonstrated by application of computerized library search (Menger *et al.*, 1972). The library contained 9-peak-spectra and search was conditioned by retention time and peak intensity. A correct identification of 9 out of 10 strains was reported.

5.8. Vibrio

An extensive Py-GC study of 57 choleralike Vibrio strains was carried out by Haddadin *et al.* (1973). A clear differentiation was obtained between *V. cholerae* and strains of *Salmonella, Aeromonas, Shigella, Proteus, Pseudomonas*, and *Vibrio parahaemoliticus*. Two characteristic peaks in the low-resolution GC patterns allowed the identification of the "classical" and "El Tor" biotypes. An intermediate group was also observed. Various Heiberg groups, serotypes, or agglutinable/nonagglutinable types could not be distinguished.

5.9. Pseudomonas

Direct probe MS was applied by Anhalt and Fenselau (1975) to differentiate species of *Pseudomonas, Proteus, Salmonella, Neisseria*, and *Staphylococcus*. Main spectral dissimilarities could be interpreted as originating from phospholipids and ubiquinones. However, the authors did not provide any indication of the pattern-reproducibility. Based on Py-GC data and discriminant analysis McFie *et al.* (1978) demonstrated the differentiation between strains of *Pseudomonas, Lactobacillus, Micrococcus, Microbacter*, and *Moraxella* (20 strains total). Using the same technique, *Ps. putida, E. coli*, and *S. aureus* were also differentiated (Gutteridge and Norris, 1980). Additionally it was shown that differences in culturing conditions, although clearly influencing individual pyrograms, did not interfere with the ability to differentiate between the three species.

Differentiation within the genus *Pseudomonas* was already reported by Reiner (1977), who tested strains of *Ps. aerugenosa* and *Ps. cepacia*. French *et al.* (1980) extended this work by analyzing 50 strains of *Ps.*

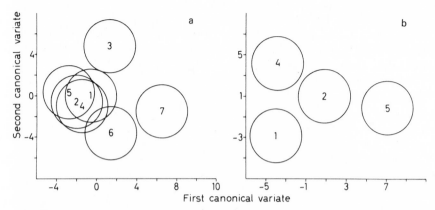

FIGURE 14. Canonical variate (discriminant) analysis of a series of *Pseudomonas* and *Acinetobacter* species. (a) Plot of the discriminant scores for the first two discriminant functions. Species means are indicated: *Pseudomonas aeruginosa* (1), *Ps. putida* (2), *Ps. maltophylia* (3), *Ps. pseudoalcaligenes* (4), *Ps. fluorescens* (5), *Acinetobacter anitratus* (6), and *Ac. Lwoffi* (7). Circles represent 95% confidence regions for the populations. (b) Reanalysis of discriminant scores after deletion of species 3, 6, and 7 from the data set. (From French *et al.*, 1980.)

aeruginosa, Ps. fluorescens, Ps. putida, Ps. maltophylia, Ps. alcaligenis, Acinetobacter anitratus, and *Ac. lwoffi* by means of Py-GC. All species could be differentiated by using two sequential runs of canonical variate analysis (DA). The results of this data analysis procedure are summarized in Figure 14. It is clearly shown that after removing the well-separated groups (i.e., species *Ps. maltophylia, Ac. anitratus,* and *Ac. lwoffi*; see Figure 14a) from the data set, the remaining initially unresolved species can be further differentiated (Figure 14b). *Ps. fluorescens* demonstrated considerable heterogeneity; dissimilarities among different strains sometimes being larger than among species.

Py-high-resolution field ionization MS of *Ps. putida* provided important information on the chemical composition of bacterial pyrolysis products (Schulten *et al.*, 1973). Close to 200 components were resolved, 60 corresponding directly to those mentioned in the Py-GC study of Simmonds (1970). In addition to the major classes of components that were identified by Simmonds, pyrolysis products originating from amines, branched hydrocarbons, and alcohols were observed. It should be noted, however, that the HRMS data do not give a decisive answer with regard to the isomeric structure of components.

5.10. Bacillus

Oxborrow *et al.* (1976, 1977a, 1977b) thoroughly investigated the influence of sample preparation and culturing conditions on low-resolution

Py-GC patterns. Reproducible pyrograms were obtained from samples that were cultured on membrane filters and directly harvested without further pretreatment. Culturing time as well as medium composition were found to be critical. Sporulation could be monitored directly. By careful standardization of all culturing and sampling procedures *B. subtilis*, *B. coagulans*, *B. firmus*, *B. cereus*, and *B. alvei* could be differentiated.

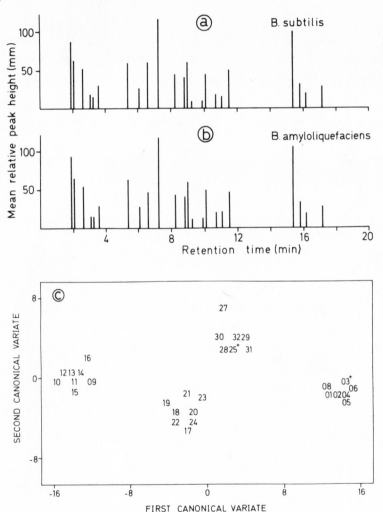

FIGURE 15. Bar-graph representation of Py–GC pyrograms of *B. subtilis* (a) and *B. amyloliquefaciens* (b) (averaged patterns). (c) Plot of the discriminant scores relative to the first two discriminant functions of 32 strains; *B. subtilus* (1–8); *B. pumilus* (9–16); *B. licheniformis* (17–24) and *B. amyloliquefaciens* (25–32). + indicates superposition of strains. (From O'Donnell *et al.*, 1980b.)

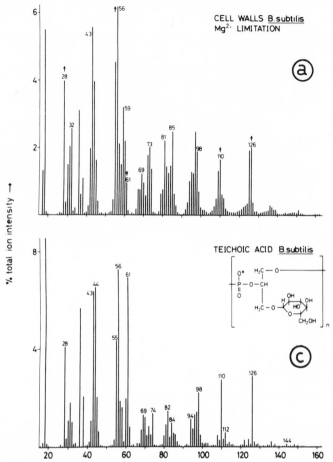

FIGURE 16. Mass pyrograms of *B. subtilus* cell walls grown under Mg^{2+}-limitation (a) and phosphate limitation (b). The presence of teichoic acid (c) in pyrogram (a) is evidenced by peaks at m/z 28, 56, 61, 110, and 126 (arrows); the presence of teichuronic acid (d) in pyrogram (b) by peaks at m/z 59, 72, 73, 102, and 125. (From Boon *et al.*, 1980.)

Canonical variate analysis (DA) was successfully used to differentiate *B. mycoides*, *B. cereus*, and *B. thuringiensis* (37 strains) by Py-GC (O'Donell *et al.*, 1980a, b). However, when only 2 peaks, traced by stepwise-DA, were used, *B. cereus* and *B. mycoides* tended to overlap. By the same data-processing technique it was also possible to differentiate between the closely related species *B. subtilis*, *B. pumilus*, *B. licheniformis*, and *B. amyloliquefaciens* (32 strains tested) (O'Donell *et al.*, 1980b). On visual inspection the pyrograms show only minor quantitative differences (see

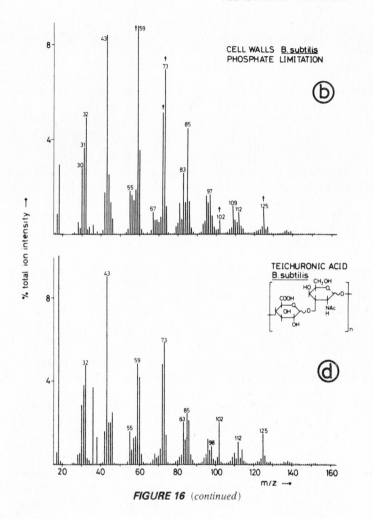

FIGURE 16 (*continued*)

Figures 15a and 15b), however, by using DA the four species can clearly be differentiated (Figure 15c). The results obtained correspond to DNA-hybridization testing and were superior to biochemical tests.

Py-GC/MS was used to give a chemical interpretation of pyrolysis products present in pyrograms of *B. subtilis* and *Micrococcus luteus* (Simmonds, 1970). Merely quantitative dissimilarities were observed, which were ascribed to predominantly carbohydrate and protein components. The most abundant product, acetamide, was postulated to originate from the *N*-acetylmuramic acid units of peptidoglycan. Propionamide represented another major product and was thought to originate from the lactylpeptide bridges of peptidoglycan. Hudson *et al.* (1982) confirmed Simmond's interpretation by analyzing peptidoglycan and a number of related model compounds. Py-GC/MS measurements revealed that peptidoglycan could be detected quantitatively in whole cell samples on the basis of the above-mentioned pyrolysis products. Clear quantitative differences were observed between gram-positive and gram-negative strains.

Boon *et al.* (1981) presented a detailed biochemical account of Py-mass spectra of *Bacillus* strains which were cultured under magnesium- or phosphate-limiting conditions. Spectra of whole cells and of cell walls (see Figures 16a and 16b) showed essential differences which could be interpreted on guidance of the analysis of purified cell wall components, i.e., peptidoglycan, teichoic acid, and teichuronic acid (see Figures 16c and 16d). In Mg^{2+}-deficient cells teichoic acid was identified, whereas in phosphate-limited cells teichuronic acid was traced.

5.11. Clostridium

By visual inspection of low-resolution mass pyrograms toxintypes A, B, and E of *Cl. botulinum* could be differentiated, irrespective of culturing media. (Cone and Lechowich, 1970). Reiner and Bayer (1978) observed dissimilarities between toxintypes B, C, and E and also between proteolytic and nonproteolytic strains. Only minor differences between toxins B and F were observed. *Cl. sporogenes* was easily differentiated from nonproteolytic type-F strains, but little or no difference could be observed relative to proteolytic strains of the same physiological group (types B and F).

A well-documented study by Gutteridge *et al.* (1980) involved Py-GC analysis of 65 *Clostridium* strains followed by canonical variate analysis (DA) of the data. Toxin types A and B overlapped but were significantly different from types E and F. Three conventional physiological groups were obvious from the pyrograms; however, differentiation between proteolytic *Cl. botulinum* and representatives of *Cl. sporogenes* could not be made.

5.12. Mycobacterium

Mycobacteria are intensively investigated with Py-GC as well as with Py-MS. In his early report Reiner (1965) could differentiate *M. avium* and *M. intracellulare*. In a later study 50 coded strains classically identified as *M. avium*, *M. scrophulaceum*, *M. fortuitum*, and *M. intracellulare* were analyzed (Reiner and Kubica, 1969). Upon visual inspection of the pyrograms a 100% accurate differentiation and classification was obtained. Duplicate samples, analyzed on a replicate column, showed different retention times and absolute peak intensities; however, the pattern of characteristic peaks remained unaltered. Reiner *et al.* (1969) stressed the importance of standardizing culturing conditions. Löwenstein medium was best suited to differentiate between species, whereas Proskauer–Beck medium revealed more of intraspecies differentiation. The influence of culture media was found to be species-dependent. Culturing time was shown not to be critical. In a third report Reiner *et al.* (1971) analyzed nine coded species, each in quadruplicate. Four species, viz., *M. tuberculosis, M. avium, M. intracellulare,* and *M. kansasii* were correctly identified by visual comparison of the pyrograms to library pyrograms. However, the pyrograms of *M. fortuitum* did not match with the library pattern of this species. Pyrograms of *M. terrae, M. bovis-BCG, M. marinum,* and *M. triviale,* which had not been analyzed before, were all correctly classified. Wickman (1977) was able to differentiate Py-GC patterns of *M. avium, M. tuberculosis, M. phlei,* and *M. gastri.* However, reproducibility was judged to be insufficient for routine identification of unknowns.

Meuzelaar *et al.* (1976) used automated Py-MS to analyze 10 batches of mycobacteria (97 different strains of 14 species). Löwenstein medium was observed to provide more characteristic pyrograms than Sauton or Middlebrook medium. Computerized data evaluation (FOMPYR package)

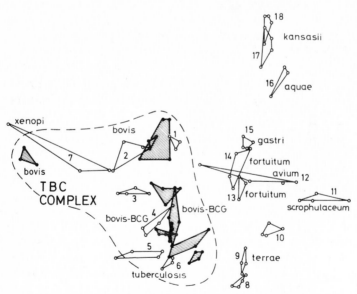

FIGURE 17. Nonlinear map of 18 clinical isolates of mycobacteria and 9 reference strains of the "tuberculosis complex" (shaded). Dissimilarities were calculated from a set of key-masses: m/z 31, 50, 58, 59, 71, 86, and 98. Strains 1–6 were identified as tuberculosis complex strains; strains 8–18 as "atypical" strains. Only strain 7 was falsely typed as a tuberculosis complex strain. (Stress: 13.8%.)

was used to compare pyrograms. Most species could be differentiated; however, some revealed considerable heterogeneity. The very heterogeneous species *M. bovis* showed some overlap with *M. avium* and *M. xenopi*. Long-term reproducibility was tested by matching the data from different batches. All *M. tuberculosis* and *M. bovis-BCG* strains were correctly identified, but for the heterogeneous species *M. bovis* only 8 out of 12 strains were correctly identified. Wieten *et al.* (1981b) identified 91 coded strains by Py-MS as either members of the "tuberculosis complex" or so-called "atypicals". Identification was based on calculation of the dissimilarity (FOMPYR routine) between an unknown and a set of reference strains. This reference set was reanalyzed with every new batch of unknowns. The references (3 for each species) were selected from the results of extensive heterogeneity studies (Wieten *et al.*, 1979). Of the 91 strains that were tested, 86 were correctly identified; 2 were incorrectly typed as atypical, and 5 were incorrectly identified as tuberculosis complex. Long-term reproducibility (measured over more than one year) proved to

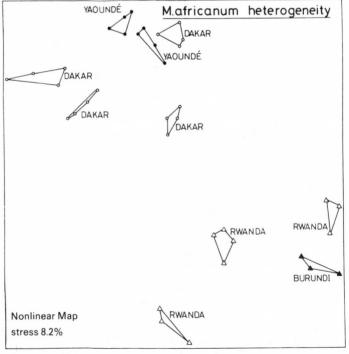

FIGURE 18. Nonlinear map representing the numerical relations among a set of mass pyrograms from 10 different *M. africanum* strains. Sites of strain isolation are indicated.

be sufficient to select a small set of key masses for the identification of the tuberculosis complex. (Wieten *et al.*, 1981b). In a computerized one-way analysis of variance procedure, the key-mass pyrograms of unknown strains were compared to a hypothetical mean tuberculosis complex pyrogram, which was calculated from the pyrograms of the reference strains. Of 125 different strains that were analyzed by this routine, 118 were correctly identified; 2 strains were incorrectly typed as atypicals and 5 were incorrectly typed as tuberculosis complex strains. The possibility of discriminating the tuberculosis complex and the atypicals is illustrated in Figure 17. Py-CADMS was indicated to be a powerful method to study the chemical structures of the ions that contribute to such key masses (Louter *et al.*, 1980b). Strains of *M. africanum* that were analyzed by this key-mass procedure were all identified as tuberculosis complex strains. Yet, dis-

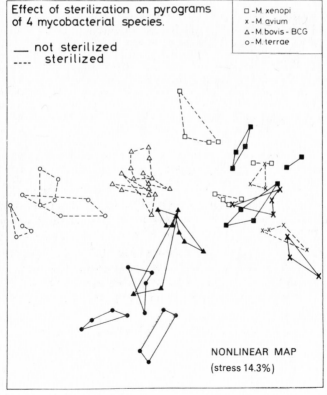

FIGURE 19. Nonlinear map representing the numerical relations among strains of four mycobacterial species that were either autoclaved prior to Py–MS analysis (○) or not autoclaved (●).

criminant analysis carried out on 100 features of the mass pyrograms of a batch of 10 *M. africanum* strains revealed evident subclusters that corresponded to the geographical origin of the strains. In Figure 18 a nonlinear map based on the four best discriminant functions illustrates this finding. Other aspects of species heterogeneity were evident from the mass pyrograms of a series of *M. kansasii* strains (Wieten *et al.*, 1979). Fresh isolates obtained from England and The Netherlands could not be differentiated by Euclidean dissimilarity calculation; however, "laboratory" strains of *M. kansasii* (over ten years old) proved to be highly dissimilar. In this study *M. marinum* was used as a reference. All strains that were analyzed by Meuzelaar *et al.* and Wieten *et al.* were autoclaved prior to MS analysis. The influence of autoclaving mycobacterial cells was shown to be systematic (Wieten, unpublished results); however differentiation between species was not affected, as can be seen in Figure 19.

Wieten *et al.* (1982) did also use Py-MS to investigate batches of *M. leprae* that were isolated from armadillo liver. A clear heterogeneity was observed. It was shown that quantitative detection of low levels of chemical impurities, e.g., polyethylene glycol, introduced during the isolation and purification procedure in the *M. leprae* samples, was possible. It was concluded that quality control of *M. leprae* was feasible.

6. CONCLUSION

At present analytical pyrolysis is not yet an established technique in clinical or pharmaceutical microbiology. However, from the results cited above the high potential for applications in this area will be evident.

A number of characteristic aspects that are of special interest for microbiological applications are apparent: (1) Analytical pyrolysis is a highly flexible technique, i.e., it can be used for the analysis of virtually any microbiological specimen. (2) Due to the high information content of pyrolysis data an extensive range of applications can be covered, provided that appropriate data evaluation procedures are used. (3) Low sample sizes are sufficient. And (4), especially for automated Py-MS, a high throughput of samples can be achieved.

In many cases it was possible to fit the results from analytical pyrolysis directly into existing taxonomical or diagnostic schemes. In other cases important additional information, or information not available by other techniques, was obtained.

From the previous sections a number of aspects are apparent that should be taken into consideration when development of routine applications of analytical pyrolysis in microbiology is envisaged:

(1) Standardization of instrumental conditions. This is essential to achieve sufficient long-term and interlaboratory reproducibility and is especially important when pyrogram libraries are to be used. However, for many applications, e.g., screening, process monitoring, and quality control, "local" references can be used as an alternative.

(2) The need for standardized culturing and sampling. Conditions should be optimized with regard to the effectiveness of the method to solve the analytical problem.

(3) Biological heterogeneity among the samples. Various aspects of heterogeneity were encountered in Section 5. It can be expected that heterogeneity can affect the consistency of results and should therefore be properly considered. This requires the analysis of relatively large numbers of samples.

(4) The use of computerized data processing routines instead of—subjective—visual procedures for the evaluation of the analytical data. Sophisticated data-handling procedures are often indispensible to obtain detailed and comprehensive information on a specific analytical problem. Whenever possible, however, an attempt should be made to deduce simple procedures from this information, that can be used for specific routine applications.

The optimal facility where all these aspects can be considered simultaneously would be a well-equipped, multidisciplinary working group within a medical or pharmaceutical research center or national institute for public health.

REFERENCES

Abbey, L. E., Highsmith, A. K., Moran, T. F., and Reiner, E. J., 1981, Differentiation and characterization of *Klebsiella pneumoniae* strains by pyrolysis–gas-liquid chromatography–mass spectrometry, *J. Clin. Microbiol.* **13**:313.

Anhalt, J. P., and Fenselau, C., 1975, Identification of bacteria using mass spectrometry, *Anal. Chem.* **47**:219.

Beuvery, E. C., Miedema, F., Van Delft, R., Haverkamp, J., Leusink, H. B., te Pas, B. J., Teppema, K. S., and Tiesjema, R. H., 1983a, Preparation, physico-chemical, and immunological characterization of polysaccharide–outer membrane protein complexes of *Neisseria meningitidis*, *Infect. Immun.* **40**:369.

Beuvery, E. C., Miedema, F., Van Delft, R., and Haverkamp, J., 1983b, Preparation and immunochemical characterization of meningococcal group C polysaccharide–tetanus toxoid conjugates as a new generation of vaccines, *Infect. Immun.* **40**:39.

Blomquist, G., Johansson, E., Söderström, B., and Wold, S., 1979, Data analysis of pyrolysis-chromatograms by means of SIMCA pattern recognition, *J. Anal. Appl. Pyrolysis* **1**:53.

Boon, J. J., De Boer, W. R., Kruyssen, F. J., and Wouters, J. T. M., 1981, Pyrolysis mass spectrometry of whole cells, cell walls and isolated cell wall polymers of *Bacillus subtilis* var. niger WM., *J. Gen. Microbiol.* **122**:119.

Borst, J., Van der Snee-Enkelaar, A. C., and Meuzelaar, H. L. C., 1978, Typing of *Neisseria genorrhoeae* by pyrolysis mass spectrometry, *Antonie van Leeuwenhoek* **44**:253.

Burns, D. T., Stretton, R. J., and Jayatilake, S. D. A. K_t, 1976, Pyrolysis gas chromatography as an aid to the identification of *Penicillium* species, *J. Chrom.* **16**:107.

Cone, R. D., and Lechowich, R. V., 1970, Differentiation of *Clostridium botulinum* types A, B, and E by pyrolysis gas–liquid chromatography, *Appl. Microbiol.* **44**;25.

Denoyer, E., Van Grieken, R. Adams, F., and Natusch, D. F. S., 1982, Laser microprobe mass spectrometry I, *Anal. Chem.* **54**:26A.

Emswiler, B. S., and Kotula, A. W., 1978, Differentiation of *Salmonella* serotypes by pyrolysis gas-liquid chromatography, *Appl. Environ. Microbiol.* **35**:97.

Eshuis, W., Kistemaker, P. G., and Meuzelaar, H. L. C., 1977, "Some Numerical Aspects of Reproducibility and Specificity," in *Analytical Pyrolysis* (C. E. R. Jones, and C. A. Cramers, eds.), pp. 151–166, Elsevier, Amsterdam.

Farré-Rius, F., and Guiochon, G., 1968, On the conditions of flash pyrolysis of polymers as used in pyrolysis gas chromatography, *Anal. Chem.* **40**:998.

Frane, J. W., 1976, The BMD and BMDP series of statistical computer programs, *Communication of the ACM* **19**:570.

French, G. L., Gutteridge, C. S., and Phillips, I., 1980, Pyrolysis gas chromatography of *Pseudomonas* and *Acinetobacter* species, *J. Appl. Bact.* **49**:505.

Gough, T. A., and Jones, C. E. R., 1975, Precision of the pyrolysis–gas chromatography of polymers, *Chromatographia* **8**:696.

Gutteridge, C. S., and Norris, J. R., 1979, The application of pyrolysis techniques to the identification of micro-organisms, *J. Appl. Bact.* **47**:5.

Gutteridge, C. S., and Norris, J. R., 1980, Effect of different growth conditions on the discrimination of three bacteria by pyrolysis gas–liquid chromatography, *Appl. Environ. Microbiol.* **40**:462.

Gutteridge, C. S., MacFie, H. J. H., and Norris, J. R., 1979, Use of principal components analysis for displaying variation between pyrograms of microorganisms, *J. Anal. Appl. Pyrolysis* **1**:67.

Gutteridge, C. S., MacKey, B. M., and Norris, J. R., 1980, A pyrolysis gas–liquid chromatography study of *Clostridium botulinum* and related organisms *J. Appl. Bacteriol* **49**:165.

Haddadin, J. M., Stirland, R. M., Preston, N. W., and Collard, P., 1973, Identification of *Vibrio cholerae* by pyrolysis gas–liquid chromatography, *Appl. Microbiol.* **25**:40.

Haverkamp, J., Kistemaker, P. G., and Eshuis, W., 1979, Pyrolysis mass spectrometry as an analytical tool in microbiology, *Anthonie van Leeuwenhoek* **45**:627.

Haverkamp, J., Eshuis, W., Boerboom, A. J. H., and Guinée, P. A. M., 1980a, "Pyrolysis Mass Spectrometry as a Rapid Screening Method of Biological Materials", in *Advances in Mass Spectrometry*, Volume 8A (A. Quayle, ed.) pp. 983–989, Heyden and Son, London.

Haverkamp, J., Meuzellaar, H. L. C., Beuvery, E. C., Boonekamp, P. M., and Tiesjema, R. H., 1980b, Characterization of *Neisseria meningitidis* capsular polysaccharides containing sialic acid by pyrolysis mass spectrometry, *Anal. Biochem.* **104**:407.

Heinen, H. J., Meier, S., Vogt, H., and Wechsung, R., 1980, "Laser-Induced Mass Spectrometry of Organic Compounds and Inorganic Compounds with a Laser Microprobe Mass Analyzer", in *Advances in Mass Spectrometry*, Vol. 8 (A. Quayle, ed.) pp. 942–953, Heyden and Son, London.

Hercules, D. M., Day, R. J., Balasanmugam, K., Dang, T. A., and Li, C. P., 1982, Laser microprobe mass spectrometry, *Anal. Chem.* **54**:280A.

Hudson, J. R., Morgan, S. L., and Fox, A., 1982, Quantitative pyrolysis GC-MS studies of bacterial cell walls, *Anal. Chem.* **120**:59.

Huff, S. M., Meuzelaar, H. L. C., Pope, D. L., and Kjeldsberg, C. R., 1982, Characterization of leukemic and normal white blood cells by Curie-point pyrolysis–mass spectrometry, I, *J. Anal. Appl. Pyrolysis* **3**:95.

Huis in't Veld, J. H. J., Meuzelaar, H. L. C., and Tom, A., 1973, Analysis of streptococcal cell wall fractions by Curie-point pyrolysis gas–liquid chromatography, *Appl. Microbiol.* **26**:92.

Irwin, W. J., 1979a, Analytical pyrolysis—An overview (Part I), *J. Anal. Pyrolysis* **1**:1.

Irwin, W. J., 1979b, Analytical pyrolysis—An overview (Part II), *J. Anal. Pyrolysis* **1**:89.

Kistemaker, P. G., Meuzelaar, H. L. C., and Posthumus, M. A., 1975, "Rapid and Automated Identification of Microorganisms by Curie-Point Pyrolysis Techniques. II. Fast Identification of Microbiological Samples by Curie-point Pyrolysis–Mass Spectrometry", in *New Approaches to the Identification of Microorganisms* (C. G. Hedén, and T. Illéni, eds.), pp. 179–191, John Wiley and Sons, London.

Kowalski, B. R., 1975, Measurement analysis by pattern recognition, *Anal. Chem.* **47**:1152A.

Kruskal, J. B., 1964, Multidimensional scaling by optimizing goodness of fit to a non-metric hypothesis, *Psychometrika* **29**:1.

Levy, R. L., 1966, Pyrolysis gas chromatography: A review of the technique, *Chrom. Rev.* **8**:48.

Levy, R. L., Fauter, D. L., and Wolf, C. J., 1972, Temperature rise time and true pyrolysis temperature in pulse mode pyrolysis gas chromatography, *Anal. Chem.* **44**:38.

Lincoln, K. A., 1965, Flash-pyrolysis of solid-fuel materials by thermal radiation, *Pyrodynamics* **2**:13.

Louter, G. J., Boerboom, A. J. H., Stalmeier, P. F. M., Tuithof, H. H., and Kistemaker, J., 1980a, A tandem mass spectrometer for collision-induced dissociation, *Int. J. Mass Spectrom. and Ion Physics* **33**:335.

Louter, G. J., Stalmeier, P. F. M., Boerboom, A. J. H., Haverkamp, J., and Kistemaker, J., 1980b, High sensitivity in CID mass spectrometry, structure analysis of pyrolysis fragments, *Z. Naturforsch.* **35c**:6.

MacFie, H. J. H., Gutteridge, C. S., and Phillips, I., 1978, Use of canonical variates analysis in differentiation of bacteria by pyrolysis gas–liquid chromatography, *J. Gen. Microbiol.* **104**:67.

McLafferty, F. W., 1981, Tandem mass spectrometry, *Science*, **214**:280.

Menger, F. M., Epstein, G. A., Goldberg, D. K., and Reiner, E., 1972, Computer matching of pyrolysis chromatograms of pathogenic microorganisms, *Anal. Chem.* **44**:423.

Meuzelaar, H. L. C., 1978, Pyrolysis mass spectrometry; prospects for interlaboratory standardization, *Proceedings 26th Annual ASMS Conference on Mass Spectrometry and Allied Topics*, St. Louis, Missouri, pp. 39–52.

Meuzelaar, H. L. C., and In't Veld, R. A., 1972, A technique for Curie point pyrolysis gas chromatography of complex biological samples, *J. Chrom. Sci.* **10**:213.

Meuzelaar, H. L. C., and Kistemaker, P. G., 1973, A technique for fast and reproducible fingerprinting of bacteria by pyrolysis mass spectrometry, *Anal. Chem.* **45**:587.

Meuzelaar, H. L. C., Posthumus, M. A., Kistemaker, P. G., and Kistemaker, J., 1973, Curie-point pyrolysis in direct combination with low voltage electron impact ionization mass spectrometry: A new method for the analysis of nonvolatile organic materials, *Anal. Chem.* **45**. Chem. **45**:1546.

Meuzelaar, H. L. C., Ficke, H. G., and Den Harink, H. C., 1975a, Fully automated Curie-point pyrolysis gas–liquid chromatography, *J. Chrom. Sci.* **13**:12.

Meuzelaar, H. L. C., Kistemaker, P. G., and Tom, A., 1975b, "Rapid and Automated Identification of Microorganisms by Curie-Point Pyrolysis Techniques. I. Differentiation of Bacterial Strains by Fully Automated Curie-Point Pyrolysis Gas–Liquid Chromatography," in *New Approaches to the Identification of Microorganisms* (C. G. Hedén, and T. Illéni, eds.), pp. 165–178, John Wiley and Sons, London.

Meuzelaar, H. L. C., Kistemaker, P. G., Eshuis, W., and Engel, H. W. B., 1976, "Progress in Automated and Computerised Characterisation of Microorganisms by Pyrolysis Mass Spectrometry", in *Rapid Methods and Automation in Microbiology* (S. W. B. Newson, and H. H. Johnston, eds.), pp. 225–230, Learned Information, Oxford.

Meuzelaar, H. L. C., Kistemaker, P. G., Eshuis, W., and Boerboom, A. J. H., 1978, "Automated Pyrolysis–Mass Spectrometry; Application to the Differentiation of Micro-organisms, in *Advances in Mass Spectrometry*, Volume 7B (N. D. Daley, Ed.) pp. 1452–1456, Heyden and Sons, London.

Meuzelaar, H. L. C., Haverkamp, J., and Hileman, F. D., 1982, *Pyrolysis Mass Spectrometry of Recent and Fossil Biomaterials*, Elsevier, Amsterdam.

Needleman, M., and Stuchbery, P., 1977, "The Identification of Microorganisms by Pyrolysis Gas–Liquid Chromatography," in *Analytical Pyrolysis* (C. E. R. Jones, and C. A. Cramers, eds.), pp. 77–88, Elsevier, Amsterdam.

Nie, N. N., Hull, C. H., Jenkins, J. G., Steinbrenner, K., and Bent, D. H., 1975, *Statistical Package for the Social Sciences* (SPSS), McGraw Hill, New York.

O'Donnell, A. G., MacFie, H. J. H., and Norris, J. R., 1980a, An investigation of the relationship between *Bacillus cereus, Bacillus thuringiensis* and *Bacillus mycoides* using pyrolysis gas–liquid chromatography, *J. Gen. Microbiol.* **119**:189.

O'Donnell, A. G., Norris, J. R., Berkeley, R. C. W., Claus, D., Kaneko, T., Logan, N. A., and Nozaki, R., 1980b, Characterization of *Bacillus subtilis, Bacillus pumilus, Bacillus licheniformis,* and *Bacillus amiloliquefaciens* by pyrolysis gas–liquid chromatography, deoxyribonucleic acid–deoxyribonucleic acid hybridization, biochemical tests, and API-systems, *Int. J. Syst. Bacteriol.* **30**:448.

Oxborrow, G. S., Fields, N. D., and Puleo, J. R., 1976, Preparation of pure microbiological samples for pyrolysis gas–liquid chromatography studies, *Appl. Environ. Microbiol.* **32**:306.

Oxborrow, G. S., Fields, N. D., and Puleo, J. R., 1977a, Pyrolysis Gas–Liquid Chromatography Studies of the Genus *Bacillus* Effect of Growth Time on Pyrochromatogram Reproduci-bility", in *Analytical Pyrolysis* (C. E. R. Jones, and C. A. Cramers, eds.), pp. 69–76, Elsevier, Amsterdam.

Oxborrow, G. S., Fields, N. D., and Puleo, J. R., 1977b, Pyrolysis gas–liquid chromatography of the genus *Bacillus*: Effect of growth media on pyrochromatogram reproducibility, *Appl. Environ. Microbiol.* **33**:865.

Oyama, V. I., 1963, Use of gas chromatography for the detection of life on Mars, *Nature* **200**:1058.

Oyama, V. I., and Carle, G. C., 1967, Pyrolysis gas chromatography application to life detection and chemotaxonomy, *J. Gas. Chrom.* **5**:151.

Quinn, P. A., 1974, Development of high-resolution pyrolysis gas chromatography for the identification of microorganisms, *J. Chrom. Sci.* **12**:796.

Reiner, E., 1965, Identification of bacterial strains by pyrolysis gas–liquid chromatography, *Nature* **206**:1272.

Reiner, E., 1977, "The Role of Pyrolysis Gas–Liquid Chromatography in Biomedical Studies," in *Analytical Pyrolysis* (C. E. R. Jones, and C. A. Cramers, eds.), pi. 49–56, Elsevier, Amsterdam.

Reiner, E., and Bayer, F. L., 1978, Botulism: A pyrolysis gas–liquid chromatographic study, *J. Chrom. Sci.* **16**:62.

Reiner, E., and Ewing, W. H., 1968, Chemotaxonomic studies of some gram-negative bacteria by means of pyrolysis gas–liquid chromatography, *Nature* **217**:191.

Reiner, E., and Kubica, G. P., 1969, Predictive value of pyrolysis gas–liquid chromatography in the differentiation of mycobacteria, *Am. Rev. Resp. Dis.* **99**:42.

Reiner, E., Beam, R. E., and Kubica, G. P., 1969, Pyrolysis gas–liquid chromatography studies for the classification of mycobacteria, *Am. Rev. Resp. Dis.* **99**:750.

Reiner, E., Hicks, J. J., Beam, R. E., and David, H. L., 1971, Recent studies on mycobacterial differentiation by means of pyrolysis gas–liquid chromatography, *Am. Rev. Resp. Dis.* **104**:656.

Reiner, E., Hicks, J. J., Ball, M. M., and Martin, W. J., 1972, Rapid characterization of *Salmonella* organisms by means of pyrolysis gas–liquid chromatography, *Anal. Chem.* **44**:1058.

Risby, T. H., and Yergey, A. L., 1976, Identification of bacteria using linear programmed thermal degradation mass spectrometry, *J. Phys. Chem.* **80**:2839.

Risby, T. H., and Yergey, A. L., 1978, Linear programmed thermal degradation mass spectrometry, *Anal. Chem.* **50**:327A.

Schulten, H. R., 1979, Biochemical, medical, and environmental applications of field-ionization and field-desorption mass spectrometry, *Int. J. Mass. Spectrom.* **32**:97.

Schulten, H. R., Beckey, H. D., Meuzelaar, H. L. C., and Boerboom, A. J. H., 1973, High-resolution field-ionization mass spectrometry of bacterial pyrolysis products, *Anal. Chem.* **45**:191.

Sekhon, A. S., and Carmichael, J. W., 1972, Pyrolysis gas–liquid chromatography of some dermatophytes, *Can. J. Microbiol.* **18**:1593.

Seydel, U., and Heinen, H. J., 1979, First results on fingerprinting of single mycobacteria cells with LAMMA, *Proceedings of 6th International Symposium on Mass Spectrometry in Biochemistry and Medicine*, Venice, Italy.

Shafizadeh, F., 1982, Introduction to pyrolysis of biomass (review), *J. Anal. Appl. Pyrolysis* **3**:283.

Simon, W., and Giacobbo, H., 1965, Thermische Fragmentierung und Strukturbestimmung organischer Verbindungen, *Chemie Ingenieur Technik* **37**:709.

Simmonds, P. G., 1970, Whole microorganisms studied by pyrolysis gas chromatography mass spectrometry: Significance for extraterrestial life detection experiments, *Appl. Microbiol.* **20**:567.

Stack, M. V., Donoghue, H. D., Tyler, J. E., and Marshall, M., 1977, "Comparison of Oral Streptococci by Pyrolysis Gas–Liquid Chromatography," in *Analytical Pyrolysis* (C. E. R. Jones, and C. A. Cramers, Eds.), pp. 57–68, Elsevier, Amsterdam.

Stack, M. V., Donoghue, H. D., and Tyler, J. E., 1978, Discrimination between oral streptococci by pyrolysis gas–liquid chromatography, *Appl. Environ. Microbiol.* **35**:45.

Stern, N. J., Kotula, A. W., and Pierson, M. D., 1979, Differentiation of selected Enterobacteriaceae by pyrolysis gas–liquid chromatography, *Appl. Environ. Microbiol.* **38**:1098.

Stern, N. J., Kotula, A. W., and Pierson, M. D., 1980, Virulence prediction of *Yersinia enterocolitica* by pyrolysis gas–liquid chromatography, *Appl. Environ. Microbiol.* **40**:646.

Stretton, R. J., Campbell, M., and Burns, D. T., 1976, Pyrolysis gas chromatography as an aid to the identification of *Aspergillus* species, *J. Chrom.* **129**:321.

Taylor, J. J., 1967, *Ex vivo* determination of pententially virulent *Sporothrix schenkii*, *Mycopathologia.* **58**:107.

Van de Meent, D., Brown, S. C., Philip, R. P., and Simoneit, B. R. T., 1980, Pyrolysis high-resolution gas chromatography and pyrolysis gas chromatography–mass spectrometry of kerogens and kerogen precursors, *Geochim. Cosmochim. Acta* **44**:999.

Van de Meent, D., De Leeuw, J. W., Schenck, P. A., Windig, W., and Haverkamp, J., 1982, Quantitative analysis of biopolymer mixtures by pyrolysis–mass spectrometry, *J. Anal. Appl. Pyrolysis* **4**:133.

Vincent, P. G., and Kulik, M. M., 1970, Pyrolysis gas–liquid chromatography of fungi: Differentiation of species and strains of several members of the *Aspergillus flavus* group, *Appl. Microbiol.* **20**:957.

Vincent, P. G., and Kulik, M. M., 1973, Pyrolysis gas–liquid chromatography of fungi: Numerical characterization of species variation among members of the *Aspergillus glaucus* group, *Mycopath. Mycologia. Appl.* **51**:251.

Walker, J. Q., 1977, Pyrolysis gas chromatograhpic correlation trials of the American Society for Testing and Materials, *J. Chrom. Sci.* **15**:267.

Wechsung, R., Hillenkamp, F., Kaufmann, R., Nitsche, R., and Vogt, H., 1978, Laser-Mikrosonden-Massen-Analysator, LAMMA: Ein neues Analysenverfahren für Forschung und Technologie, *Mikroskopie* **34**:47.

Wells, G., Voorhees, K. J., and Futrell, J. H., 1980, Heating profile curves for resistively heated filament pyrolysers, *Anal. Chem.* **52**:1782.

Wickman, K., 1977, Pyrolysis gas–liquid chromatography of mycobacteria, *Acta Path. Microbiol. Scand.* **259**:49.

Wieten, G., Haverkamp, J., Engel, H. W. B., and Tarnok, I., 1979, Pyrolysis Mass Spectrometry in Mycobacterial Taxonomy and Identification," in *Twenty-Five Years of Mycobacterial Taxonomy* (G. P. Kubica, L. G. Wayne, and L. S. Good, eds.), pp. 171–189, CDC. Press, Atlanta.

Wieten, G., Haverkamp, J., Engel, H. W. B., and Berwald, L. G., 1981a, Application of pyrolysis mass spectrometry to the classification and identification of mycobacteria, *Ref. Infect. Dis.* **3**:871.

Wieten, G., Haverkamp, J., Meuzelaar, H. L. C., Engel, H. W. B., and Berwald, L. G., 1981b, Pyrolysis mass spectrometry: A new method to differentiate between the mycobacteria of the "Tuberculosis complex" and other mycobacteria, *J. Gen. Microbiol.* **122**:109.

Wieten, G., Haverkamp, J., Berwald, L. G., Groothuis, D. G., and Draper, P., 1982, Pyrolysis mass spectrometry: its applicability to mycobacteriology, including *M. leprae*, *Ann. Microbiol.* **133B**:15.

Windig, W., and De Hoog, C. S., 1982, Pyrolysis mass spectrometry of selected yeast species, II, *Sporidiobolus* and relationships, *Stud. Mycol.* **22**:60.

Windig, W., Kistemaker, P. G., Haverkamp, J., and Meuzelaar, H. L. C., 1979, The effects of sample preparation, pyrolysis, and pyrolyzate transfer conditions on pyrolysis mass spectra, *J. Anal. Appl. Pyrolysis* **1**:39.

Windig, W., Kistemaker, P. G., Haverkamp, J., and Meuzelaar, H. L. C., 1980, Factor analysis of the influence of changes in experimental conditions in pyrolysis–mass spectrometry, *J. Anal. Appl. Pyrolysis* **2**:7.

Windig, W., De Hoog, G. S., and Haverkamp, J., 1981a, Chemical characterization of yeasts and yeastlike fungi by factor analysis of their pyrolysis mass spectra, *J. Anal. Appl. Pyrolysis* **3**:213.

Windig, W., Kistemaker, P. G., and Haverkamp, J., 1981b, Chemical interpretation of differences in pyrolysis mass spectra of simulated mixtures of biopolymers by factor analysis with graphical rotation, *J. Anal. Appl. Pyrolysis* **3**:199.

Windig, W., Haverkamp, J., and Kistemaker, P. G., 1983, Chemical interpretation of sets of pyrolysis mass spectra by discriminant analysis and graphical rotation, *Anal. Chem.* **55**:88.

Wishart, D., 1978, Users manual CLUSTAN, Inter-University Res. Council Series, Rep. no. 47.

Zemany, P. D., 1952, Identification of complex organic materials by mass spectrometric analysis of their pyrolysis products, *Anal. Chem.* **24**:1709.

11

VOLATILE C^1–C^8 COMPOUNDS IN MARINE SEDIMENTS

Jean K. Whelan, Chemistry Department, Clark
Building, Woods Hole Oceanographic Institution,
Woods Hole, Massachusetts 02543

1. INTRODUCTION

Volatile organic compounds in both recent and ancient marine sediments can provide information about the sediment history (Hunt, 1975; Hunt and Whelan, 1979; Whelan *et al.*, 1980a and references cited therein). In recent sediments (subbottom depths less than 3 m), these compounds are related to factors such as depositional environment, organisms (including microorganisms) living in or which have lived in the sediment, and surface oil seeps. In more deeply buried sediments, these compounds can give an indication of past geothermal history and petroleum generation. Investigations of these compounds with respect to petroleum generation have been carried out in a number of laboratories (see, for example, Thompson, 1979; Leythauser *et al.*, 1982; Kvenvolden and Claypool, 1980; Hunt, 1979, Chap. 5 and references cited therein; Tissot and Welte, 1978, Chap. 6 and references cited therein).

The purpose of this chapter is to present the methodology currently used in obtaining GC profiles and mass spectra of these compounds in sediments and to summarize some of the results found to date using these methods. One of the most important elements in any good GC/MS analysis is first obtaining a GC in which peak separation and resolution are satisfactory. Because few laboratories have full-time use of a GC/MS system, this discussion will assume that the MS part of the system is to be used primarily for qualitative identification of compounds present in a particular set of samples. Once the MS identification is accomplished, quantitative analysis can usually be carried out more inexpensively and reliably on the GC rather than on the GC/MS system.

Because volatile compounds tend to escape from sediments to varying degrees, it is very difficult to obtain good interlaboratory GC intercomparisons of data as will be shown later by some examples. However, it is possible for a particular laboratory to obtain internally consistent data over a period of time because all of the samples are handled and analyzed in the same way. For this reason, examples used here will be taken primarily from data obtained in this laboratory. However, because the changes in levels of compounds are often not subtle (changes of 3–4 orders of magnitude are common), it is often possible to see trends even with data obtained in different laboratories.

It will be assumed that most readers of this book will be most interested in shallow sediments where the biology is still active. In spite of this, some examples will be used from more deeply buried sediments where compounds undoubtedly result primarily from chemical (thermal) processes. The experimental methodology is the same no matter what the source of the volatile compounds. In addition, if studies of compounds produced by microorganisms are to be correctly interpreted, a knowledge of other potential compound sources is also necessary.

It might be assumed that the volatility of these compounds would cause large scatter in the data so that no meaningful conclusions are possible. However, several laboratories including our own (see, for example, Thompson 1979, and Leythauser et al., 1979; 1982) have observed a surprisingly tight binding of these compounds in low levels in frozen or wet sediments. This point will be discussed in more detail later after the methodology and reliability of the measurements have been evaluated. For now, the views of most investigators working in this area can be summarized by stating that low levels (less than 1500 ng/g) of these compounds in many sediments represent adsorbed molecules which were probably generated very close to the site where they are detected—an important point if these molecules are to be used as indicators of past sediment history.

2. METHODS

2.1. Sample Collection and Storage

Recent sediments are collected with gravity corers, box corers, or grab samplers. The samples are sectioned on shipboard if possible, or, in the case of gravity cores, are kept frozen and brought back to the laboratory where they are sectioned. Each sample (generally 10–100 g) is placed in a heavy duty Kapak Bag (Kapak Corp., Bloomington, Minnesota) which is heat-sealed and frozen until the time of analysis. It has been found that samples can be thawed briefly without affecting results of analysis for compounds larger than C_2 as long as the samples are not allowed to dry out.

Precautions are taken during sampling and analysis to avoid organic vapors such as exhaust fumes, solvent fumes, cigarette smoke, paint fumes, or any other compounds which produce an odor. All tools and glassware are soap and water washed, distilled water rinsed, and where possible, heated overnight at 120°C before use. (No organic solvents should be used in the cleaning process.) All helium used either in sample preparation or in GC analyses is purified by passage through a 12-in. × 3/8-in. alumina U-trap immersed in dry ice. This trap removes all organic contaminants except for traces of methane. All water used in sample preparation is deionized, charcoal filtered, distilled in an all glass still, and bubbled with helium before use.

All solvent handling required to prepare standards or to clean equipment is carried out in a good fume hood. Blanks are run frequently—at least once a week when hydrocarbon-rich samples, such as petroleum source rocks, are being analyzed. When samples with low levels of compounds are being analyzed, a blank is run on each headspace vessel (to be described below) before the actual analysis is carried out. From experience, it has been found that routine ultrasonic cleaning of the head-space vessels with soap and water after each use prevents intersample contamination. That no systematic source of contamination occurs in either sampling (including heat sealing of the Kapak bags) or analysis is shown by samples from many areas which show little or no C_1–C_8 compounds.

2.2. GC Analysis

2.2.1. Head-Space Sample Preparation

Sediment C_1–C_8 volatile organic compound analyses are currently obtained by a modification of methods described previously (Whelan, 1979;

Whelan *et al.*, 1980a). The procedure described here was adopted to ensure that lithified sediments and cuttings are completely broken up during analysis. For recent sediments, the procedure using cans described previously (Whelan, 1979) works almost as well except that some of the early eluting capillary GC peaks are obscured by a higher blank in the "can" procedure.

The technique involves placing a weighed quantity (5–20 g) of frozen wet sediment into a stainless steel vessel equipped with a screw cap, two stainless steel ball bearings (about 4–8 mm in diameter), a silicone rubber gasket, and a silicone rubber septum as shown in Figure 1. Silicone is used

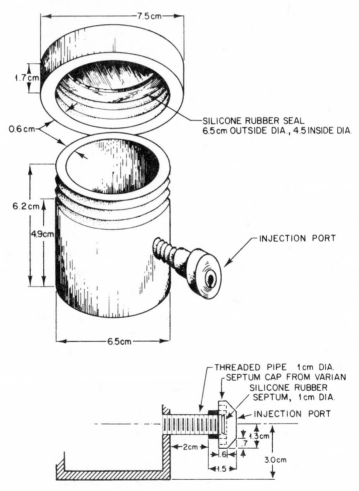

FIGURE 1. Diagram of stainless-steel vessel used in head-space analysis.

when any rubber gaskets or seals are needed because it can be heated to get rid of absorbed organic compound prior to analysis (and because the characteristic $M + 1$ and $M + 2$ silicone isotope peaks are easily detected in the GC/MS if any rubber contamination occurs). The vessel containing sediment is placed in a helium-filled glove bag (Figure 2; IR4 Company, Cheltenham, Pennsylvania) along with a flask of helium-bubbled distilled water. The water is added to the vessel until sediment plus water reach a level indicated by a calibration mark engraved into the inside of the vessel. The mark indicates the level of water needed to obtain a 50 ml gas head-space after the vessel is sealed. The cap is screwed on the vessel, which is then removed from the glove bag. The cap is tightened further with a strap wrench and the vessel is allowed to stand at room temperature overnight so the sediment can thaw. It is shaken vigorously for five minutes on a commercial paint shaker (such as available from Red Devil, Inc., Union, New Jersey) followed by heating for 30 min in a 95°C water bath. The head-space gas is then analyzed by GC as described in Section 2.2.2.

The use of a glove bag as shown in Figure 2 is a modification of techniques used by chemists who work with oxygen-sensitive compounds (for a review of general methodology, see Shriver, 1969). The purpose of the glove bag is to allow sealing of the head-space vessel in a helium

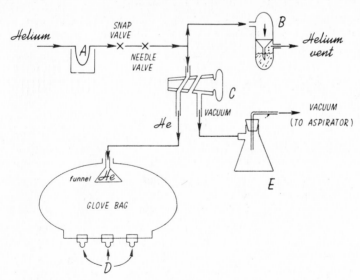

FIGURE 2. Glove bag for preparation of head-space vessels. (A) Alumina trap immersed in dry ice; (B) glass bubble tube (Kontes Glass, Vineland, New Jersey)—liquid used in bottom is water; (C) three-way stopcock; (D) clamps to seal front opening of glove bag; (E) aspirator trap-filter flask equipped with rubber stopper and glass evacuation tube. All tubing is made of silicone rubber (A. H. Thomas Company).

atmosphere away from any contaminants which might be present in the laboratory air so that air does not need to be rigorously excluded. It is sufficient to make sure the bag contains mostly helium and that whenever the bag is opened (by removing the clamps, D in Figure 2, and unfolding the edge of the bag) helium is blowing out of the opening so that very little air is diffusing in. Thus, before the head-space vessel and the water are put in the glove bag, the bag is sealed at D and evacuated (an aspirator is used in this laboratory) and then filled with helium by slowly turning three-way stopcock C after a vigorous flow of helium is flowing through the gas bubbler B. After the bag fills with helium, as noted by its inflation, the clamps at D are removed so that the helium flow is diverted from B through stopcock C into the glove bag. The head-space vessel containing frozen sediment and the water to be added are quickly placed inside and the bag is reclamped at D so that the helium flow is diverted completely through gas bubbler B. The vessel is then prepared as described above. This procedure has several advantages over the can procedure described previously (Whelan, 1979). It is much easier to prepare the head-space this way than by the syringe method (Whelan, 1979). The bags are relatively inexpensive and can be replaced when they start to leak. Furthermore, leaks in the bag, which are easily detected by the slowdown or disappearance of bubbles through B, are not a big problem because helium blowing out of the leak (instead of through B) prevents air from passing back into the bag. Finally, the stainless steel vessels described here do not leak (unlike some cans) once they are sealed.

2.2.2. GC Analysis of Head-Space Gas

GC analysis of head-space gas is carried out in two stages. A packed column analysis is first done to determine amounts of C_1–C_5 hydrocarbons. A capillary column analysis is then carried out to determine the distribution of C_6–C_8 components. There are definite advantages to carrying out these analyses in two stages as will be pointed out in the following discussion.

A 5–10 ml aliquot of the head-space gas is first analyzed for C_1–C_5 hydrocarbons using a Hewlett–Packard 5710A GC equipped with a gas sampling valve (F in Figure 3) attached to a 1/8 in. × 8 in. loop (G) packed with 60/80 mesh alumina (gas chromatographic grade) as described by Swinnerton and Linnenbom. The gas sample to be analyzed is injected through a silicone rubber septum (D) into a helium stream (flow rate 15 cm^3/min) which then passes through the alumina-filled loop (G) chilled in liquid nitrogen. After the sample is frozen in the loop, the helium flow is stopped via two toggle valves (J) on each side of the loop and the loop is heated in a 95°C water bath for one minute. The sampling valve (F) is

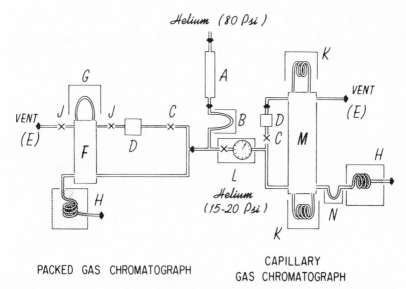

PACKED GAS CHROMATOGRAPH

CAPILLARY
GAS CHROMATOGRAPH

FIGURE 3. Valving system for injection into packed and capillary gas chromatographs. (A) Drying trap; (B) alumina trap immersed in dry ice; (C) needle value; (D) injection port; (E) vent (flow rate 15 ml/min); (F) gas sampling value (Carle, 6 port, mini-volume valve, model 2818); (G) alumina-filled gas sampling loop; (H) gas chromatograph; (I) carrier gas through valve to GC; (J) snap (on/off) valves; (K) capillary gas sampling loops (cooled in liquid nitrogen during filling); (L) pressure regulating valve and gauge to capillary GC; (M) gas sampling valve (Carle 8-port microvolume valve No. 2014) to capillary GC; (N) liquid-nitrogen-cooled loop made from front of capillary column.

turned, which injects the sample into the helium stream of the GC column (1/8 in. × 6 ft spherosil, 40–100 mesh, Supelco, Inc. Attached to 1/8 in. × 12 ft 20% OV-101 on Analabs AS, 100/110 mesh as described by Durand and Espitalie, 1971). The purpose of the valve F, which is not shown in Figure 3, is to interchange the two helium streams. Analysis is carried out via temperature programming from 60 to 200°C at 8°C/min after holding for 4 min at the initial temperature. Peak measurements are carried out with an electronic integrator. Some loss of methane occurs during this analysis. However, losses also occur during core collection so that this measurement is presumed to be qualitative only. Figure 4 shows a typical GC pattern obtained on a hydrocarbon-rich sediment with this column. Some halogenated compounds can also be detected. For example, traces of Freon (chlorodifluoromethane) have been observed as a contaminant in many Deep Sea Drilling Project (DSDP) samples. The freon appears as a very small peak, always of about the same size, about halfway between ethane and propane. Its detection via GC/MS was very clear—both from

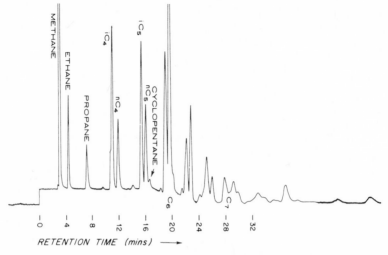

FIGURE 4. Typical packed-column GC analysis–petroleum source rock sample.

the characteristic chlorine isotopic peaks (at m/z of 35 and 37) and fragmentation via loss of chlorine producing the strongest peak in the spectrum at m/z of 51.

This column was chosen for this work because it could be used at ordinary GC temperatures rather than cryogenic ones, which often produce loss of GC peak sharpness and separation due to solidification of the liquid phase. Furthermore, the column gives good separation between ethylene and ethane and between propylene and propane—separations not easily accomplished on several other columns tested including Porapak Q. In addition, this column gives good separation between methane and ethane—an important consideration for marine sediments which often contain large amounts of methane and only traces of C_2-C_5 components. Even on this column, the methane peak often becomes so large that the much smaller C_2-C_5 peaks cannot be quantitated because of methane tailing. In these cases, a second analysis is carried out by chilling the alumina loop (G in Figure 3) in dry ice rather than liquid nitrogen. The methane is not retained by the dry-ice-filled loop while the C_2-C_5 compounds are.

Compounds larger than C_5 have not been analyzed routinely with this column because of periodic partial decomposition of C_6 plus compounds on the alumina loop and because the capillary column described below gives much better separation of isomeric C_6-C_8 components. The alumina loop decomposition problem also periodically plagues C_4-C_5 hydrocarbons. It has been found that the problem can be eliminated by periodically (about once a month) deactivating the alumina by injection of 50–100 μl of water into the loop G and allowing helium to pass through the loop overnight. This will sometimes cause a huge unresolved complex mixture of compounds to elute from the packed column for a day or so. Usually baking the column out overnight at 200°C solves the problem.

Commercially available gas standards are used for periodic calibrations. Scott gas working Standard No. 1 (C_1–C_6 n-alkane mixture) is measured against these fairly precise mixtures and is then used in obtaining response factors in day-to-day work. Once calibrated, this analysis is very stable over long periods of time (response factors often change very little over several months, see Table 1) and over a range of several orders of magnitude (see Figure 5). Similar results have been obtained by injecting the standards into head-space vessels prepared as described above and then heated for 30 min in a 90°C water bath.

The stability of this system is demonstrated by its successful use over a period of several years aboard the DSDP drilling ship, the Glomar Challenger. The analysis is used to monitor all gassy cores as they are brought on deck as a safety precaution. Thus, at some sites, the system must operate continuously for an eight-week period with a maximum sample analysis rate of about one analysis every 30 min. At other times, when gassy sediments are not encountered, the system will be idle for

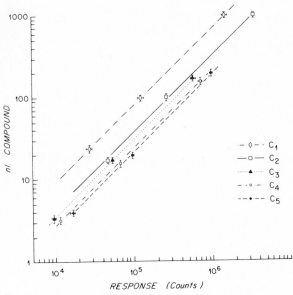

FIGURE 5. Linearity of packed-column GC analysis.

TABLE 1. Precision Packed-Column Analysis: Five Replicate Analyses

Compound	1 ml Big Shot calibration gas[a]			1 ml Scott calibration gas[b]		
	nl Hydrocarbon in gas	Counts per nl	Coefficient variation (%)	nl[c] found	Coefficient variation	nl gas[d] from specs.
Methane	36.9	1321 ± 177	13.4	(a) 22.2 ± 1.9[a]	8.5%	18 ± 1.8
				(b) 21.4 ± 8.2[e]	38%	
Ethane	40.0	2520 ± 58	2.3	(a) 18.2 ± 0.02[e]	0.1%	17 ± 1.7
				(b) 18.3 ± 0.2	1%	
Propane	38.5	3685 ± 85	2.3	(a) 18.2 ± 0.2	1%	17 ± 1.7
				(b) 17.9 ± 0.1[e]	0.5%	
Butane	30.8	4158 ± 104	2.5	(a) 20.1 ± 0.2	1%	16 ± 1.6
				(b) 20.1 ± 0.2[e]	1%	
Pentane	32.3	4892 ± 171	3.5	(a) 22.8 ± 0.3	1.3%	20 ± 2.0
				(b) 22.3 ± 0.9[e]	4%	
Hexane	33.8	5318 ± 148	2.8	(a) 23.9 ± 1.6	6.7%	21 ± 2.1
				(b) 24.0 ± 3.0[e]	12.5%	

[a] Five replicate analyses.
[b] Reponse factors from B.S. standard shown here used in calculation.
[c] nl gas from specifications on can-blend tolerance (maximum deviation from the desired concentrations) is advertised as ±10%.
[d] Amounts determined using response factors calculated from Big Shot standard.
[e] Triplicate analyses [run about two months after (a)].

several months—a condition that often causes more problems for instruments than running continuously. The system has now run reliably for several years with multiple operators with only minimal maintenance being required. The only difference between the drilling ship system and the Woods Hole system is that a mechanical refrigeration unit is used with propanol as coolant for the sample loop because cryogenic liquids are not readily available at sea. The coolant reaches a minimum temperature of −65°C so that methane is lost. Thus methane is measured by direct injection of a small (about 50–100 μl) sample into a second GC unit (Claypool, *et al.*, 1973).

There are many column packing materials now available which might accomplish the above analysis as well and without the periodic alumina loop decomposition problems. For example, in this laboratory we have been using a column (1/8 in. × 8 ft) and gas sampling loop (1/8 in. × 8 in.) packed with *n*-octane-Porasil C (Alltech Associates, Deerfield, Illinois) in another application (Whelan *et al.*, 1980b). This column gives good separation and no decomposition of C_1–C_6 alkanes and alkenes. Because the *n*-octane is covalently bonded to the Porasil, the column is thermally stable

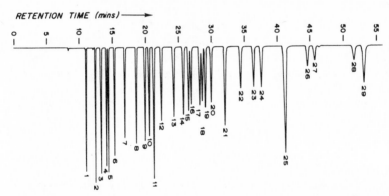

FIGURE 6. Capillary gas chromatogram of standard hydrocarbon mixture. Compounds present are as follows: (1) *n*-pentane (*n*C₅); (2) 2,2-dimethylbutane (2,2DMB); (3) cyclopentane (CP); (4) 2,3-dimethylbutane (2,3DMB); (5) 2-methylpentane (2MP); (6) 3-methylpentane (3MP); (7) *n*-hexane (*n*C₆); (8) methylcyclopentane (MCP); (9) 2,2-dimethylpentane (2,2DMP); (10) 2,4-dimethylpentane (2,4DMP); (11) benzene (Benz) plus 2,2,3-trimethylbutane (2,2,3TMB); (12) cyclohexane (CH); (13) 3,3-dimethylpentane (3,3DMP); (14) 1,1-dimethylcyclopentane (1,1DMCP); (15) 2-methylpentane (2MP); (16) 2,3-dimethylpentane (2,3DMP); (17) 3-methylhexane (3MH); (18) 1-*t*-3-dimethylcyclopentane (1T3DMCP); (19) 1-*t*-2-dimethylcyclopentane (1T2DMCP); (20) 3-ethylpentane (3EP); (21) 2,2,4-trimethylpentane (2,2,4TMP); (22) *n*-heptane (*n*C₇); (23) 1-*c*-2-dimethyliycyclopentane (1C2DMCP); (24) methylcyclohexane (MCH); (25) (A) ethylcyclopentane (ECP) plus (B) 2,2-dimethylhexane (2,2DMH); (26) 2,5-dimethylhexane (2,5DMH); (27) 2,4-dimethylhexane (2,4DMH); (28) toluene (Tol); (29) 2,3,4-trimethylpentane (2,3,4TMP).

to 150°C and has a very low bleed rate. The reason for not replacing the column described here with one that does not cause decomposition is that the decomposition of many functional group compounds (excluding C_2 and C_3 alkenes) has turned out to provide useful information on compound structures. Thus, the capillary column, to be described below, gives very rapid elution of a whole series of C_5 and C_6 alkenes between $n\,C_5$ and $n\,C_6$ (as shown in Figures 6 and 8) so that it is sometimes difficult to distinguish the alkenes from the alkanes in this region. If a peak appears in both analyses, it can be confidently identified as an alkane. However, if it appears in the capillary analysis, but not in the packed analysis, it can be identified as an alkene or some other low-molecular-weight organic functionalized compound.

Depending on the quantity of material detected on the packed column, a 1–15 ml sample of head-space gas is then analyzed by capillary gas chromatography for C_6–C_8 volatile compounds using a hexadecene–hexadecane–KEL F (HHK) column (0.01 in. × 150 ft stainless steel, run at 30°C) as described by Schwartz and Brasseaux (1963). Before injection, the sample is concentrated first on a 10 in. × 0.01 in. i.d. stainless steel loop (K in Figure 3) cooled in liquid nitrogen connected to an 8-port microvolume gas sampling valve (M in Figure 3). After the loop is filled, the loop is heated in a 90°C water bath to inject the sample into a 3-in. section of the front of the column frozen in liquid nitrogen (N in Figure 3). The liquid-nitrogen bath is removed and the loop, N, is heated briefly in a warm water bath to start the analysis. The second freezing step improves resolution by minimizing injection of dead volumes. Peak areas are measured with an electronic integrator.

This column separates almost all of the isomeric C_6–C_8 saturated and aromatic hydrocarbons as shown in Figure 6 as well as most of the other volatile functionalized compounds encountered in recent sediments (Figures 7 and 8). These include aldehydes, ketones, furans, alkenes, and thioethers as shown in Figure 7. The degree of separation of one class of compounds from another depends on the precise proportions of three coating components in the liquid phase so that "problematical" separations will vary from one column to another. A sporadic difficulty has been that benzene often overlaps either 2,4-dimethylpentane or 2,2,3-trimethylbutane. Fortunately, most samples have either predominantly aromatics or alkanes so that GC/MS analyses will show the presence of one or the other.

The biggest problem with this column is its very high bleed rate, which means that it must be recoated about every 1 to 2 months. Furthermore, the bleed is very deleterious to the mass spectrometer source. For this reason, other capillary columns with lower bleed rates have been tested including OV-101, Squalane, DC-550, and UCON-LB-550. All of these phases remain liquids at the low temperatures required for the analysis and, therefore, give excellent resolution. However, they all have problems in separating cycloalkanes and methylcycloalkanes from other components. Further testing of other stationary phases is planned. In the meantime, HHK is still the column of choice due to its superior separating capabilities. However, an OV-101 column is now being used in GC/MS analyses because that analysis does not require complete separation of all components as will be shown later.

The standard mixture for calibration of the capillary GC head-space analysis is a mixture of all of the C_5–C_7 and most C_8 saturated and aromatic hydrocarbons which is injected into a "blank" stainless steel vessel prepared with water as described above. The amount of each standard injected is

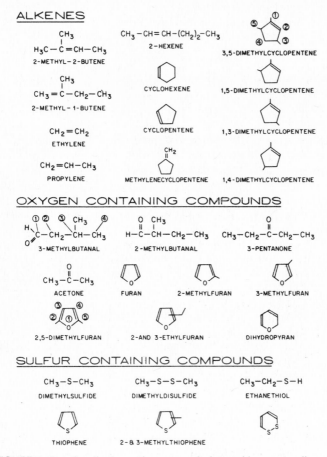

FIGURE 7. Functionalized organic compounds detected in recent sediments.

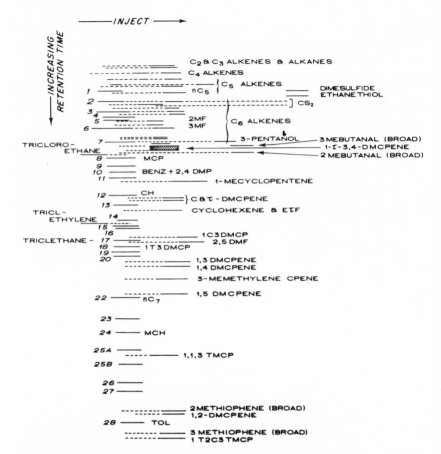

FIGURE 8. Capillary gas chromatogram—retention times of functionalized compounds (dashed lines) with respect to hydrocarbons (solid lines). HHK capillary column. For key to numbers, see Figure 6. Other abbreviations: Me, methyl; 2MF and 3MF, 2- and 3-methylfuran; DMCPene, dimethylcyclopentene; EtF, ethylfuran; 2,5DMF, 2,5-dimethylfuran; CPene, cyclopentene; TriCl, trichloro. Trichloro compounds shown are common contaminants when polyvinylchloride core liners are used.

adjusted so that 20–40 ng of compound will be injected into the GC during analysis. Standard solutions are prepared in methanol and stored in portions in glass ampules until needed.

Response factors for other compounds are obtained in a similar way except that more compound must usually be injected into the head-space vessel to obtain the same GC response because of the greater water solubility of most of these molecules. Since the volume of head-space gas injected into the GC is adjusted so that the amount of compound is always approximately the same (within an order of magnitude, based on packed column GC results), nonlinearity of the FID signal at high compound concentrations does not become a problem.

Precision of the capillary gas chromatography head-space analyses is 5% (coefficient of variation) for amounts of compound in the range 100 to 0.1 ng compound/g dry weight of sediment. Precision of absolute amounts of compound in duplicate sediment samples is 25%–30% (coefficient of variation) primarily due to sample variation. However, precision of peak ratios in duplicate samples is generally 5% or less on duplicate samples analyzed within a short time. Duplicate samples run a year apart on two samples of an organic-rich surface sediment (from the upwelling region of S.W. Africa at Walvis Bay, Whelan *et al.*, 1980a) showed the same compounds with the same proportions (within 10%–15%), which indicates that both hydrocarbons and functionalized organic compounds in this sediment were stable in the frozen core sample and that the analytical procedure gives internally consistent results.

The absolute sensitivity of the methods described above is about 2.0×10^{-11} moles for ethane and 7×10^{-12} moles for pentane. Assuming use of a 20-g sediment sample and a 20-ml gas sample used in the GC analysis, this is equivalent to 0.08 ng ethane per gram/dry weight sediment and 0.06 ng pentane per gram dry weight. Qualitative analysis is possible to about an order of magnitude less. Sensitivity for aromatic and functionalized compounds is less because of their greater water solubility in the head-space vessel. The aromatic hydrocarbons, benzene and toluene, have responses of about half of those found for the alkanes. Some of the oxygenated compounds, particularly the aldehydes and furans, have responses as much as 1000 times less than for the hydrocarbons in the head-space analysis because of their greater water solubility. This response can be greatly varied by addition of inorganic salts and sediments to the water. Thus, levels of oxygenated molecules such as the furans in Figure 11 should be assumed to show relative changes within a series of samples only.

2.2.3. Stability of Samples to Analysis

Experiments have been carried out on changes caused in hydrocarbon compositions due to sediment heating. A sample was analyzed via the can procedure (Whelan, 1979). Table 2 shows that for heating periods longer than 1.5 h at 90°C, the amounts of several C_6 and C_7 components generally increased. For heating periods of 1.5 h or less, no change occurs (outside of analytical error) with the possible exception of toluene. This experiment shows that the normal short heating times of 30 min routinely used in these

TABLE 2. Amounts of Compound Found by Heating a Marine Sediment Sample (DSDP, Leg 42A, 381–54–0) for Varying Lengths of Time[a]

Compound	Time (h)							
	0.5	1.5	2.5	3.5	4.5	5.5	6.5	7.5
n-Hexane	1.0	1.0	1.2	1.4	2.0	1.3	3.0	2.3
MCP	0.42	0.43	0.50	0.50	0.59	0.43	1.0	0.71
MCH	0.59	0.55	0.64	n.d.	0.67	0.67	1.6	1.4
CH	0.30	0.25	0.40	0.44	0.75	0.36	0.63	0.38
nC_7	0.28	0.29	0.38	0.42	0.51	0.41	1.1	1.1
1C2DMCP	0.14	0.13	0.13	n.d	0.16	0.13	0.35	n.d
Toluene	1.6	2.5	2.8	3.5	3.2	3.2	n.d	2.9
CH/MCP	0.7	0.6	0.8	0.9	1.3	0.8	0.6	0.5
MCP/MCH	0.7	0.8	0.8	n.d	0.9	0.6	0.6	0.5
Tol/MCH	2.8	4.6	4.4	n.d	4.8	4.7	n.d	2.4

[a] Amounts of compound are given in ng hydrocarbon/g dry weight sediment.
[b] n.d. means not determined.

analyses are not producing artifacts with respect to hydrocarbon analyses. Similar experiments carried out with diatomaceous sediments containing functionalized compounds indicate that the alkenes, ethers, ketones, and even (for the minimal heating periods shown in Table 2) the aldehydes are stable to the analysis. Their stability in more clay- and carbonate-rich sediments has not been tested.

Hydrocarbon recovery experiments were carried out by injecting a standard mixture of all of the C_5–C_8 saturated and aromatic hydrocarbons into the head-space vessel containing distilled water only, distilled water plus a claystone sediment, and distilled water plus a chalky sediment. Recoveries from the two sediments were 94% and 100%, respectively, as compared to the distilled water sample. Similar experiments have not yet been carried out for all of the functionalized compounds.

2.2.4. Interlaboratory Comparisons of Volatile Hydrocarbon Data

These comparisons can be confusing because of differences in storing and analyzing samples as mentioned earlier. As a note of caution to the uninitiated, it should be mentioned that all areas of organic geochemistry suffer from problems similar to those described here (see, for example, Hilpert et al., 1978 for a summary of results on a laboratory intercomparison involving the C_{15} + hydrocarbons). In the case of volatile compounds, it is possible to obtain good internal consistency within any one laboratory as well as comparable ratios of compounds between laboratories. In addition, many geologically interesting changes in these volatile compounds are not subtle—increases and decreases of several orders of magnitude are common.

Table 3 shows results from an interlaboratory comparison carried out on a recent gravity core sample taken from the Guaymas Basin in the Gulf of California. It can be seen that isomer ratios show generally good agreement between the two laboratories. Because alternating depth sections of the core were analyzed, some of the variation in amounts shown between the two laboratories is real. The shallowest section at 1.23 m shows low levels of C_1–C_6 typical of organic rich young sediments. Methane levels are consistently higher in the California analyses—probably because methane has the most tendency to escape with storage. The California results show higher methane levels because analyses were carried out closer to the time of sediment collection. The high value found by the California group for hexane in the 2.51–2.58 section is peculiar, but probably real. The deeper sections of the core show an increase in light hydrocarbons in results from both laboratories. The reason for this increase is volcanic activity causing localized sediment heating and hydrocarbon production (Einsele *et al.*, 1980).

A second laboratory intercomparison was carried out between our laboratory at Woods Hole and that of Schaefer and Leythaeuser and

TABLE 3. Comparison of Light Hydrocarbon Data[a]

Sample depth (m)	C_1	$C_2 + C_3$	nC_6	$C_4 - C_6$	$\dfrac{C_1}{(C_2 + C_3)}$	$\dfrac{C_1}{(C_1 - C_4)}$
1.23–1.29*	9	1.0	0.52	1.1	8.9	0.90
2.51–2.58	2190	15	1760	1944	149	0.99
2.59–2.65*	72	53	1.8	669	1.4	0.85
2.94–2.99	7520	154	328	7290	48	0.82
3.31–3.37*	4104	235	1080	14,884	17	0.56
3.38–3.45	6350	153	162	5116	41	0.76
	iC_4/nC_4	iC_5/iC_5	NeoC$_5$/C$_5$	CP/C$_5$	2,3DNB/2MP	
1.23–1.29*	No C_4	No C_5	—	—	1	
2.51–2.58	7.4	12.7	0.37	0.008	40	
2.59–2.65*	5.4	9.4	0.27	0.003	132	
2.94–2.99	1.8	2.2	0.005	0.02	0.76	
3.31–3.37*	1.4	2.0	0.004	0.04	0.63	
3.38–3.45	1.8	2.2	0.0001	0.001	0.41	

[a] Samples with asterisks were analyzed at Woods Hole by Whelan and co-workers (1981); samples without asterisks were analyzed in the laboratory of Simoneit and co-workers (Simoneit *et al.*, 1979).

co-workers in Germany. In this case the same samples were examined but very different sample preparation and analyses procedures were used in the two laboratories (Schaefer *et al.*, 1978). The Woods Hole methodology at the time was set up to look at unlithified surface sediments (Whelan, 1979) while these samples were fairly lithified petroleum source rock cuttings. Thus the Woods Hole C_2–C_6 levels are generally much lower than the German values, sometimes by as much as 4–5 times. These low recoveries by the Woods Hole group caused a change to the methods described in this paper which are capable of better breaking up the lithified sediment particles before analysis. In spite of the large differences in compound recoveries, the agreement between *ratios* of compounds between the two laboratories was generally within experimental error.

2.3. GC/MS Analysis

2.3.1. Instrumentation

Compounds were identified by GC/MS using a Varian Aerograph 1400 gas chromatograph (modified for use with capillary columns) coupled via a glass capillary interface to a Finnigan 1015C quadrupole mass spectrometer. The total eluant from the capillary column goes into the mass spectrometer. The electron energy was 70 eV, the ionization current was 350 MA and the preamp gain was 10^{-7} A/V. The electron multiplier was set at 1.7 KV and the scan conditions were 40–200 amu at 15 msec/amu. The mass spectrometer was interfaced to a Riber 150 data system with DEC PDP 8-E (32K memory) computer. More recently, this computer system has been replaced with a Finnigan INCOS 2300 Data System.

GC/MS analyses of standards and samples were carried out using the loop injection system and capillary columns described earlier (Figure 3) attached to the Varian 1400 GC interfaced with the mass spectrometer as detector. Standards, either purchased or synthesized, were injected into a head-space vessel and analyzed as described previously. After identification of a particular compound in two or three representative samples from a particular suite of about 10 (such as sections from particular sediment core) via MS, routine analyses were carried out via GC retention times if the GC peak patterns of other samples appeared to be similar. GC–MS analyses are not carried out on all samples due to the expense and scarcity of mass spectral time. Chemical ionization was not used in these analyses because electron impact mass spectrometry generally gives simple and characteristic fragmentation patterns for these small molecules.

The GC injection system, including sample loops and valves, are mounted so that detaching at one point (just after L in Figure 3) allows convenient transfer from the GC to the GC/MS system. Analyses by GC/MS are then fairly rapid—about five or six samples a day can be analyzed. The most time-consuming part of the process must then be carried out—retrieval of mass spectra of all GC peaks, comparison to reference spectra, and identification of compounds present as described in the next section.

2.3.2. Scheme for Recovery of GC/MS Data and Peak Identification

Outlined below is a "search" scheme which has been found to be useful in detecting frequently encountered classes of compounds as well as compounds peculiar to one sample. It is probable that some version of this scheme will eventually be used even on the automatic mode of the new INCOS data system in collecting data from our analyses.

Two typical GC/MS analyses are shown in Figure 9. The Peru shelf sample is typical of recent sediments containing primarily biological debris. This particular sample comes from an upwelling region of the western South American coast where the rapidly deposited fine-grained sediments are primarily of marine origin. It can be seen that the GC/MS total ion scan is fairly simple. The simplicity is typical of fine-grained gravity cores recovered from many areas of the world including the Persian Gulf/Arabian Sea, Walvis Bay, and the Gulf of Maine. In contrast, a second sample is shown from offshore California. The compound mixture is much more complex and consists almost exclusively of a mixture of many of the possible isomers of C_1–C_8 saturated and aromatic hydrocarbons. This young sediment (Pleistocene) was recovered from a more deeply buried (587 m subbottom) sediment (DSDP Site 467, Leg 63). Due to the relatively high geothermal gradient throughout the area, extensive hydrocarbon generation has occurred in the sediment. The complex series of hydrocarbons is typical of the presence of petroleum hydrocarbons.

The general scheme for gathering scan numbers for all compounds present is the same for either type of sample. The scan numbers of all peaks obviously larger than background noise are collected from the total ion scans (Figure 9) along with scan numbers of the base line just before each peak which is used in base-line subtraction. The subtraction is particularly important if the GC column used has a high bleed rate (such as HHK). Next, mass scans (see examples in Figure 10) characteristic of various classes of compounds previously found in sediments (shown in Table 4) are examined for the presence of any peaks not previously found

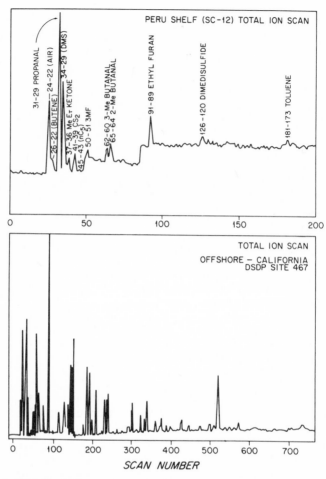

FIGURE 9. GC/MS—Sediments from the Peru Shelf and offshore California (DSDP Site 467, Sample 467-63-2, subbottom depth 587 m)—total ion scans.

in the total ion scan. These mass scans also sometimes allow a more precise determination of the scan numbers, particularly in the case of noisy runs or in cases where peaks of two compounds overlap. The mass scans which have been found by experience to be diagnostic of particular groups of compounds (Table 4) are printed out along with the total ion scan (total in Figure 10). In this way, it is fairly easy to rapidly see what type of compound each peak represents. For example, in the Peru sample in Figure 10, the strong peaks in mass scans 96, 81, and 82 indicate that the peak centered at scan 91 is either an alkylfuran or a dimethylcyclo-pentene. A second much smaller amount of a second isomer can be seen in the m/z 96 mass scan (scan number 98). These compounds are absent in the offshore California sample also shown in Figure 11.

It can be seen in Table 4 that some masses used in mass scans are diagnostic of more than one type of compound. However, the masses chosen have been found to be particularly strong in or unique to the classes of compound indicated with some obvious exceptions, such as 57. Compound identification is done after collecting mass spectra for all of the GC peaks of interest detected in the mass scans. The mass spectra are then

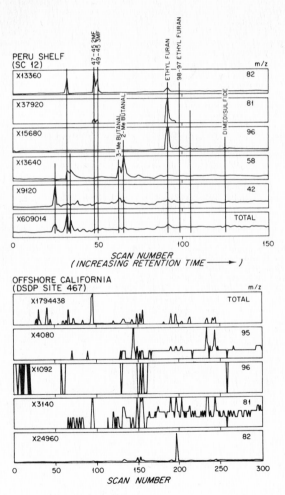

FIGURE 10. GC/MS—mass scans (m/z) of sediments from the Peru Shelf and offshore California.

TABLE 4. Mass Scans Which Have Been Found to be Useful in Detecting Particular Groups of Volatile Compounds in Sediments[a]

m/z	Compound types
57	Alkanes; alkyl groups
55, (70), ㊷	Alkenes and cycloalkanes
(84)	Cyclohexane
68	Isoprene
91	Toluene, alkylbenzenes, phenyl alkyl groups
78	Benzene
105, 106	Xylene, ethylbenzene, dialkylbenzenes
67	Cyclohexene, dialkylfurans, pyrrole, some dienes
82	Cyclohexene, dimethylcyclopentene, methylfurans
81	Dimethylcyclopentane, alkyl furans
96, (95)	Methylcyclohexene, dimethylcyclopentene, dimethylfurans, furfural
58, (70), ㊷	Aldehydes; 2- and 3-methylbutanal
62, 64[b]	Dimethylsulfide
(94), 96[b], 98[b]	Dimethyldisulfide
(94), 66	Phenol
76, 78[b], 80[b]	Carbon disulfide
84, 86[b]	Thiophene
(97), 98, 100[b]	Methylthiophenes
(95), 110	C_8 dienes
(97), 112	Dimethylcyclohexanes

[a] Circles indicate masses typical of more than one type of compound in this scheme.
[b] Peaks characteristic of sulfur M + 2 isotope

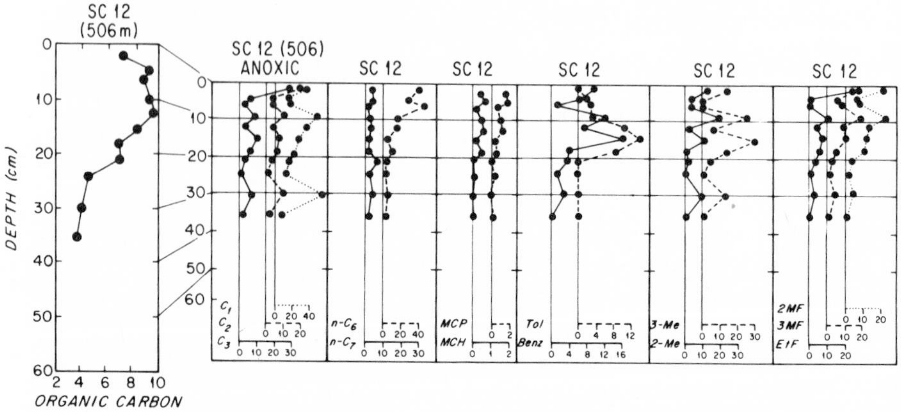

FIGURE 11. C_1–C_7 volatile compounds in sediment cores from Peru Shelf (gravity core). Abbreviations: org C, organic carbon; C_1, methane; C_2, ethane; C_3, propane; n-C_6-n, hexane; n-C_7-n, heptane; MCP, methylcyclopentane; MCH, methylcyclohexane; Tol, toluene; Benz, benzene; 3-Me and 2-Me are 1-methyl- and 2-methylbutanal; 2MF, 3MF, and EtF are 2-methyl-, 3-methyl-, and ethylfuran.

compared to reference spectra—preferably run on the same instrument. Mass spectra can vary somewhat from one instrument to another, particularly with respect to exact peak intensities. However, if a compound is known, it is usually possible to eliminate all except two or three possible structures via comparison to library spectra. At this point, standards are purchased or synthesized and final compound identifications done by comparing GC retention times and mass spectra obtained on our own instruments. Only if a mass spectrum is very weak or very noisy are mass scans alone used for tentative compound identification.

The comparison of sample mass spectra with reference spectra is a very time-consuming process. The operation promises to be greatly speeded up in the future by the recent addition of the Finnigan INCOS data system to the Woods Hole GC/MS facility. Preliminary results on analyses of standards indicate a very high percentage (>95%) of "hits" in identifying the compounds present in these volatile mixtures. This result was obtained using the NBS/EPA/NIH reference library (31,000 spectra) and the forward search algorithm included with the INCOS system. Because of variability of mass spectra from one system to another, which was mentioned above, this reliability should improve even more as we build a library of reference mass spectra run in our own laboratory.

Occasionally, examination of several mass chromatograms or a particular mass spectrum shows the presence of two compounds under a particular GC peak. A good example is the overlapping of benzene and thiophene in many analyses of DSDP sediments. Benzene and thiophene are characterized by strong molecular ions at 78 and 84 respectively. The mass scans of these two masses indicates that thiophene is concentrated a little more in the tail of the GC peak while benzene is concentrated more in the front. Thus, by scanning across the peak, it is possible by subtraction to obtain spectra of the pure compounds. The INCOS system includes programs which will do this subtraction and reconstruction automatically if the two mass spectra each have their own characteristic strong fragment or molecular ion peaks (as is true in the benzene–thiophene case). Ions which maximize simultaneously (within one scan) are assumed to belong to a single compound. A new "enhanced" spectrum is then constructed which contains little or no contributions from background or overlapping components.

Background subtraction has a strong influence on the mass spectrum obtained. With a low bleed liquid phases (such as OV101) on a WCOT capillary column, often no background subtraction is necessary. If a background subtraction is needed, often one scan from almost any "clean" part of the GC can be used for all of the mass spectra as long as no GC base-line drift is observed. In cases where a liquid phase is known to bleed a lot (such as HHK), a background scan must be selected from the area just before the GC peak elutes. The reason for picking an area just before rather than just after the GC peak is to avoid subtraction of the peak tail which usually occurs on the rear of the GC peak on this particular column. Subtracting the peak tail produces the effect of subtracting the peak from itself which can produce some strange results with overlapping peaks. This "tailing" of GC peaks can cause problems for the INCOS automatic data collecting software so that processing of mass spectra via the manual subtraction scheme outlined previously seems to work better for these samples.

The overall search scheme outlined above works well for our particular analyses. It could undoubtedly be streamlined and modified to suit the needs of other laboratories. However, the most important idea to be taken away from this section is the absolute necessity of having *some* organized data retrieval scheme which has been optimized for a particular analysis. The GC/MS computer system gathers an enormous amount of information in a very short time. A great deal of time can be wasted in attempting mass spectral retrieval in an unorganized way.

3. TYPICAL RESULTS AND INTERPRETATION

In this section, some typical results obtained from various depositional settings will be presented in summary form. For more detail, the reader is referred to the original references.

One general comment about investigation of low-molecular-weight compounds in sediments (and probably in other materials) should be made because it may not be obvious to those who have worked with larger and more complex molecular structures. Because of the compound structural simplicity, many volatile compounds have more than one source. For example, the thiophenes, toluene, and the gem-dimethylalkanes all appear to have biological–low-temperature chemical ($<20°C$) sources in surface gravity cores (Whelan *et al.*, 1980a). All of these compounds are also known to have thermal sources in more deeply buried geothermally hot ($>100°C$) sediments (Hunt, 1979; Tissot and Welte, 1978). Thus, it is generally more useful to rely on the presence of a group of compounds using all available information about the sample rather than trying to get information from the presence of only one particular compound.

3.1. Potential Sources of Volatile Organic Compounds in Marine Sediments

Potential sources of volatile organic compounds in recent marine sediments are the same as those which must be considered for the higher-molecular-weight molecules. In addition, the lower molecular weight and higher volatility of these compounds causes the potential for greater movement of these molecules in the sediments, water, and in air. Thus, the major sources which must be considered are (1) transport either by bottom currents or by slumping of older diagenetically transformed sediments, (2) planktonic or benthic fauna and flora, (3) transport from terrestrial sources either by the atmosphere or rivers, (4) diagenesis (via biological and/or low-temperature chemical alteration) of biologically produced precursors, and (5) migration from more deeply buried sediments. Sometimes, it is possible to decide which of these might be important in a particular area.

It is well known that methane is produced microbiologically in anaerobic recent marine sediments. In addition, preliminary data have shown that microorganisms can make traces of C_4–C_7 branched alkanes and alkenes plus toluene as degradation products when using terpenoids as food sources (Hunt et al., 1980b) These compounds probably do not represent major metabolic paths in the organism. The trace amounts of substrate transformed argue that these paths probably represent reactions which are incidental to the baceria. However, the small (ppm to ppb) levels of compounds found are comparable to those of recent sediment analyses so that these microbial process can explain one source of the compounds in recent sediments. In addition to microbiological processes there may be low-temperature (<20°C) chemical processes, possibly aided by the presence of water and silical/organic matrix catalysis via the types of processes postulated earlier (Whelan et al., 1980a) Future work may elucidate specific sets of these compounds associated with particular communities of microorganisms or depositional environments in marine sediments.

Vertical migration in the sediment column of many of these compounds in recent fine-grained sediments seems unlikely although it may be important for the lightest compounds methane and ethane. A number of investigators have pointed out the remarkable correlation that exists in many fine-grained sediments between light hydrocarbon composition and lithology (see, for example, Thompson, 1979; Leythaeser et al., 1979; Whelan, 1979). The correlation often holds for very abrupt changes in composition over very short distances even in unconsolidated gravity cores (Whelan et al., 1980a; 1983). In addition, significant quantities of functionalized organic molecules are often found in surface (1 m) gravity cores. These compounds generally disappear with depth leaving 90% or more saturated and aromatic hydrocarbons in deeper sediments. Thus a deep source for the alkenes and oxygen-containing compounds detected in surface gravity cores is not possible.

Most of the surface gravity cores examined in this laboratory contain compound levels low enough so that they can be dissolved in water. Thus, a limit on the amount of vertical or horizontal movement possible would be equivalent to that of interstitial water inorganic ions. This has been shown to be controlled only by diffusional processes in very young sediments (Sayles, 1979). The movement of organic molecules should be even slower in organic-rich sediments because the distribution of these molecules between the water and the organic matrix of the sediment decreases the diffusion coefficients even of methane and ethane about 100 times, as compared to diffusion in water without sediments (Kartsev *et al.*, 1959). These considerations together with the abrupt changes found in depth profiles in many cores suggest that organic compounds remain close to the place where the molecule was generated in many shallow fine-grained sediments. Similar arguments have also been made for more deeply buried sediments in the absence of fracturing, interlayered sand beds, and geothermal temperatures in excess of 50°C (Leythaeser *et al.*, 1979 and 1982; Thompson, 1979).

3.2. Examples

Table 5 summarizes some patterns of volatile organic compounds found in marine sediments, current ideas on their meaning, and appropriate references. These patterns are discussed briefly below. Because this book is concerned with the analysis of compounds produced by microorganisms, all of the various light hydrocarbon patterns associated with petroleum (or thermogenic) hydrocarbon generation have not been included. For more information on this subject, the reader is directed to several excellent recent reviews (Thompson, 1979; Leythaeser, 1979 and 1982; Hunt, 1979, Chap. 5).

3.2.1. Presence of Functionalized Organic Molecules

The presence of functionalized organic molecules such as ethers, alkenes, aldehydes, or ketones is indicative of biogenic debris. These compounds are typically found in the top 10 m of sediment, usually with the highest levels being encountered near the sediment–water interface where the bacterial community is particularly intense. Most of these compounds disappear in sediments which have been exposed to even small rises in geothermal temperature (>20°C) although traces can sometimes be found in deeper sediments. In more deeply buried DSDP sediments, saturated and aromatic hydrocarbons are generally the major (>95%) compounds detected.

3.2.2. Presence of a Limited Number of Compounds

The presence of a limited number of compounds (in the range of about 2 to 20) in many organic geochemical analyses is indicative of biologically produced compounds (Blumer *et al.*, 1971). In contrast, petroleum consists of hundreds of compounds—primarily saturated and aromatic hydrocarbons. Thus, it is often possible to get a rough qualitative idea of biogenic as opposed to petrogenic or anthropogenic organic compound input by looking at the complexity of the GC or GC/MS. Figure 9 shows a typical example where a Peru shelf gravity core (containing only a few compounds) is contrasted to a DSDP sample (Site 467, offshore California) containing a complex hydrocarbon mixture. The latter is from an area where petroleum hydrocarbon generation in sediments is a common occurrence.

3.2.3. Presence of Sulfur Compounds

The presence of sulfur compounds including dimethylsulfide, dimethyl-disulfide, thiophene, methylthiophene, and carbon disulfide is characteristic of sediments which contain or have contained a community of sulfate-reducing microorganisms following the initiation of anoxic conditions in the sediments. Unambiguous identifications of even traces of these compounds is possible using GC/MS. There are also occasional occurrences of more exotic structures such as a compound tentatively identified (via GC/MS) as the heterocyclic disulfide shown in Figure 7. These sulfur compounds have been seen in many surface sediments including Walvis Bay (Whelan *et al.*, 1980) and the Peru Shelf (Whelan and Hunt, 1983) as well as in deeper sections of cores deposited under more oxic conditions such as the Gulf of Maine and the Arabian Sea/Persian Gulf (Hunt and Whelan, 1979). A number of possible low-temperature mechanisms for production of these compounds in surface sediments have been proposed (Whelan *et al.*, 1980a).

In deeper sediments (for example DSDP sediments), the aliphatic thioethers, such as dimethylsulfide and dimethyldisulfide, are generally not found, probably because of their thermal instability. These compounds are not as stable as the thiophenes which can also result from higher-temperature processes, and are well-known petroleum constituents.

3.2.4. Methane Levels Qualitatively High

Methane levels qualitatively high (typically 50–300 ng/g) and C_2–C_5 levels lower (typically 10–1000 times less) indicates a sediment which has been anoxic enough at some time to support the activities of anaerobic

TABLE 5. Patterns of Volatile Organic Compounds Found in Marine Sediments

Trend	Indicators of	Examples (references)
I. Presence of functionalized organic compounds such as ethers, alkenes, aldehydes, and ketones (Figure 7)	Biogenic debris (a) Generated in the water column or near the sediment–water interface; compounds typically disappear in the top 10 m of sediment (b) Found in deeply buried fractured geothermally cold sediments; may represent: (1) Biodegradation of sediment organic matter via percolation of oxygenated waters through fractures. (2) Geothermally cold sediments (<20°C) where biogenic debris has survived to depth.	Walvis Bay (Whelan, Hunt, and Berman, 1980) Persian Gulf/Arabian Sea (Hunt and Whelan, 1979; Ross, 1978) Peru Shelf (Whelan, Hunt, and Tarafa, *Organic Geochemistry*, in press) Japan Trench—DSDP Sites 436/438 (Whelan and Hunt, 1980) (1) DSDP Site 438 (2) DSDP Site 436
II. Limited number of compounds	Biogenic or low temperature (<20°C) chemical source	Walvis Bay; Gulf of Maine (Scranton and Whelan, in press)
III. Presence of sulfur compounds—dimethylsulfide, dimethyldisulfide, thiophene, methylthiophenes, and carbon disulfide	(a) Sediment which was anoxic at or shortly after time of deposition; compounds believed to be debris from community of sulfate-reducing bacteria (b) Thiophene and methylthiophenes are also common petroleum constituents where source is believed to be	Walvis Bay Persian Gulf Peru Shelf Hunt, 1979; Tissot and Welte, 1978

IV. Methane levels high; Levels of C_2–C_5 much lower	Anaerobic methanogenic bacteria present in sediment which was anoxic or became anoxic shortly after time of deposition	Walvis Bay; Peru Shelf
V. Traces of alkenes and alkane near surface–water interface; decrease with depth in gravity cores	Oxic sediment–water interface	Gulf of Maine; Persian Gulf
VI. Presence of neopentane and other gemdimethylalkanes	(a) Sediment contained terpenes derived predominantly from terrigenous sources which have undergone aerobic bacterial degradation (b) Also found in deeply buried sediment exposed to temperatures in excess of 150°C	(a) Black Sea DSDP Site 381 (Hunt and Whelan, 1978) Japan Trench DSDP Site 438 (Whelan and Sato, 1980) Canary Islands DSDP Site 397 (Whelan, 1979) (b) Hunt, Huc, and Whelan, 1980
VII. Shallow subsurface toluene maximum—often accompanied by no or very small quantities of benzene	Anoxic sediment water interface at time of deposition	Walvis Bay; Peru Shelf
VIII. Exponential increase in variety and amount of C_4–C_8 saturated and aromatic hydrocarbons	Petroleum generation	For reviews, see references cited at the beginning of Section 3.2

methanogenic bacteria. These methane levels are lower (with respect to C_2–C_5) than reported elsewhere (see, for example, Claypool and Kaplan, 1974; Claypool, 1976; Bernard, 1978; and Kvenvolden and Redden, 1980). However, it should be remembered that values measured in this work represent *adsorbed* C_1–C_5 components with the lighter methane preferentially escaping when frozen cores are stored for long periods of time. Even in frozen cores stored for several years, enough methane remains so that qualitative information is still available (Whelan *et al.*, 1980a).

Much smaller amounts of C_2–C_5 hydrocarbons are also usually present in these anoxic sediments so that C_1/C_2 or $C_1/(C_2 + C_3)$ ratios are often used along with other data such as δC^{13} measurements to distinguish biogenic from petrogenic methane. Petroleum generation causes higher levels of saturated C_2–C_5 hydrocarbons, and low $C_1/(C_2 + C_3)$ ratios (generally less than 10 and often less than 1).

Oxic sediments generally show lower levels of methane. The values of 0–5 ng/g for ten Gulf of Maine cores are fairly typical (Scranton and Whelan, in press). However, the C_2–C_5 levels are often in the same range in these sediments (0–2 ng/g for the Gulf of Maine) so that the $C_1/(C_2 + C_3)$ ratio must be considered together with total levels of hydrocarbon in trying to decide whether or not the ratio is diagnostic of petroleum.

Excellent correlations (correlation coefficients of 0.7–0.9 are common) have been observed between concentrations of C_2 and C_3 as well as for several C_4–C_7 compounds in many gravity cores. The Peru Basin (see Figure 11) and Walvis Bay sediments provide typical examples. The correlation often seems to hold for more deeply buried DSDP sediments (see, for example, Whelan, 1979; Whelan and Hunt, 1980a, 1983) as well as for some oxic surface gravity cores where only traces of these compounds are present. In very young organic-rich surface gravity cores, it is likely that this correlation results from either biological or a combination of biological-low-temperature chemical processes. The specific mechanisms of production are unknown, although some possibilities have been proposed (Whelan *et al.*, 1980a). It is known that such a high degree of specificity is more typical of biological processes than of chemical ones.

3.2.5. Traces of Alkene and Alkanes at Sediment–Water Interface

Traces of alkene and alkanes at a sediment–water interface are a common pattern seen in a number of areas which have an oxic sediment–water interface as found in many oceanic and continental shelf sediments. It is very common for these compounds to decrease within the top meter. In ten cores recovered from the Gulf of Maine, typical surface concentrations of C_2 to C_7 alkanes are 1 to 10 ng compound/g dry weight of sediment. These levels decrease to 0 to 0.1 ng/g within the top meter. Total alkene levels in these cores are 10–100 ng/g at the surface and decrease to 0.1–1 ng/g within the top 100 cm. Similar behavior has been observed for surface cores recovered from other areas such as the Arabian Sea/Persian Gulf, and the California Bight (Whelan and Hunt, 1982; Hunt and Whelan, 1979; Ross and Stoffer, 1978). The general trend is for alkanes and aromatics to decrease in the first few meters while organic carbon levels remain constant. The exact source of the alkanes in these surface sediments is not completely clear. The most reasonable possibilities are microbial production, air transport via prevailing offshore breezes in this area, or deposition of land-derived sediments via continental runoff. Oil seeps and large sources of pollution are unlikely in the Gulf of Maine. They could be minor contributors in areas such as the Arabian Sea/Persian Gulf area, which carry a lot of oil tanker traffic.

3.2.6. Presence of Neopentane and Other Gem-Dimethylalkanes

The presence of neopentane and other gem-dimethylalkanes in recent sediments has been found to be associated with terrigenous source material and sediments exposed to oxygenated waters. Possible reasons for this correlation have been presented (Whelan, 1979; Whelan and Sato, 1980). Neopentane is a minor constituent of petroleum although it is the major C_5 hydrocarbon found in the Athabasca Asphalt Sands where petroleum initially present has been subjected to partial aerobic bacterial degradation (Strausz *et al.*, 1977). Microbial activity may concentrate other gem-dimethylalkanes such as 2,2-dimethylbutane; 2,2- and 3,3-dimethylalkanes; and 1,1-dimethylcyclopentane. For example 2,2-dimethylbutane was produced in small amounts when the terpene, 2-pinene (which contains a gem dimethyl group), was exposed to marine microorganisms (Hunt *et al.*, 1980b).

Sediments examined to date from many areas suggest that two factors may be necessary to cause formation of neopentane in immature marine sediments: (a) terpene precursors derived predominantly from terrigenous organic matter and (b) exposure of sediments containing these precursors to oxygenated waters and aerobic microorganisms—either at the time of deposition or after burial.

One note of caution—the gem dimethyl compounds can also be formed by petrogenic (thermal) processes. However, evidence has been presented that this process begins only in the geothermally hottest (deepest) sediments after generation of most other petroleum constituents is well underway (Hunt *et al.*, 1980a).

3.2.7. Shallow Subsurface Toluene Maximum

Shallow subsurface toluene maximum often accompanied by little or no benzene is a common pattern observed in sediment deposited under anoxic bottom water conditions. The sediment data fit well with the terpene microbial degradation studies mentioned previously (Hunt *et al.*, 1980b). Thus, toluene was produced when marine bacteria were exposed to β-carotene (a common terpene found in green plant matter) under anaerobic conditions. The compound was not produced in experiments carried out under aerobic conditions. It is also common for toluene and benzene to be either absent or present in only very small proportions in sediments deposited in oxygenated-bottom-water environments (for example, in the Gulf of Maine and Persian Gulf sediments). In fact, levels of benzene and toluene are generally so low in shallow oxic sediments that their presence can only be detected via mass scans of m/z 78 and 91, respectively during GC/MS analyses.

3.2.8. Thermogenic (Petroleum) Hydrocarbon Generation

Some excellent recent reviews on the subject are cited at the beginning of Section 3.2. The presence of petroleum in sediments differs from the cases cited above in exhibiting an exponential increase in the variety and amount of C_2–C_8 saturated and aromatic hydrocarbons along with a disappearance (or swamping out) of functionalized molecules. This "main phase" of hydrocarbon generation occurs as the sediments reach temperatures of 50–150°C, where petroleum generation reaches its maximum rate (see, for example, Hunt, 1979; Huc and Hunt, 1980). Levels of hydrocarbons present typically reach microgram per gram levels (rather than the ng/g levels discussed here in earlier sections). Extreme heating of sediments (to temperatures greater than 200°C) can produce further hydrocarbon cracking to give primarily the lightest alkanes, methane, ethane, propane, along with the aromatic hydrocarbons benzene and toluene.

ACKNOWLEDGMENTS. Special thanks to Nelson Frew for his help in obtaining GC/MS data. This work was supported by National Science Foundation Grants Nos. 79-19861 and 81-19508. Contribution No. 5045 from the Woods Hole Oceanographic Institution.

REFERENCES

Bernard, B. B., 1978, Light hydrocarbons in marine sediments, Ph.D. Thesis, Texas A & M University, Department of Oceanography, College Station, Texas 77843.

Blumer, M., Guillard, R. R. L., and Chase, T., 1971, Hydrocarbons of marine phytoplankton, *Marine Biology* **8**:183.

Claypool, G. E., 1976, Manual on pollution prevention and safety, *JOIDES J.* **1**: Special Issue No. 4.

Claypool, G. E., and Kaplan, I., 1974, "The Origin and Disbribution of Methane in Marine Sediments," in *Natural Gases in Marine Sediments* (I. Kaplan, ed.), pp. 99–139, Plenum Press, New York.

Claypool, G. E., Presley, B. J., and Kaplan, I. R., 1973, "Gas Analysis in Sediment Samples from Legs 10, 11, 13, 14, 15, 18, and 19," in *Initial Reports of the Deep Sea Drilling Project*, Vol. 19, pp. 879–884, U.S. Government Printing Office, Washington, D.C.

Durand, B., and Espitalie, J., 1971, Formation et evolution des hydrocarbures de C_1 a C_{15} et des gaz permanents dans les argiles du Toarcien du Bassin de Paris, *Advances in Organic Geochemistry* **1971**:455.

Einsele, G., Gieskes, J. M., Curray, J., et al., 1980. Intrusion of basaltic sills into highly porous sediments, and resulting hydrothermal activity, *Nature* **283**:441.

Hilpert, L. R., May, W. E., Wise, S. A., Chesler, S. N., and Hertz, H. S., 1978, Interlaboratory comparison of determinations of trace level petroleum hydrocarbons in marine sediments, *Anal. Chem.* **50**:458.

Huc, A. Y., and Hunt, J. M., 1980, Generation and migration of hydrocarbon in offshore South Texas Gulf Coast sediments, *Geochim. Cosmochim. Acta* **44**:1081.

Hunt, J. M., 1975, Origin of gasoline range alkanes in the deep sea, *Nature* **254**:411.

Hunt, J. M., 1979, *Petroleum Geochemistry and Geology*, Freeman, San Francisco.

Hunt, J. M., and Whelan, J. K., 1978, "Dissolved Gases in Black Sea sediments," in *Initial Reports of the Deep Sea Drilling Project*, Vol. 42B, pp. 661–665, U.S. Govt. Printing Office, Washington, D.C.

Hunt, J. M., and Whelan, J. K., 1979, Volatile organic compounds in Quaternary sediments, *Org. Geochem.* **1**:219.

Hunt, J. M., Huc, A. Y., and Whelan, J. K., 1980a, Generation of light hydrocarbons in sedimentary rocks, *Nature* **288**:688.

Hunt, J. M., Miller, R. J., and Whelan, J. K., 1980b, Formation of C_4–C_7 hydrocarbons from bacterial degradation of naturally occurring terpenoids, *Nature* **288**:577.

Kartsev, A. A., Tabasaranskii, Z. A., Subbota, M. I., and Mogilevskii, G. A., 1959, *Geochemical Methods of Prospecting and Exploration for Petroleum and Natural Gas*, University of California Press, Berkeley, Chaps. 4 and 5.

Kvenvolden, K. A., and Claypool, G. E., 1980, Origin of gasoline-range hydrocarbons and their migration by solution in carbon dioxide in Norton Basin, Alaska, *Am. Ass. Pet. Geol. Bull.* **64**:1078.

Kvenvolden, K. A., and Redden, G. D., 1980, Hydrocarbon gas in sediment from the shelf, slope, and basin of the Bering Sea, *Geochim. Cosmochim. Acta* **44**:1145.

Leythaeuser, D., Schaefer, R. G., and Weiner, B., 1979, Generation of low molecular weight hydrocarbons from organic matter in source beds as a function of temperature and facies, *Chemical Geology* **25**:95.

Leythaeuser, D., Schaefer, R. G., and Yukler, X., 1982, Role of diffusions in primary migration of hydrocarbons, *Am. Assoc. of Petr. Geol. Bull.* **66**:408.

Ross, D. A., and Stoffers, P., 1978, Report C: General data on bottom sediments including concentration of various elements and hydrocarbons in the Persian Gulf and Gulf of Oman, Woods Hole Oceanographic Institution Technical Report No. WHOI-78-39, Woods Hole, Massachusetts, pp. 62–74.

Sayles, F. L., 1979, The composition and diagenesis of interstitial solutions—I. Fluxes across the seawater–sediment interface in the Atlantic Ocean, Geochim. Cosmochim. Acta **43**:527.

Schaefer, R. G., Weiner, B., and Leythaeuser, D., 1978, Determination of sub-nanogram per gram quantities of light hydrocarbons (C_2–C_9) in rock samples by hydrogen stripping in the flow system of a capillary gas chromatograph, *Anal. Chem.* **50**:1848.

Schwartz, R. D., and Brasseaux, D. J., 1963, Resolution of complex hydrocarbon mixtures by capillary column gas liquid chromatography. Composition of the 28° to 114°C portion of petroleum, *Anal. Chem.* **35**:1375.

Scranton, M. I., and Whelan, J. K., 1983, Dissolved gases in water and sediments from the Gulf of Maine, in *Georges Bank*, R. Backus, ed., MIT Press, Cambridge, Massachusetts, in press.

Shriver, D. F., 1969, *The Manipulation of Air-Sensitive Compounds*, McGraw-Hill, New York, Chaps. 7–9.

Simoneit, B. R. T., Mazurek, M. A., Brenner, S., Crisp, P. T., and Kaplan, I. R., 1979, Organic geochemistry of recent sediment from Guaymas Basin, Gulf of California, *Deep-Sea Res.* **26A**:879.

Strausz, O. P., Jha, K. N., and Montgomery, D. S., 1977, Chemical composition of gases in Athabasca bitumen and in low-temperature thermolysis of oil, sand, asphaltene and maltene, *Fuel* **56**:114.

Swinnerton, J. W., and Linnenbom, V. J., 1967, Determination of C_1–C_4 hydrocarbons in sea water by gas chromatography, *J. Gas Chromatogr.* **5**:570.

Thompson, K. F. M., 1979, Light hydrocarbons in subsurface sediments, 1979, *Geochim. Cosmochim. Acta* **43**:657.

Tissot, B. P., and Welte, D. H., 1978, *Petroleum Formation and Occurrence*, Springer-Verlag, Berlin.

Whelan, J. K., 1979, "C_1–C_7 hydrocarbons from IPOD Holes 397 and 397A", in *Initial Reports of the Deep Sea Drilling Project*, Vol. 47A, pp. 531–539, U.S. Government Printing Office, Washington, D.C.

Whelan, J. K., and Hunt, J. M., 1980, "C_1–C_7 Volatile Organic Compounds in Sediments from Deep Sea Drilling Project Legs 56 and 57: The Japan Trench, in *Initial Reports of the Deep Sea Drilling Project*, Vol. 56–57, pp. 1349–1365, U.S. Government Printing Office, Washington, D.C.

Whelan, J. K., and Sato, S., 1980, "C_1–C_5 Hydrocarbons from Core Gas Pockets, Deep Sea Drilling Project Legs 56 and 57, Japan Trench Transect", in *Initial Reports of the Deep Sea Drilling Project*, Vol. 56–57, pp. 1335–1347, U.S. Government Printing Office, Washington, D.C.

Whelan, J. K., Hunt, J. M., and Berman J., 1980a, Volatile C_1–C_7 organic compounds in surface sediments from Walvis Bay, *Geochim. Cosmochim. Acta* **44**:1767.

Whelan, J. K., Hunt, J. M., and Huc, A. Y., 1980b, Applications of thermal distillation-pyrolysis to petroleum source rock studies and marine pollution, *J. Anal. Appl. Pyr.* **2**:79.

Whelan, J. K., and Hunt, J. M., 1982, "C_1–C_8 Hydrocarbons in Leg 64 Sediments, Gulf of California", in *Initial Reports of the Deep Sea Drilling Project*, Vol. 64B, pp. 763–779, U.S. Government Printing Office, Washington, D.C.

Whelan, J. K., and Hunt, J. M., 1983, "Volatile C_1–C_8 Organic Compounds in Sediments from the Peru Upwelling Region", *Org. Geochem.* in press.

12

MASS SPECTROMETRY OF NITROGEN COMPOUNDS IN ECOLOGICAL MICROBIOLOGY

Göran Bengtsson, Laboratory of Ecological
Chemistry, University of Lund, Ecology Building,
Helgonavägen 5, S-223 62 Lund, Sweden

1. INTRODUCTION

New ionization techniques recently introduced in mass spectrometry (MS) have been successfully applied in determination of the structure and molecular weight of amino acids, amines, and other nitrogen-containing compounds. Fragmentation pathways of derivatives of these compounds can now be obtained with EI, CI, radionuclide fission fragmentation, and negative ion MS. Applications of MS also benefit from the development of higher accuracy in isotopic measurements and new techniques for derivatization and vaporization. Three areas of research in ecological microbiology are particularly interesting in the light of advances in mass spectrometric techniques: (1) Structure elucidation of metabolites; (2) quantitative MS; (3) isotope ratio measurements. Although the power of modern MS in research on nitrogen compounds has been demonstrated in some fields of application, such as biochemistry and medicine, comparatively few papers have reported applications of MS in ecological microbiology. The following is an attempt to describe some efforts to use MS to elucidate information on microbial activity of ecological significance.

2. STRUCTURE DETERMINATION

2.1. Cell Wall Components

Cell wall composition has become a well-established and widely used criterion in the classification and identification of gram-positive bacteria (Keddie and Bousfield, 1980). Rapid GC methods have been developed to determine the occurrence of cell wall components of value in classification and identification, viz., diaminopimelic acid, N-acetylmuramic acid, and certain sugars and fatty acids. Work is in progress to test the significance of these compounds as a measure of bacterial standing crop. Diaminopimelic and muramic acids originate from wall peptidoglycans, which show a remarkable consistency throughout the bacterial world. The mass spectrum of muramic acid recovered from tissues of rats experimentally infected with streptococcal cell walls is given in Chapter 8. The carboxyl groups of the muramic acid residues are linked to peptide units, which, in turn, are cross-linked through bridges. Wietzerbin *et al.* (1974) found that the cross-linking in mycobacteria was mediated through D-alanyl-(D)-*meso*-diaminopimelic acid and *meso*-diaminopimelyl-*meso*-diaminopimelic acid linkages. The occurrence of the latter was established by partial hydrolysis and mass spectrometric analysis. Hydrolysis was carried out after enzymatic degradation of the peptidoglycans and the peptides were acetylated in acetic anhydride and esterified in methanol-HCl. The resulting dipeptide derivative had a molecular ion at m/z 530 (Figure 1) and corresponded to the structure in Figure 2. The presence of diaminopimelic acid in the cell wall of the Legionnaires' disease bacterium (*Legionella pneumophila*), an agent in outbreaks of upper respiratory disease, was also confirmed by GC/MS (Guerrant *et al.*, 1979; see also Chapter 8).

The formation of certain microbial cell constituents can be examined by MS. It is known that the C-2 carbon of the thizole ring of thiamine in

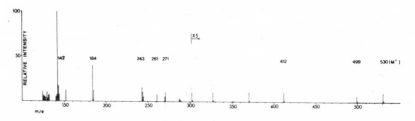

FIGURE 1. Mass spectrum of the N-acetyl methyl ester derivative of the dipeptide *meso*-diaminopimelyl-*meso*-diaminopimelic acid (from Wietzerbin *et al.*, 1974).

FIGURE 2. Mass fragmentation of the dipeptide *meso*-diaminopimelyl-*meso*-diaminopimelic acid. The molecular ion at m/z 539 and the peaks at m/z 508, 277, and 249 are observed in the mass spectrum of its N-trideuterioacetyl methyl ester (from Wietzerbin et al., 1974).

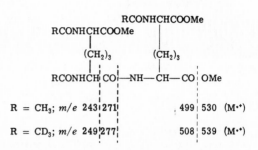

RCONHCHCOOMe
        |
     (CH₂)₃
        |
RCONHCH CO —NH— CH — CO OMe

R = CH₃; m/e 243 271   , 499 530 (M·⁺)

R = CD₃; m/e 249 277   508 539 (M·⁺)

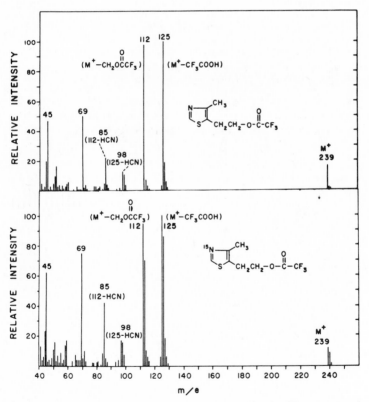

FIGURE 3. Mass spectrum of 4-methyl-5-β-hydroxyethyl thiazole trifluoroacetate obtained from cells grown on L-(^{14}N)tyrosine (above) and L-(^{15}N)tyrosine (below) (from White and Rudolph, 1978).

Escherichia coli is derived from the C-2 carbon of L-tyrosine. White and Rudolph (1978) used N-15 labeled tyrosine to test whether the nitrogen of the ring also originated from L-tyrosine. Thiamine was extracted from cells grown in stock cultures supplied with N-15 tyrosine. The phosphates were enzymatically removed from the thiamine pyrophosphate and the thiamine subsequently cleaved into its thiazole and pyrimidine moieties with bisulfite. The thiazole was converted into its trifluoroacetic acid ester and the isotopic incorporation measured directly by GC/MS. The mass spectrum obtained from cells grown on N-14 tyrosine (Figure 3a) and N-15 tyrosine (Figure 3b) are shown. The protein amino acids were assayed for their N-15 content to eliminate the possibility that the incorporated N-15 could have been derived from another amino acid by transamination. It was shown that the N in the thiazole was derived solely from L-tyrosine.

2.2. Uptake and Exudation of Amino Acids

Microbial conversion of specific amino-compounds can be determined by a combination of different analytical techniques, viz., isotope labeling, electrophoresis, NMR, and GC/MS. Confirmatory tests can be made of hypotheses about specific biochemical pathways. Switching in absorption of one nitrogen source to another can be monitored, and metabolites of importance in reproductive behavior, for attraction and avoidance of parasites and predators, may be found. Mass spectrometric monitoring of the formation of microbial metabolites of amino acids is convenient in studies of factors that influence, e.g., palatability and nutritional quality of a product.

The formation of L-(2,3-dihydroxyphenyl)alanine from L-phenylalanine is an example of a biosynthetic pathway in bacteria, established by MS analyses. The metabolism of phenylalanine is connected to the catabolism of chloridazone (5-amino-4-chloro-2-phenyl-2H-pyridazine-3-on), the active substance in the herbicide "Pyramin," which can be used as an exclusive carbon source by certain soil bacteria. L-(2,3-dihydroxyphenyl)alanine was isolated from a culture medium supplied with phenylalanine (Figure 4) by exclusion chromatography and identified by MS (Buck *et al.*, 1979). $(M-46)^+$ and m/z 151, representing fragmentation of HCOOH from the carboxyl group of the alanine moiety and the *o*-hydroxy group, were abundant.

Wegst *et al.* (1981) isolated and characterized L-phenylalanine dihydrodiol as a metabolite in the formation of L-(2,3-dihydroxyphenyl)alanine from L-phenylalanine. The experiment was aimed to test whether the catechol compound was the product of coupled dioxygenase and dehydrogenase reactions or whether it was formed by the action of two mono-

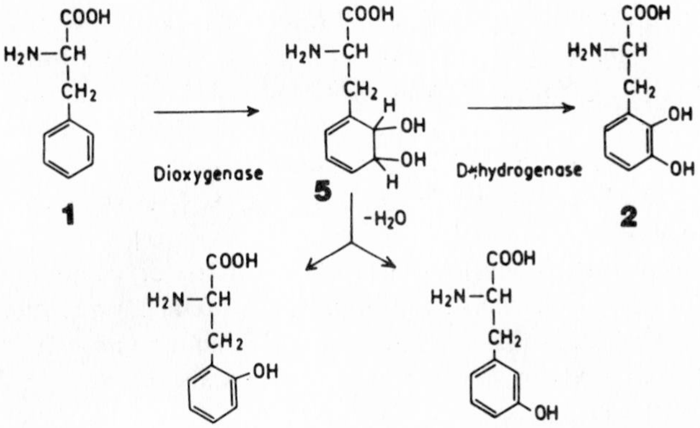

FIGURE 4. Formation of L-(2,3-dihydroxyphenyl)alanine (2) by chloridazon-degrading bacteria incubated with L-phenylalanine (1). Small amounts of L-o-hydroxyphenylalanine (3) and L-m-hydroxyphenylalanine (4) can also be detected in the culture medium (from Buck et al., 1979).

oxygenases. The characterization of the dihydrodiol confirmed the presence of the first mentioned reaction. The dihydrodiol was isolated from the growth medium by exclusion chromatography on Sephadex, HPLC, and TLC and identified by a combination of uv spectrometry and MS. It was difficult to obtain mass spectra of the L-phenylalanine dihydrodiol, so the authors used t-butoxycarbonyl-L-phenylalanine and N-acetyl-L-phenylalanine, which showed M^+ peaks in field desorption MS.

The calcium salt of the hydroxy-analog of methionine, 2-hydroxy-4(methylthio)butanoic acid (M), is used as a nutrient supplement in commercial broiler feeds. There has been considerable interest in improving the protein nutritional response of ruminant animals to maximize meat and milk production by supplementation of the M-analog. However, data indicate that the M-analog is more resistant to rumen microbial degradation than methionine, and the M-analog has occasionally been found in control milk samples. It was thought that the natural microflora of milk and fermentation products, such as bread and beer, would be capable of biochemical formation of the M-analog from free methionine. An experiment was undertaken by Belasco et al. (1978) to evaluate this possibility. L-C-14-methionine was added to synthetic nutrient media of several Bacillus and Saccharomyces species and to sterile skim milk that was inoculated with bacteria. Samples were withdrawn from the incubation mixtures for TLC (autoradiography) and GC analysis. Confirmation of the identity of the M-analog in cultured milk products was made as a silyl derivative and in bread, beer and bacterial cultures as ethyl ester derivatives. The M-analog was conclusively found in spoiled milk. Monitoring at m/z 294 (M^+) showed a large peak at the correct retention time of the derivative while no peak was evident with extracts of fresh milk. A prominent molecular ion was obtained at m/z 178 for the ester of standard and derivatized extracts of pure cultures (Figure 5). Other fragments were found at m/z 104 ($CH_3SCH_2CH=O^+$) and m/z 76 ($CH_3SCH_2CH_3^+$) resulting from hydrogen rearrangements.

Variations in the composition of organic nitrogen compounds in the exudates from microorganisms are widespread. Uncommon amino acids or amines can be identified in exudates by MS, and the absence of certain common amino acids in exudates can reveal some interesting details in metabolic pathways.

It has been known for some time that microorganisms belonging to the anaerobically growing genus *Bifidobacterium* release into culture broths significant amounts of amino acids except for tryptophan, which has not been found to accumulate. This was the basis for an experiment by Aragozzini *et al.* (1979) with 51 strains of *Bifidobacterium*. After extraction of

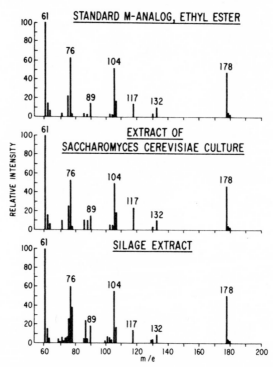

FIGURE 5. Mass spectra of ethyl ester derivatives of the M analog from a *Saccharomyces cerevisiae* culture and in corn silage (from Belasco *et al.*, 1978).

cultures incubated under anaerobic conditions in trypticase phytone-glucose broth they found an unknown metabolite, visible on TLC. The product was isolated and derivatized as a methyl ester prior to MS analysis. The product was identified as indole-3-lactic acid (M^+ at m/z 219). Tryptophane was found to deaminate to indole-3-pyruvic acid, which was converted to indole-3-lactic acid.

Herrmann and Jüttner (1977) employed the method of trifluoroacetylation for the separation of biogenic amines excreted by algae. Different species of algae were grown axenically and the amines isolated from the medium by vacuum sublimation. The amine hydrochlorides were derivatized with trifluoroacetic anhydride in ethyl acetate. Thirteen biogenic amines were identified by MS and most of them showed a fragment ion at m/z 114 corresponding to a loss of CF_3CONH_3. Molecular ions were either of low intensity or nondetectable.

Onodera and Kaudatsu (1969) showed that about 30% of endogenous nonprotein nitrogen of rumen ciliate protozoa was present in the form of free amino acids. Analysis by paper chromatography indicated that alanine, glutamic acid, proline, and lysine were major components. In addition, an unknown spot was discovered. The compound was isolated by column chromatography and ion exchange, washed by ethanol and recrystallized. Molecular structure was determined by MS using the ethyl ester. A parent peak was indicated at m/z 157, so the molecular weight was estimated as 129. The most intense peak was at m/z 84, an amine peak. The absence of m/z 102, which would indicate an ester fragment, led the authors to the presumption of a cyclic structure. The compound was determined as L-pipecolic acid (2-piperidinecarboxylic acid).

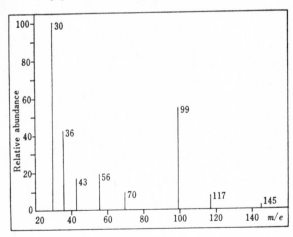

FIGURE 6. Mass spectrum of the ethyl ester of δ-aminovaleric acid (Tsutsumi et al., 1975).

In 1975, Tsutsumi *et al.* identified δ-aminovaleric acid in the metabolites of rumen ciliates. The growth medium was deprotonated by trichloroacetic acid and the compound isolated by ion-exchange and silica gel column chromatography, and then derivatized with an ethanol–HCl mixture. The molecular weight was estimated to 117 (Figure 6) as the peak found at m/z 145 seemed to be a parent peak. The peak at m/z 30 was thought to be an amine peak and since there was no peak at m/z 102 (ester peak), the compound seemed to have an amino group at a terminal carbon atom. The steric configuration was verified by optical rotation measurements (ORD) and ir-spectroscopy.

2.3. Stimulatory and Inhibitory N-Compounds

Some nitrogen-containing compounds are known to exert a stimulatory influence on certain microbial behavior. Thus, Sakurai *et al.* (1976) purified by partition chromatography and gel filtration a peptide that induced sexual agglutinability in *Saccharomyces cerevisiae*. The peptide was treated with 1,1,3,3-tetramethoxypropane in methanol–HCl to convert the terminal arginine residue, confirmed by Edman-dansyl degradation, into pyrimidylornithine, whose α-amino group was acetylated with Ac_2O–AcONa in methanol. The resulting hexapeptide methyl ester was analyzed by MS whereby the molecular ion was observed at m/z 777. It was found that the pyrimidyl-ornithine residue was eliminated upon electron impact in the ion source. The fragmentation pattern elucidated a methyl ester of a pentapeptide having an acetylglycine residue at the N-terminus.

Nordbring-Hertz and Odham (1980) used a mass spectrometer in EI mode to confirm the presence of ammonia in the exudate of the nematode *Panagrellus redivivus*. An average of 2 ng of ammonia was produced per nematode during a 24-h period. In the mass spectrum, NH_3^+ and NH_2^+ ions were superimposed on the background (Figure 7). Ammonia was found to stimulate trap formation in the nematode trapping fungus *Arthrobotrys oligospora*.

Allelopathy, i.e., chemical inhibition of one population by another, is represented in microorganisms by a variety of chemical compounds, viz., oxygen, ammonia, short-chain fatty acids, and alcohols. The role of antibiotics, toxic substances specifically produced by microorganisms, has been difficult to demonstrate in soil and aquatic habitats, but several papers have been devoted to identification of antibiotics by mass spectrometry.

Actinomycin is a commonly occurring antibiotic being produced by different species of actinomycetes belonging largely to the genus Streptomyces. Actinomycins constitute a family of chromopeptide antibiotics which differ solely in the peptide portion of the molecule. The actinomycin Z

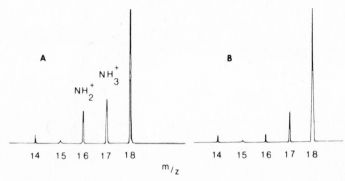

FIGURE 7. Mass spectra of ammonia from an aqueous medium with *Panagrellus* nematodes (A) and background from ion isource (B) (from Nordbring-Hertz and Odham, 1980).

complex produced by *Streptomyces fradiae* consists of several compounds designated Z_0 to Z_5 which differ markedly in amino acid content from other actinomycins described in the literature. Katz *et al.* (1973) investigated hydrolysates of actinomycin Z_5 by paper, gas, and ion-exchange chromatography and MS. An hitherto unknown natural amino acid, *cis*-5-methylproline, was identified as its *N*-trifluoroacetyl methyl ester.

In 1974 Katz *et al.* confirmed the presence of hydroxythreonine (α-amino-β,γ-dihydroxybutyric acid) in the hydrolysate of the actinomycin Z_1. Katz *et al.* (1975) compared by EI-MS a derivatized synthetic 3-hydroxy-5-methyl-proline and the corresponding peak in the hydrolysate of actinomycin Z_1 and found that the spectra of the first isomer of the synthetic compound was identical with the unknown (Figure 8). The other three isomers of the amino acid gave slightly different spectra, the molecular ion (m/z 351) being more pronounced. The spectra showed fragmentation peaks representing loss of $COOCH_3$ (292), CF_3COOH (237), $COOCH_3$ –CF_3COOH (178), and CF_3^+ (69). The amino acid composition of the antibiotic was then elucidated and threonine, hydroxy-threonine, D-valine, 4-oxo-5-methylproline, sarcosine, *N*-methylalanine, and *N*-methylvaline identified.

Longicatenamycin, an antibiotic isolated from the strain S-520 of *Streptomyces diasticus*, is active against gram-positive bacteria. It has been suggested that the antibiotic is a complex mixture composed of several closely related peptides, which have three exchangeable positions for amino acid residues in their molecules. The presence of four new amino acids was reconfirmed by Shibu *et al.* (1975) by comparison with the synthetic specimens. 5-chloro-D-tryptophan was isolated from the hydrolysate and its structure assigned by a combination of NMR, MS, and ORD and by comparison with a synthetic moiety.

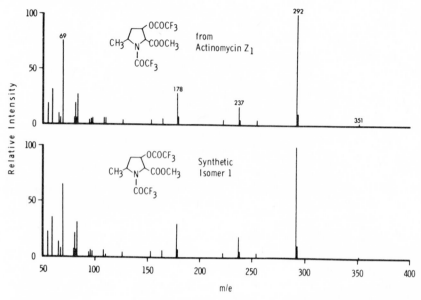

FIGURE 8. Mass spectra of derivatized 3-hydroxy-5-methylproline (from Katz *et al.*, 1975).

The structures of two sulfur-containing amino acids, lanthionine and β-methyl-lanthionine, were established by MS of a purified hydrolysate of gardimycin, a peptide antibiotic produced by species of the genus *Actino-planes* and active *in vitro* against gram-positive bacteria, gram-negative cocci, and obligate anaerobes (Zerilli *et al.*, 1977). The mass spectrum of the diethyl ester hydrochloride of lanthionine (Figure 9) shows M^+ and $(M + H)^+$ at m/z 264 and 265, and, as observed for amino acid ethyl esters, the second is more intense than the first. The peak at m/z 102 corresponds

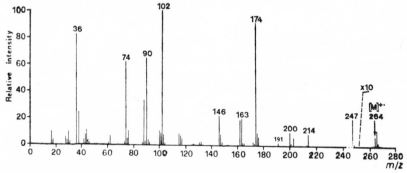

FIGURE 9. Mass spectrum of lanthionine diethyl ester hydrochloride (from Zerilli *et al.*, 1977).

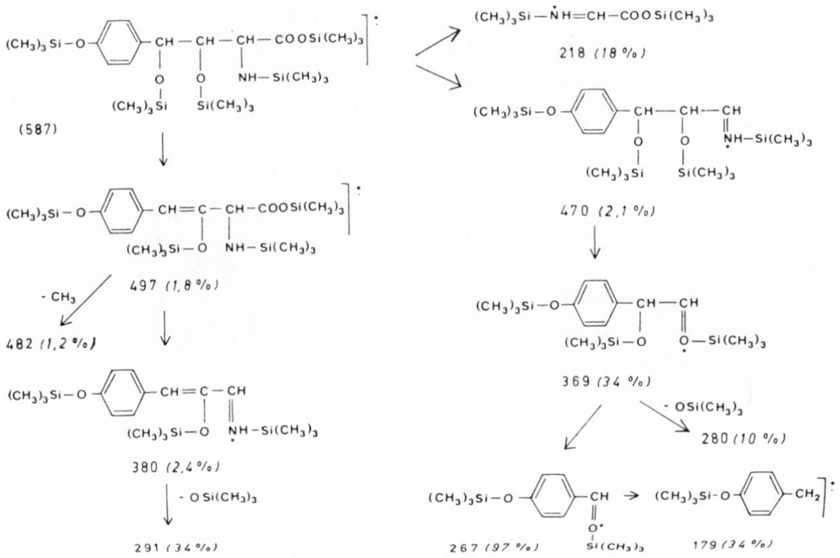

FIGURE 10. Mass fragmentation of penta(trimethylsilyl)-3,4-di-hydroxyhomotyrosine (from Keller-Schierlein and Widmer, 1976).

to the typical ester fragment, which is abundant in this case because of the influence of a sulfur atom in this symmetric molecule.

The tetrahydroderivative of the polypeptide antifungal antibioticum Echinocandin B was treated with trifluoroacetic acid and methanol–HCl to yield the tripeptide threonyl-(4-hydroxyprolyl)-(4-oxohomotyrosine)-methyl ester, which could by hydrolyzed to 4-oxohomotyrosine (Keller-Schierlein and Widmer, 1976). This and some other facts suggested the presence of 3,4-dihydroxyhomotyrosine in Echinocandin B. This was confirmed by isolation of the compound by mild acidic treatment of the polypeptide and a combination of NMR and MS analyses. The fragmentation pattern of the TMS derivative did not show any molecular peak, but two distinct fragmentation pathways were recognized in elucidating the structure of the molecule (Figure 10).

3. ISOTOPE MASS SPECTROMETRY

3.1. Analytical Methods

Traditionally, N-containing samples in nitrogen-15 tracer studies have been converted into N_2 for measurement of the isotopic abundance. The superior accuracy of the mass spectrometer is one of its main advantages in work to determine the isotopic composition of samples, whereas the alternative method, of optic emission spectroscopy, is simpler and easier to maintain. The precision of MS in analysis of biological samples is typically in the range of 0.1%–0.5% and the accuracy of a modern instrument is about one δ unit, which is equivalent to 0.0004 at. % N-15.

The different methods in preparing N_2 for isotopic analyses have been reviewed in detail by Fiedler and Proksch (1975). The most common techniques are as follows:

1. The Rittenberg technique. Bound nitrogen is converted to ammonium by the Kjeldahl method followed by oxidation to N_2 by the action of hypobromite:

$$2NH_3 + 3BrO^- \rightarrow N_2 + 3H_2O + 3Br^-$$

2. The Dumas technique. The nitrogen is oxidized and/or reduced, as the case may be, to N_2 in the presence of copper or nickel oxide, viz.:

$$CO(NH_2)_2 + 3CuO \rightarrow N_2 + CO_2 + 2H_2O + 3Cu$$

H_2O and CO_2, which are evolved simultaneously with N_2, are usually absorbed by a liquid trap, CaO/Al_2O_3 or by molecular sieves.

3.2. Isotope Fractionation

Some concern has been expressed that the abundance of N-15 relative to N-14 would be significantly altered during nitrogen metabolism. The small increase in abundance of ^{15}N normally found in nitrogenous compounds of biological origin is due to mass discrimination and not to analytical errors (Gaebler et al., 1963). Wellman et al. (1968) estimated the isotope abundance during denitrification experiments in synthetic media with *Pseudomonas stutzeri*. On an average, $^{14}NO_3$ and $^{14}NO_2$ were utilized 2% faster than the labeled species. The isotope fractionation was smaller during the initial stages of nitrate reduction, i.e., during the population growth phase, than during the stationary phase, a behavior that is analogous to the microbial reduction of SO_4. Nitrogen fixation by *Azotobacter vinelandii*, the oxidation of ammonium ion to nitrite by *Nitrosomonas europaea*, denitrification by *Pseudomonas denitrificans*, and assimilation of ammonium by *Azotobacter vinelandii* and three soil yeasts all showed a discrimination in favor of the lighter isotope (Delwiche and Steyn, 1970). All soils examined showed a ^{15}N-abundance that was never less than that of atmospheric nitrogen, and it was shown that, in addition to microbial processes, chromatographic processes on soil particles played a strong role in the isotope discrimination. The faster utilization of the N-14 isotope relative to the N-15 isotope in denitrification has been emphasized in a theoretical model (Focht, 1973).

Differences in the range of isotope ratios for nitrate from soils of different land-use environments permit identification of the predominant source of nitrate contamination of ground water (Kreitler and Jones, 1975). Nitrate in samples was converted to N_2 and analyzed by mass spectrometry by comparing the ratio of m/z 29 to m/z 28 with a ±1‰ experimental error. The $\delta^{15}N$ range, that is the ratio of the $^{15}N/^{14}N$ of a sample to the $^{15}N/^{14}N$ of a standard, of natural soil nitrate was 2–8‰ whereas the $\delta^{15}N$ range of animal or human waste nitrate was 10–20‰. More than 66% of ground water nitrates analyzed in Runnels County, Texas, an area highly contaminated with nitrate, were in the $\delta^{15}N$ range of natural soil nitrates that is, increased concentrations of nitrate with a low $\delta^{15}N$ indicate contamination from fertilizer. Extensive terracing of the farmland had raised the water table and leached the nitrate into the ground water. The same technique was used by Kreitler et al. (1978) on ground water samples from Long Island, New York. A shift was demonstrated from lighter values in ^{15}N (3–9‰) in the eastern part of the island, where land was used predominantly for agriculture, to heavier values towards New York City (12–21‰), where septic systems and sewers are more common.

3.3. Assimilation Studies

Factors that influence the uptake and transformation of N-containing compounds in microbial environments can be conveniently studied with N-15 tracer technique. Usually, the N-15 content has to be determined with high sensitivity and a limited amount of sample, because of some inherent problems associated with the tracer experiments (Wada et al., 1977): (1) The metabolic activity is generally low; the rate of utilization of inorganic nitrogen is 1–10 ng atoms N per liter and hour. (2) Measurements of activity should be completed within a few hours to minimize error during the experiment. (3) The amount of N-15 labeled compounds to be added must be small.

There has been some interest in studying the relative importance of fixation of N_2 and assimilation of soil nitrogen in legumes. Measurement of the enzymatic activity that is responsible for the nitrogen fixation is an instantaneous determination making extrapolation of fixation rate over a growing season delicate. Amarger et al. (1979) approached the problem by adding a small amount of labeled N-15 fertilizer to the soil and then determining the ratio of N-15 content in legumes inoculated with

Rhizobium japonicum and in noninoculated control plants. In order to have different levels of fixation, nitrogen fertilizer was added to half of the treatments. Measurements were made on aerial parts sampled 91 days after planting. After mineralization and distillation, ammonium-nitrogen was oxidized *in vacuo* and the dried N_2 purified and analyzed by MS. The reproducibility of the measures was ±0.4‰. Each measure necessitated about 5 ng of nitrogen. The proportion of fixed nitrogen determined from the natural N-15 abundance in plants was directly proportional to the measures of reducing activity. Between 15% and 35% of the nitrogen was fixed by *Rhizobium*. Nitrogen fixation was lower in fertilized plants and the uptake of soil nitrogen was less in inoculated plants than in control plants. Inoculation thus seemed to inhibit to some extent the uptake of mineral nitrogen.

The stimulation of microbial activity in soils and the uptake of N by plants by addition of organic residues was studied by Vaacob and Blair (1980). [15]N labeled residues of soybean (*Glycine max*), a grain legume, and Siratro (*Macroptillium atropurpureum*) were added to a soil which was sown to Rhodes grass (*Chloris gayna*). Plants were harvested and analyzed for nitrogen, which was extracted by 2 *M* KCl followed by steam distillation and MS. Nitrogen uptake was higher in grass on Siratro amended soil than on soybean soil. Recovery of incorporated N-15 was affected by the number of crops that preceded the experiment; 13.7%, 42.4%, and 55.5% on soils that had grown 1, 3 and 6 previous Siratro crops. In all treatments, the plants growing in amended soil took up more N.

That mycorrhizal infection improves nitrogen nutrition and growth of the infected plant is well established. Using N-15 labeling of a soil, Stribley and Read (1974) showed that the greater nitrogen content of mycorrhizal *Vaccinium macrocappon* Ait. did not originate from ammonium nitrogen in the soil but that the endophyte more likely was able to assimilate nitrogen from organic compounds. The ^{15}N excess of shoots of nonmycorrhizal plants was significantly greater than that of shoots from mycorrhizal plants after 6 months' incubation in a sterile soil amended with [15]$(NH_4)_2SO_4$.

Several aspects of nitrogen assimilation by phytoplankton in natural waters, viz., primary production in a nitrogen limited environment, preferential uptake of different nitrogen sources, and the relatedness of nitrogen and carbon assimilation, have been subject to ^{15}N tracer experiments. The seasonal variation in nitrogen assimilation by phytoplankton was studied by Dugdale and Dugdale (1965). Known amounts, corresponding to about 10% of the total amount present in the water, of [15]NH_4 and [15]NO_3 were added to water samples that were filtered after a short incubation. The organic N was converted to N_2 in a modified Coleman nitrogen analyzer prior to MS analysis. The spring bloom of phytoplankton was correlated to a strong uptake of NH_3-N, NO_3-N and N_2-N, and a simultaneous

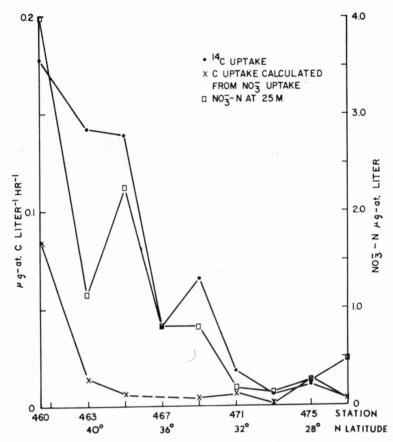

FIGURE 11. Comparison of primary productivity measured by the ^{14}C method and new productivity measured by the ^{15}N method (from Dugdale and Goering, 1967).

drop in phytoplankton standing crop and N-assimilation occurred in the midsummer period.

While ^{14}C measurements of primary production in the sea reflect a total primary production, uptake of nitrogen can be used to calculate new production while uptake of ammonia represents regenerated primary production (Dugdale and Goering, 1967). Fractions of primary production that corresponded to nitrate and ammonia were separated in a ^{15}N tracer experiment in the Sargasso Sea at Bermuda, where the nitrate uptake accounted for between 8.3% and 39.5% of ammonia plus nitrate uptake. The nitrogen in incubated and filtered samples was converted to N_2 by the Dumas method and the isotope ratio determined by MS installed on board the ship for immediate analysis of the water samples. The precision of the instrument was 0.01 at. % for replicate samples. The average uptake of ammonia and nitrate in light and dark was well correlated with the carbon uptake, and the computed ratio of carbon to nitrogen uptake, 6.1:1 was similar to the expected ratio 7:1. The difference between total primary production, derived by ^{14}C measurements, and primary production based on NO_3 uptake, was considerable, especially at higher latitudes (Figure 11).

The regeneration of ammonia for primary production was studied in a ^{15}N tracer experiment in an estuary near Hawaii (Caperon et al., 1979). Incubated water samples were separated into macrophytoplankton and nanoplankton, including bacteria, by filtering and ammonia was converted to N_2 prior to MS analysis. The excretion for nanoplankton populations made up 70% of excretion for all organisms that passed a 0.333-mm mesh sieve. Mean uptake rate for ammonium was 0.213 μg/Lgh while daytime excretion rate was 0.035 μg/Lgh, i.e. 16% of uptake. Over a daily cycle, the uptake and regeneration processes were in balance.

3.4. Mineralization Studies

^{15}N tracer techniques have been used to estimate denitrification rates both in aquatic and terrestrial environments. Goering and Dugdale (1966) collected water samples from under ice in a subarctic lake and incubated the samples with $^{15}NO_3$. Evolved gases were analyzed with MS. CO_2 was differentiated from N_2O by determining the proportion of the sum of m/z 44, 45, and 46 from the magnitude of m/z 22, which is only produced from CO_2^{++} ions. Ammonia was distilled and analyzed separately. Free nitrogen was the only significant product of denitrification at pH 7.7 in the lake water (evolution of N_2O and NO would not be expected at pH above 7.0). In Lake Mendota, one of the most extensively examined lakes in the USA, rates of nitrate reduction to ammonia and organic nitrogen was estimated by in situ incubation of $^{15}NO_3$-dosed samples for several days (Brezonik and Lee, 1968). Nitrate reduction ranged from 1.4 to 13.4 μg of N per liter and day in the hypolimnion. The relative importance of denitrification, immobilization, and nitrification in lake sediments under anaerobic and aerobic conditions was studied by Chen et al. (1972). Sediment samples were amended with $^{15}NO_3$ and incubated under anaerobic(He) and aerobic conditions. The ^{15}N content of the sediment organic N and NH_4-N was determined by MS. Under anaerobic conditions, added nitrate disappeared from the sediment within 48 h with one portion present as organic N and another as NH_4-N. In aerobic sediments, the rate of NO_3-N loss was less than half that in the anaerobic system and subsequent mineralization was not detected. The same approach was extended by Tirén *et al.* (1976), who estimated the nitrate consumption rate in two eutrophic lakes in central Sweden. Water in contact with the sediment was enclosed in a cylinder equipped with a magnetic stirrer and PVC tubes for sampling (Figure 12). $^{15}NO_3$ was added to the free water and water samples were analyzed daily for nitrogen gas formation by a double collector MS. Nitrate consumption rates varied between 0.50 and 0.65 mg/lit day. At the end of the experiment between 80% and 90% of the labeled nitrate was found in the form of nitrogen gas, about 4% in the ammonium fraction in the free water, and less than 1% in the form of organic nitrogen.

The rate of denitrification in a soil is influenced by the amount of dissolved oxygen and the pH. The formation of nitrogen gases from a forest soil that was amended with labeled nitrite and nitrate was examined in a special incubation unit that was connected to the inlet system of a MS by a capillary tube (Nömmik and Thorin, 1972). Gas samples of 1.0 to 1.5 ml were removed from the unit and the mass spectrum obtained analyzed using the peak height for m/z 20 (neon) as an internal standard. When nitrite was added to the soil under anaerobic atmosphere, NO, nitric acid, was the predominant nitrogen gas during the initial stages of incubation, independent of pH. As incubation proceeded, the proportion of N_2O and N_2 increased. When nitrate was added to the soil, only about 6% of added

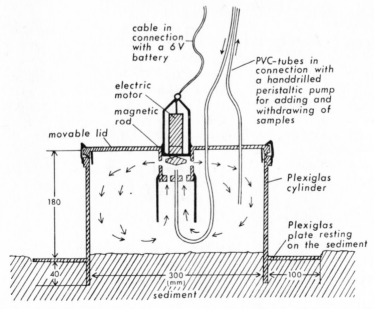

FIGURE 12. The cylinder for denitrification experiments resting on the bottom sediment, with closed lid (from Tirén *et al.*, 1976).

nitrate-N was recovered in gaseous form. When pH was increased, denitrification gases increased and as incubation proceeded, N_2 increased and N_2O decreased.

The effects of different concentration of salts and $^{15}NH_4$-N on mineralization, nitrification, and immobilization were studied by Westerman and Tucker (1974). Soils were incubated with various concentrations of $^{15}NH_4Cl$, NaCl, $CuCl_2$, and $CaCl_2$ and the isotope ratio determined by MS. Dilute concentrations of salts and $^{15}NH_4Cl$ were found to stimulate mineralization of soil N. Nitrification decreased with increasing concentration of salts, especially copper and calcium salts, while immobilization of $^{15}NH_4$-N was decreased by high concentrations of salts.

A nitrogen deficit is commonly observed during incubation of soils, i.e., more $^{15}NO_3$ is lost from a soil than can be accounted for by gaseous losses. This phenomenon has been explored by Letey et al. (1980), who incubated soils in cylinders with solutions of $^{15}NO_3$ and carbon in the form of glucose. Gas samples were periodically removed and analyzed for $^{46}N_2O$, $^{45}N_2O$, and $^{30}N_2$ by a GC–MS system. Soils were extracted with $2\,M$ KCl after incubation and distillates converted to N_2 by the hypobromite method. Under saturated soil conditions only 30% of the produced N_2O was evolved, the fraction diffusing increasing as the air-filled porosity increased. Clearly, N_2O could be retained in the soil and reduced to N_2 if the soil was sufficiently wet.

McGill et al. (1975) demonstrated that N-transformations were dependent on C-transformations in a soil and that both N and C were incorporated into organic fractions solely by microbial activity. Soils were incubated with amended $^{15}(NH_4)_2SO_4$ and ^{14}C-acetate and fractionated into humic and fulvic acids and a residue that was sonicated and further separated into two size fractions. The fulvic acid fraction contained initially large amounts of labeled extracellular microbial metabolites, that were rapidly transformed by successive populations of bacteria and actinomycetes. About 5% of added labeled nitrogen was recovered from the human acid fraction. Very fine ($<0.04\ \mu m$) dissolved and suspended material, presumably cytoplasmic constituents, contained about 20% of labeled N.

3.5. Microbial Synthesis of ^{15}N Amino Acid

In addition to their use as tracers in metabolic experiments, ^{15}N labeled NH_4 and NO_3 can be used as precursors in microbial biosynthesis of ^{15}N-labeled organic compounds. A method was developed for obtaining high-purity $L(^{15}N)$-aspartic acid under the action of aspartase utilizing whole bacteria as a source of enzyme (Ivanof et al., 1981):

$$\text{Fumaric acid} + {}^{15}NH_3 \xrightleftharpoons{\text{aspartase}} L({}^{15}N)\text{-aspartic acid}$$

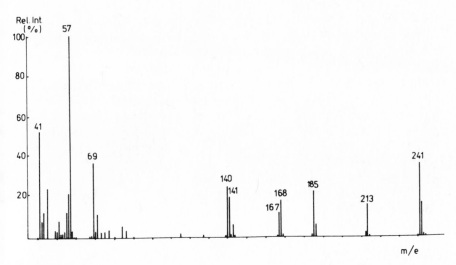

FIGURE 13. Mass spectrum of the $N(O)$-trifluoroacetyl n-butyl ester of L-(^{15}N)aspartic acid (from Ivanof *et al.*, 1981).

Citrobacter freundi was incubated with fumaric acid and $^{15}NH_4Cl$ at pH 10, and 70% of the fumaric acid was converted to L(^{15}N)-aspartic acid. The compound was precipitated, washed, and recrystallized and the purity and isotope content measured by TLC and MS. The sample was derivatized as N-trifluoroacetyl-n-butylester, and the mass spectrum recorded with EI MS. The percentage of ^{15}N was at least 99% as obtained by the intensities of the fragment ions at m/z 240 and 241 (Figure 13).

4. CONCLUSION

Although mass spectrometry has opened up no new area of research on nitrogen compounds in ecological microbiology, the technique has developed into a first-hand alternative for studies of microbial nitrogen metabolism. The examples presented in this chapter demonstrate the usefulness of MS for structure elucidation and isotope ratio measurements, but the analytical method has found no widespread use in ecological microbiology. This is probably due to the high costs involved in purchase and maintenance of mass spectrometric equipment and the limited experience with MS in laboratories of ecological microbiology. A closer cooperation between microbiologists and analytical chemists would be beneficial for research in this area. It would open the field for several desirable advances in the applications of MS in ecological microbiology. For example, it would be advantageous to replace EI as an ionization method for chemical ionization, field ionization, field desorption, and secondary ion mass spectrometry, which reduce fragmentation and evaporation of the substances under investigation. The development of derivatives with simple fragmentation patterns and pronounced molecular ions would also be feasible, and combinations of capillary GC and MS for confirmatory identification of a variety of microbial metabolites and for determination of ^{15}N enrichment in microbial end-products should be encouraged in ecological microbiology because of the small sample size requirement and high accuracy of MS. So far, very few papers on quantitative MS of N-compounds have been published within the area of ecological microbiology. Numerous applications of selected ion monitoring may be expected in the near future, since the high selectivity and sensitivity offered by this technique would be most useful in microsampling procedures.

REFERENCES

Amarger, N., Mariotti, A., Mariotti, F., Durr, J. C., Bourguignon, C., and Lagacherie, B., 1979, Estimate of symbiotically fixed nitrogen in field grown soybeans using variations in ^{15}N natural abundance, *Plant and Soil* **52**:269–280.

Aragozzini, F., Ferrari, A., Pacini, N., and Gualandris, R., 1979, Indole-3-lactic acid as a tryptophan metabolite produced by *Bifidobacterium* spp., *Appl. Environ. Microbiol.* **38**:544–546.

Belasco, I. J., Pease, H. L., and Reiser, R. W., 1978, Microbial conversion of methionine hydroxy analogue and its natural occurrence in various foods and feed products, *J. Agric. Food Chem.* **26**:327–330.

Brezonik, P. L., and Lee, G. F., 1968, Denitrification as a nitrogen sink in Lake Mendota, Wisconsin, *Environ. Sci. Tech.* **2**:120–125.

Buck, R., Eberspächer, J., and Lingens, F., 1979, L-(2,3-Dihydroxyphenyl)alanin, eine neue natürliche Aminosäure, *Liebigs Ann. Chem.* **4**:564–571.

Caperon, J., Schell, D., Hirota, J., and Laws, E., 1979, Ammonium excretion rates in Kaneohe Bay, Hawaii, measured by a ^{15}N isotope dilution technique, *Mar. Biol.* **54**:33–40.

Chen, R. L., Keeney, D. R., Graetz, D. A., and Holding, A. J., 1972, Denitrification and nitrate reduction in Wisconsin lake sediments, *J. Environ. Quality* **1**:158–162.

Delwiche, C. C., and Steyn, P. L., 1970, Nitrogen isotope fractionation in soils and microbial reactions, *Environ. Sci. Tech.* **4**:929–935.

Dugdale, V. A., and Dugdale, R. C., 1965, Nitrogen metabolism in lakes III. Tracer studies of the assimilation of inorganic nitrogen sources, *Limnol. Oceanog.* **10**:53–57.

Dugdale, R. C., and Goering, J. J., 1967, Uptake of new and regenerated forms of nitrogen in primary productivity, *Limnol. Oceanog.* **12**:197–206.

Focht, D. D., 1973, Isotope fractionation of ^{15}N and ^{14}N in microbiological nitrogen transformations: A theoretical model, *J. Environ. Quality* **2**:247–252.

Gaebler, O. H., Choitz, H. C., Vitti, T. G., and Vukmirovich, R., 1963, Significance of N^{15} excess in nitrogenous compounds of biological origin, *Can. J. Biochem. Physiol.* **41**:1089–1097.

Goering, J. J., and Dugdale, V. A., 1966, Estimates of the rates of denitrification in a subarctic lake, *Limnol. Oceanog.* **11**:113–117.

Guerrant, G. O., Lambert, M. S., and Moss, C. W., 1979, Identification of diaminopimelic acid in the Legionnaires disease bacterium. *J. Clin. Microbiol.* **10**:815–818.

Herrmann, V., and Jüttner, F., 1977, Excretion products of algae. Identification of biogenic amines by gas–liquid chromatography and mass spectrometry of their trifluoroacetamides, *Anal. Biochem.* **78**:365–373.

Ivanof, A., Muresan, L., Quai, L., Bologa, M., Palibroda, N., Mocanu, A., Vargha, E., and Barzu, O., 1981, Preparation of ^{15}N-labeled L-aspartic acid using whole bacteria as enzyme source, *Anal. Biochem.* **110**:267–269.

Katz, E., Mason, K. T., and Mauger, A. B., 1973, Identification of *cis*-5-methylproline in hydrolysis of actinomycin Z_5, *Biochem. Biophys. Res. Comm.* **52**:819–826.

Katz, E., Mason, K. T., and Mauger, A. B., 1974, The presence of α-amino-β,γ-dihydroxybutyric acid in hydrolysates of actinomycin Z_1, *J. Antiobiotics* **27**:952–955.

Katz, E., Mason, K. T., and Mauger, A. B., 1975, 3-hydroxy-5-methylproline, a new amino acid identified as a component of actinomycin Z_1, *Biochem. Biophys. Res. Comm.* **63**:502–508.

Keddie, R. M., and Bonsfield, I. J., 1980, Cell-Wall Composition in the Classification and Identification of Coryneform Bacteria," in *Microbial Classification and Identification* (M. Goodfellow, and R. G. Board, eds.), pp. 167–188, Academic Press, New York.

Keller-Schierlein, W., and Widmer, J., 1976, Stoffwechselprodukte von Mikroorganismen. Über die aromatische Aminosäure des Echinocandins B: 3,4-Dihydroxyhomotyrosin, *Helv. Chim. Acta* **59**:2021–2031.

Kreitler, C. W., and Jones, D. C., 1975, Natural soil nitrate: The cause of the nitrate contamination of ground water in Runnels County, Texas, *Ground Water* **13**:53–61.

Kreitler, C. W., Ragone, S. E., and Katz, B. G., 1978, N^{15}/N^{14} ratios of ground-water nitrate, Long Island, New York, *Ground Water* **16**:404–409.

Letey, J., Jury, W. A., Hadas, A., and Valoras, N., 1980, Gas diffusion as a factor in laboratory incubation studies on denitrification, *J. Environ. Qual.* **9**:223–227.

Lewis, O. A. M., 1975, An ^{15}N-^{14}C study of the role of the leaf in the nitrogen nutrition of the seed of *Datura stramonium* L., *J. Exp. Bot.* **26**:361–366.

McGill, W. B., Shields, J. A., and Paul, E. A., 1975, Relation between carbon and nitrogen turnover in soil organic fractions of microbial origin, *Soil Biol. Biochem.* **7**:57–63.

Nordbring-Hertz, B., and Odham, G., 1980, Determination of volatile nematode exudates and their effects on a nematode-trapping fungus, *Microb. Ecol.* **6**:241–251.

Nömmik, H., and Thorin, J., 1972, Transformations of ^{15}N-labeled nitrite and nitrate in forest raw humus during anaerobic incubation, *Proc. International Atomic Energy Agency, Vienna*, pp. 369–382.

Onodera, R., and Kandatsu, M., 1969, Occurrence of L-(−)-pipecolic acid in the culture medium of rumen ciliate protozoa, *Agr. Biol. Chem.* **33**:113–115.

Pavlou, S. P., Firederich, G. E., and Macisaac, J. J., 1974, Quantitative determination of total organic nitrogen and isotope enrichment in marine phytoplankton, *Anal. Biochem.* **61**:16–24.

Sakurai, A., Sakata, K., Tamura, S., Aizawa, K., Yanagishima, N., and Shimoda, C., 1976, Isolation and structure elucidation of α substance-I_B, a hexapeptide inducing sexual agglutination in *Saccharomyces cerevisiae*, *Agr. Biol. Chem.* **40**:1451–1452.

Shiba, T., Mukunoki, Y., and Akiyama, H., 1975, Component amino acids of the antibiotic longicatenamycin. Isolation of 5-chloro-D-tryptophan, *Bull. Chem. Soc. Jpn* **48**:1902–1906.

Stribley, D. P., and Read, D. J., 1974, The biology of mycorrhiza in the ericaceae. IV. The effect of mycorrhizal infection on uptake of ^{15}N from labeled soil by *Vaccinium macrocarpon* Ait., *New Phytol.* **73**:1149–1155.

Tirén, T., Thorin, J., and Nömmik, H., 1976, Denitrification measurements in lakes, *Acta Agr. Scand.* **26**:175–184.

Tsutsumi, W., Onodera, R., and Kandatsu, M., 1975, Occurrence of δ-aminovaleric acid in the culture medium of rumen ciliate protozoa, *Agr. Biol. Chem.* **39**:711–714.

Volk, R. J., Pearson, C. J., and Jackson, W. A., 1979, Reduction of plant tissue nitrate to nitric oxide for mass spectrometric ^{15}N analysis, *Anal. Biochem.* **97**:131–135.

Wada, E., Tsuji, T., Saino, T., and Hattori, A., 1977, A simple procedure for mass spectrometric microanalysis of ^{15}N in particulate organic matter with special reference to ^{15}N-tracer experiments, *Anal. Biochem.* **80**:312–318.

Wegst, W., Tittmann, U., Ebersprächer, J., and Lingens, F., 1981, Bacterial conversion of phenylalanine and aromatic carboxylic acids into dihydrodiols, *Biochem. J.* **194**:679–684.

Wellman, R. P., Cook, F. D., and Krouse, H. R., 1968, Nitrogen-15: Microbiological alteration of abundance, *Science* **161**:269–270.

Westerman, R. L., and Tucker, T. C., 1974, Effect of salts and salts plus nitrogen-15-labeled ammonium chloride on mineralization of soil nitrogen, nitrification, and immobilization, *Soil Sci. Soc. Am. Proc.* **38**:602–605.

White, R. H., and Rudolph, F. B., 1978, The origin of the nitrogen atom in the thiazole ring of thiamine in *Escherichia coli, Biochim. Biophys. Acta* **542**:340–347.

Wietzerbin, J., Das, B. C., Petit, J. F., Lederer, E., Leyh-Bouille, M., and Ghuysen, J. M., 1974, Occurrence of D-alanyl-(D)-*meso*-diaminopimelic acid and *meso*-diaminopimelyl-*meso*-diaminopimelic acid interpeptide linkages in the peptidoglycan of *Mycobacteria, Biochemistry* **13**:3471–3476.

Yaacob, O., and Blair, G. J., 1980, Mineralization of ^{15}N-labeled legume residues in soils with different nitrogen contents and its uptake by rhodes grass, *Plant and Soil* **57**:237–248.

Zerilli, L. F., Tuan, G., Turconi, M., and Coronelli, C., 1977, Mass spectra of lanthionine and β-methyllanthionine isolated from gardimycin, *Annal. Chim.* **67**:691–697.

INDEX

437

Amino acids (*cont.*)
  in aqueous surface microlayers, 179
  aromatic, 173, 174
  in biological fluids, 177
  characteristic of ions, 163
  cyclic α-, 159
  D-, 178, 190
  derivatives, 165, 168, 169, 171
  derivatization of, 160
  deuterated, 175, 177
  dicarboxylic, 165
  enantiomers, 178
  free, 158, 159, 175
  hydroxy, 165, 169
  lipophilic, 192
  mass spectrometric studies of, 158
  methylthiohydantoins, 182
  in microbial products, 179
  N-acetyl esters, 162
  N-HFB amino acid isobutyl esters, 168
  N-perfluoroacyl alkyl esters, 162
  n-TFA-n-butyl esters, 168
  nonprotein, 158
  oxazolidionones, 174
  perdeuterated, 175
  perfluorinated alkyl esters, 174
  phenylthiohydantoin (PTH) derivatives, 182
  quantitative analysis of, 162
  racemization of, 178
  release by anaerobic microorganisms, 179
  resolution of enantiomers by GLC, 158
  in soil extracts, 177
  trimethylsilylated, 172
  ultramicro determination of 175, 178
  uncommon, 420
  unusual, 158
  α-, 176, 178, 194
  β-, 167, 173
  ω-, 159, 173, 178
Amino sugars, 114, 118, 146
Amoxycillin, 253
Anabaena variabilis, 86
Anacystis montane, 79
Anaerobic infections
  diagnosis of, 230
  rapid diagnosis of, 229
Antibiotics, 149, 158, 179, 180, 193, 227, 250, 422
  structure of oligosidic moiety of, 149
Archaebacteria, 57, 82

Arthritis, 229
  rapid detection of, 290
Arthrobacter simplex, 75
Arthrobotrys oligospora, 179, 422
Aspergillus oryzae, 87
Azotobacter, 363

Bacillus, 91, 219, 222, 274, 359, 363, 366, 369
Bacillus acidocaldarius, 85
Bacillus alvei, 367
Bacillus amyloliquefaciens, 367, 369
Bacillus cereus, 367, 369
Bacillus coagulans, 367
Bacillus firmus, 367
Bacillus licheniformes, 367, 369
Bacillus mycoides, 369
Bacillus pumilus, 367, 369
Bacillus subtilis, 92, 94, 193, 339, 367, 369
Bacillus thuringiensis, 369
Bacteria, 73, 82, 225, 241, 242, 251, 270, 281, 292, 359, 418
  acid-fast 260, 270
  acido-thermophilic, 74, 82
  aerobic, 240
  anaerobic, 240, 269
  anaerobic methanogenic, 410
  antibiotic susceptibility of, 250
  cell wall composition for identification of, 416
  characteristic LPS constituents of some, 278
  characterization of, 282
  chemical composition of, 280
  classification of, 275
  detection in serum, 290
  detection of, 330
  fatty acid composition of, 271, 274
  glucose nonfermenters, 286
  gram-negative, 91, 260, 265, 270, 275, 290
  gram-negative rods, 252
  gram-positive, 260, 270, 275
  halophilic, 82
  identifying anaerobic, 218
  lipids of, 73
  obligate anaerobic, 217, 218
  small gram-negative rod-shaped, 281
Bacterial biomass, 293, 330
Bacterial cell walls, 257
Bacterial classification, 258, 260
  cell wall analysis in, 275
Bacterial dextrans, structure of, 133
Bacterial fingerprinting, 359

DATE

OCT 16 1984
APR 13 85
MAY 12 1987
JUL 01 1987
DEC 18 1987
APR 20 1988
6-28-91
DEC 0 3
APR 0 4 2005
DEC 0 4 2014

DEMCO 38-297